Biochemical and Clinical Aspects of Pteridines
Volume 2
Cancer · Immunology · Metabolic Diseases

Biochemical and Clinical Aspects of Pteridines Volume 2

Cancer · Immunology Metabolic Diseases

Proceedings
Second Winter Workshop on Pteridines
March 6–9, 1983
St. Christoph, Arlberg, Austria

Editors
H. Ch. Curtius · W. Pfleiderer · H. Wachter

W DE G

Walter de Gruyter · Berlin · New York 1983

Editor

Prof. Dr. Hans Christoph Curtius
Division of Clinical Chemistry
Department of Pediatrics
University of Zurich
Steinwiesstrasse 75
CH-8032 Zurich, Switzerland

Prof. Dr. Dr. h. c. Wolfgang Pfleiderer
Department of Chemistry
University of Konstanz
P.O.B. 5560
D-7730 Konstanz, F. R. of Germany

Prof. Dr. Helmut Wachter
Institute of Medical Chemistry and Biochemistry
University of Innsbruck
Fritz-Pregl-Strasse 3
A-6020 Innsbruck, Austria

CIP-Kurztitelaufnahme der Deutschen Bibliothek

Cancer, immunology, metabolic diseases: proceedings /
... Winter Workshop on Pteridines. - Berlin; New York: de Gruyter

NE: Winter Workshop on Pteridines

2. March 6 - 9, 1983, St. Christoph, Arlberg, Austria. - 1983.
(Biochemical and clinical aspects of pteridines; Vol. 2)
ISBN 3-11-009813-X

NE: GT

Library of Congress Cataloging in Publication Data

Winter Workshop on Pteridines (2nd: 1983: St. Christoph,
Arlberg, Austria)
Proceedings / Second Winter Workshop on Pteridines,
March 6 - 9, 1983, St. Christoph, Arlberg, Austria.

(Biochemical and clinical aspects of pteridines; v. 2)
Includes indexes.
1. Pteridines--Congresses. 2. Pteridines--Metabolism--Congresses.
3. Pteridines--Diagnostic use--Congresses.
4. Pteridines--Physiological effect--Congresses.
I. Curtius, H. Ch. (Hans-Christoph). II. Pfleiderer, Wolfgang.
III. Wachter, H. (Helmut). IV. Series.
QP801.P69W56 1983 616.07'56 83-24079

ISBN 3-11-009813-X (U.S.)

 Printing: Gerike GmbH, Berlin. Binding: Dieter Mikolai, Berlin. Printed in Germany.

PREFACE

This volume contains the proceedings of the Second Winterworkshop on pteridines held in St. Christoph, Arlberg (Austria), from March 6 to 9, 1983, centred around the development and growing interest in clinical application of pteridines.

The Second Winterworkshop brought together scientists involved in the many different aspects of pteridine chemistry, biochemistry, immunology and application in medicine; the Winterworkshop being a tradition established in St. Christoph by the first meeting in 1982.

This meeting was organised to discuss recent developments in the field of pteridine research which resulted in the presentation of 31 papers from 51 participants. In addition to the participants from Austria, France, Switzerland and West Germany, this time scientists from Canada, England, Holland, Japan and the USA also reported on their most recent research.

The discovery of different forms of tetrahydrobiopterin deficiencies in atypical PKU during the last years has shed new light on the understanding of the biochemistry of pteridines. Consequently, efforts were also made towards the treatment of atypical PKU and neurological and psychiatric diseases with tetrahydrobiopterin.

Furthermore the neopterin determinations in the urine of patients with different diseases lead to their application as a diagnostic tool and initiated investigations on the involvement of pteridines in immunology.

We wish to express our thanks to all those who helped directly and indirectly in producing this volume. Also special thanks are due to all guest speakers, session chairpersons and participants especially to those who travelled so far to join us on this occasion. We look forward to seeing them all again at the next workshop.

Finally, we wish to thank de Gruyter Publishers for their helpful, efficient and professional assistance in publishing this book.

H.-Ch. Curtius
W. Pfleiderer
H. Wachter

LIST OF PARTICIPANTS

Abita, J.P.
Service de Médecine Nucléaire, Hôpital St. Louis, 2 Place du Dr. Fournier, F-75475 Paris Cédex 10, France

Aulitzky, W.
Landeskrankenhaus Salzburg, Urologische Abteilung, Müllner-Hauptstrasse 48, A-5020 Salzburg, Austria

Ayling, J.E.
Dept. of Pharmacology, College of Medicine, University of South Alabama, Mobile, AL 36688, USA

Bailey, S.W.
Dept. of Pharmacology, College of Medicine, University of South Alabama, Mobile, AL 36688, USA

Barford, Patricia
Dept. of Chemistry, University of Aston in Birmingham, Gosta Green, Birmingham B4 7ET, U.K.

Bichler, A.
Dept. of Obstetrics and Gynaecology, University of Innsbruck, Anichstrasse 35, A-6020 Innsbruck, Austria

Blau, N.
Division of Clinical Chemistry, Dept. of Pediatrics, University of Zurich, Steinwiesstrasse 75, CH-8032 Zurich, Switzerland

Cooper, B.A.
Division of Hematology, Royal Victoria Hospital, 687, Ave. des Pins Ouest, Montreal, Quebec H3A 1A1, Canada

Curtius, H.-Ch.
Division of Clinical Chemistry, Dept. of Pediatrics, University of Zurich, Steinwiesstrasse 75, CH-8032 Zurich, Switzerland

De Wit, R.
Cell Biology & Morphogenesis Unit, Subfaculteit Biologie der Rijksuniversiteit Leiden, Kaiserstraat 63, Postbus 9516, NL-2300 RA Leiden, Netherlands

Dhondt, J.L.
Laboratoire de Biochimie Structurale, Faculté de Médicine, 56 Rue du Port, F-59046 Lille Cédex, France

Ebels, Ietskina
Organisch-chemisch Laboratorium, Rijksuniversiteit Utrecht, Croesestraat 79, NL-3522 AD Utrecht, Netherlands

Fink, M.
III. Medizinische Klinik der Universität München, Klinikum Grosshadern, D-8000 München 70, F.R. of Germany

Frick, J.
Urologische Abteilung, Landeskrankenhaus Salzburg, Müllner-Hauptstrasse 48, A-5020 Salzburg, Austria

Fuchs, D.
Institut für Med. Chemie und Biochemie, Universität Innsbruck, Fritz-Pregl-Strasse 3, A-6020 Innsbruck, Austria

Ghisla, S.
Fakultät für Biochemie der Universität Konstanz, Postfach 5560, D-7750 Konstanz 1, F.R. of Germany

Harzer, W.
Milupa AG, Wissenschaftliche Abteilung Ausland, D-6382 Friedrichsdorf/Taunus, F.R. of Germany

Hausen, A.
Institut für Med. Chemie und Biochemie, Universität Innsbruck, Fritz-Pregl-Strasse 3, A-6020 Innsbruck, Austria

Heintel, D.
Division of Clinical Chemistry, Dept. of Pediatrics, University of Zurich, Steinwiesstrasse 75, CH-8032 Zurich, Switzerland

Hetzel, H.
Universitätsklinik für Frauenheilkunde, Anichstrasse 35, A-6020 Innsbruck, Austria

Hölzer, P.
Milupa AG, Wissenschaftliche Abteilung Ausland, D-6382 Friedrichsdorf/Taunus, F.R. of Germany

Huber, Ch.
Universitätsklinik für Innere Medizin, Anichstrasse 35, A-6020 Innsbruck, Austria

Huber, F.J.K.
Universität Wien, Institut für Analytische Chemie, Währingerstrasse 38, A-1090 Wien IX, Austria

Joller, P.
Division of Immunology, Dept. of Pediatrics, University of Zurich, Steinwiesstrasse 75, CH-8032 Zurich, Switzerland

Kern, H.
Varian AG, Steinhauserstrasse, CH-6300 Zug, Switzerland

Kersten, W.
Institut für Physiologische Chemie, Universität Erlangen-Nürnberg, Fahrstrasse 17, D-8520 Erlangen, F.R. of Germany

Kersten, Helga
Institut für Physiologische Chemie, Universität Erlangen-Nürnberg, Fahrstrasse 17, D-8520 Erlangen, F.R. of Germany

Körber, F.
Zeitschrift für Klinische Chemie und Klinische Biochemie, Biochemisches Institut, Horazweg 31, D-1000 Berlin 42, F.R. of Germany

Leupold, Dorothea
Universitäts-Kinderklinik, Prittwitzstrasse 43, D-7900 Ulm-Donau, F.R. of Germany

Levine, R.A.
Section on Biochemical Pharmacology, Hypertension-Endocrine Branch, NHLBI, National Institutes of Health, Bethesda, MD 20205, USA

Margreiter, R.
Universitätsklinik für Chirurgie I, Anichstrasse 35, A-6020 Innsbruck, Austria

Milstien, S.
Laboratory of Neurochemistry, National Institute of Mental Health, Bethesda, MD 20205, USA

Niederwieser, D.
Universitätsklinik für Chirurgie I, Anichstrasse 35, A-6020 Innsbruck, Austria

Pfleiderer, W.
Universität Konstanz, Fakultät für Chemie, Postfach 5560, D-7750 Konstanz 1, F.R. of Germany

Reibnegger, G.
Institut für Med. Chemie und Biochemie, Universität Innsbruck, Fritz-Pregl-Strasse 3, A-6020 Innsbruck, Austria

Rembold, H.
Max-Planck-Institut für Biochemie
D-8033 Martinsried b. München, F.R. of Germany

Renz, Berta
Abteilung für Klinische Hämatologie der GSF, Landwehrstrasse 61, D-8000 München, F.R. of Germany

Rey, Françoise
Hôpital des Enfants-Malades, Département de Pédiatrie, 149, Rue de Sèvres, F-75730 Paris Cédex 15, France

Rey, J.
Hôpital des Enfants-Malades, Département de Pédiatrie, 149, Rue de Sèvres, F-75730 Paris Cédex 15, France

Rokos, H.
Research Laboratory, Hanning Berlin GmbH, Komturstrasse 19-20, D-1000 Berlin 42, F.R. of Germany

Smith, G.K.
The Wellcome Research Laboratories, Burroughs Wellcome Co, 3030 Cornwallis Road, Research Triangle Park, NC 27709, USA

Spielberger, M.
Universitätsklinik für Chirurgie I, Anichstrasse 35, A-6020 Innsbruck, Austria

Sugimoto, T.
Laboratory of Cell Physiology, Dept. of Life Chemistry, Graduate School at Nagatsuta, Tokyo Institute of Technology, 4259 Nagatsuta, Midori-Ku, Yokohama 227, Japan

Scheibenreiter, Susan
Universitäts-Kinderklinik, Währinger Gürtel 74-76, A-1090 Wien, Austria

Stokstad, E.L.R.
Dept. of Nutritional Sciences, University of California, Berkeley, CA 94720, USA

Viscontini, M.
Institut für Organische Chemie, Winterthurerstrasse 190, CH-8057 Zurich, Switzerland

Wachter, H.
Institut für Med. Chemie und Biochemie, Universität Innsbruck, Fritz-Pregl-Strasse 3, A-6020 Innsbruck, Austria

Wachter, Ilse
Institut für Med. Chemie und Biochemie, Universität Innsbruck, Fritz-Pregl-Strasse 3, A-6020 Innsbruck, Austria

Weber, R.
Walter de Gruyter Verlag, Genthiner Strasse 13, D-1000 Berlin 30, F.R. of Germany

Zeitler, H.-J.
Abteilung für Klinische Hämatologie der Gesellschaft für Strahlen- und Umweltforschung, Landwehrstrasse 61, D-8000 München 2, F.R. of Germany

Ziegler, Irmgard
Institut für Toxikologie und Biochemie, Abteilung für Zellchemie, Arcisstrasse 16, D-8000 München 2, F.R. of Germany

CONTENTS

1 CHEMISTRY AND ANALYSIS OF PTERIDINES

2 BIOCHEMISTRY AND METABOLISM OF PTERIDINES

3 PTERIDINES IN IMMUNOLOGY

4 Pteridines in Cancer and Other Diseases

5 Tetrahydrobiopterin Deficiencies, Diagnosis and Therapy

1

CHEMISTRY AND ANALYSIS OF PTERIDINES

SYNTHESES OF PTERIDINES

Wolfgang Pfleiderer
Fakultät für Chemie, Universität Konstanz
Postfach 5560, D-7750 Konstanz

Introduction

Examination of the pyrazino[2,3-d]pyrimidine structure of pteridines reveals two principal pathways for the synthesis of this ring system, namely fusion of a pyrazine ring to a pyrimidine derivative, and annelation of a pyrimidine ring to a suitably substituted pyrazine derivative.

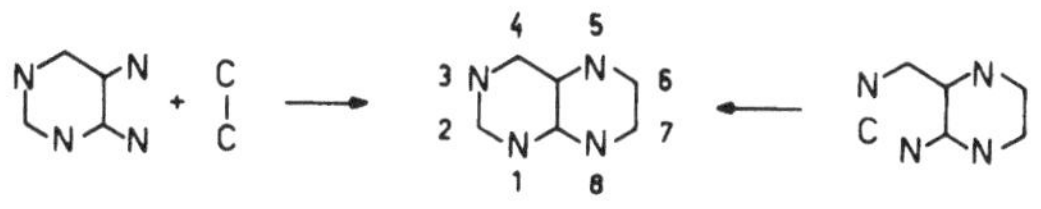

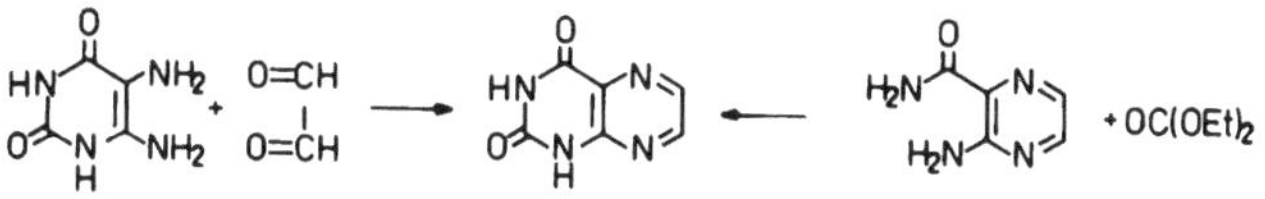

1   2   3

The synthesis of lumazine (2) may therefore be achieved either by condensation of 5,6-diaminouracil (1) and glyoxal or by the reaction of 2-aminopyrazine-3-carboxamide (3) with diethyl carbonate.
Since pyrimidines are more easily accessible the former pathway is of major importance and of more general use. Less important methods include degradations of more complex molecules and ring transformations of structurally related bicyclic nitrogen heterocycles.

Biochemical and Clinical Aspects of Pteridines, Vol. 2

Syntheses from Pyrimidines

The most straightforward synthesis of a pteridine is the Gabriel-Isay reaction (1) involving the condensation of a 5,6-diaminopyrimidine with a 1,2-dicarbonyl compound, as in Gabriel's preparation of 4-methyl-(4) (2) and Isay's preparation of 6,7-diphenylpteridine (5) (3) by this route at the beginning of this century. A further study of the reaction was published two years later (4) and it has since become the synthetic method most frequently used.

GABRIEL-ISAY-REACTION

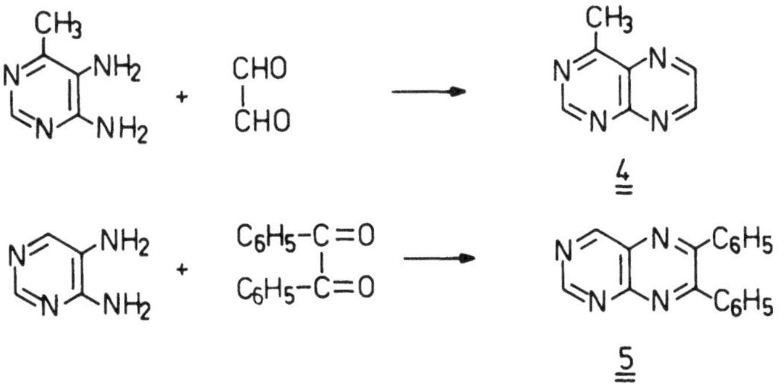

Syntheses with symmetrical 1,2-dicarbonyl compounds (glyoxal, diketones, oxalic acid derivatives) present no problems, but unsymmetrical dicarbonyl components such as α-ketoaldehydes, α-ketoacids etc. give rise to mixtures of substituted pteridines isomeric at position 6 (8) and 7 (7).

Since separation and purification of such mixtures involves losses of time and yield and is often extremely difficult, syntheses which selectively produce a single isomer are sought. Isolation of intermediates (9) as in the reaction of 2,5,6-triamino-4-pyrimidone (6) with methylglyoxal may help in the regioselective orientation, whereas blocking of the most reactive function (10) can lead to the other isomer (5).
The principle of pH-dependent condensation, first recognized

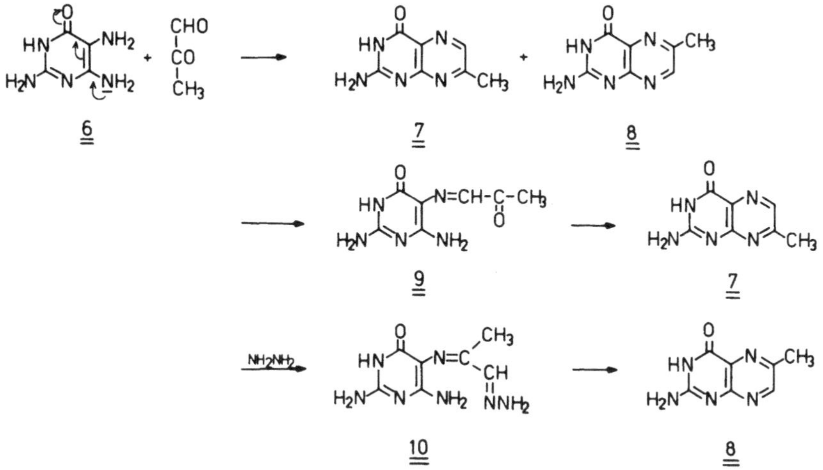

by Purrmann (6) and later applied generally (7), offers the possibility of influencing the direction of the reaction using the differences in the basicity and nucleophilic reactivity of the amino groups in the 5- and 6-positions of the pyrimidine nucleus. Particularly in condensations with 5,6-diaminouracil (8), 2,5,6-triamino-4-pyrimidone (6), 2,4,5,6-tetraaminopyrimidine and their corresponding substituted analogs with α-keto-acids and -esters, neutral or weakly acid media as well as organic solvents (9) favour the formation of pteridin-7-ones, whereas in a strongly acidic medium protonation of the more basic 5-amino groups leads to the opposite orientation of the substituents, to give pteridin-6-ones.

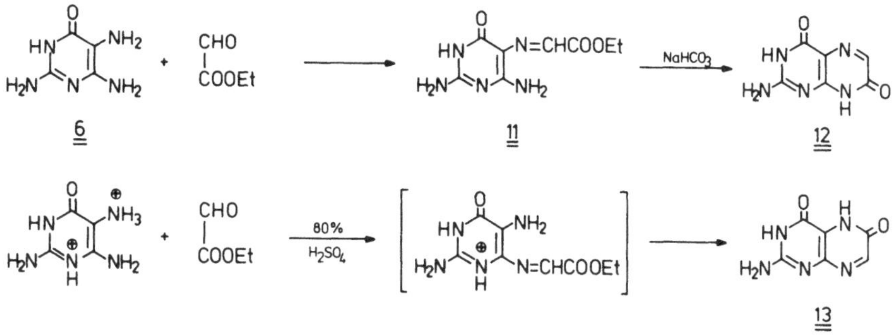

Condensation of 6 with ethyl glyoxylate hemiacetal proceeds particularly smoothly to isoxanthopterin (12) if the alkyl 5-azomethinecarboxylate (11) is isolated first and then cyclized under mild basic conditions (8,10,11). In direct condensation reactions much less specific behaviour is found depending to a large extent on the nature of the pyrimidine substituents (12). Very strong acidic medium usually initiates a regioselective orientation towards pteridin-6-ones as demonstrated by the xanthopterin (13) synthesis. Less drastic pH-differences can be applied on the other hand to condense 6 with ethyl acetonyloxalate to 6-acetonylisoxanthopterin (14) and 7-acetonylxanthopterin (15) respectively.

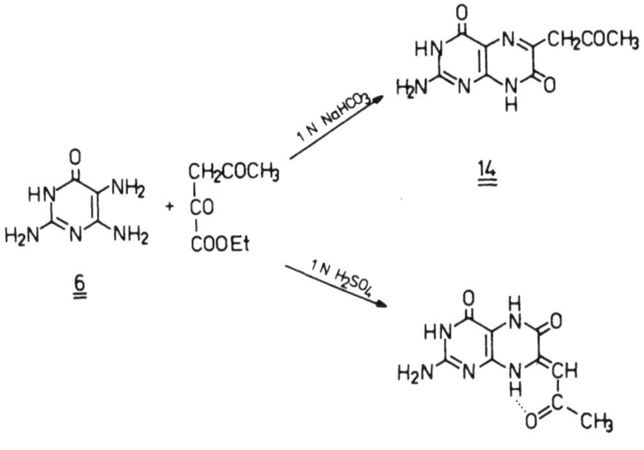

Another pH-dependent condensation is recognized between 6 and benzoin, which leads in neutral solution to 6,7-diphenyl-5,6-dihydropterin (16) and under acidic conditions to the isomeric 6,7-diphenyl-7,8-dihydropterin (17) (13).

The importance of 6-substituted pterins, which constitute most naturally-occurring pteridine derivatives, affords also regioselective approaches to this type of compounds. Condensations of 6 with α-haloaldehydes followed by subsequent oxidation

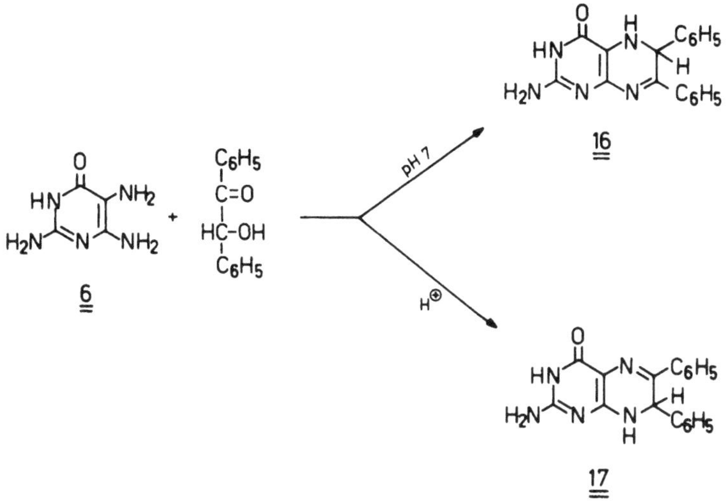

offers such a possibility and leads with 1-bromo-2,2-diethoxy-propionaldehyde (18) to 6-(diethoxymethyl)-pterin (19) (14) and with 1-(2-bromo-2-formylethyl)-pyridinium iodide (20) to 6-pterinylmethyl-pyridinium iodide (21) (15) respectively.

An interesting approach to 6-subst. polyhydroxyalkylpterins has been found in the condensation of 6 with sugar-phenyl-hydrazones in weak acidic medium. In order to explain the high

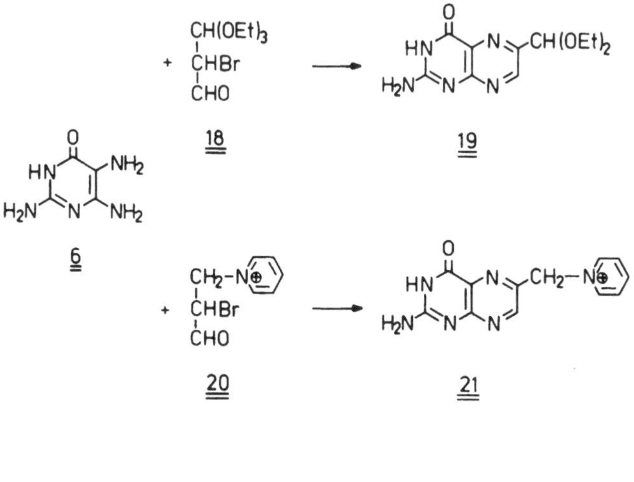

regioselectivity of this reaction it has to be assumed that the synthesis of neopterin (27) (16), for example, from 6 and the D-arabinose-phenylhydrazone (22) involves a complex series of reactions of oxido-reductions, condensations and additions. In the first step will 22 be transfered in an acid-catalysed Amadori-type rearrangement to the corresponding 1-phenylhydrazino-1-deoxy-D-ribulose (23), which forms then a Schiff's base (24) between the keto function of the sugar derivative and the most nucleophilic 5-amino group of the pyrimidine part. 24 undergoes then another intramolecular oxido-reduction to form 25 which cyclizes to 26, a tetrahydropterin derivative, and leading on oxidation finally to D-neopterin (27).

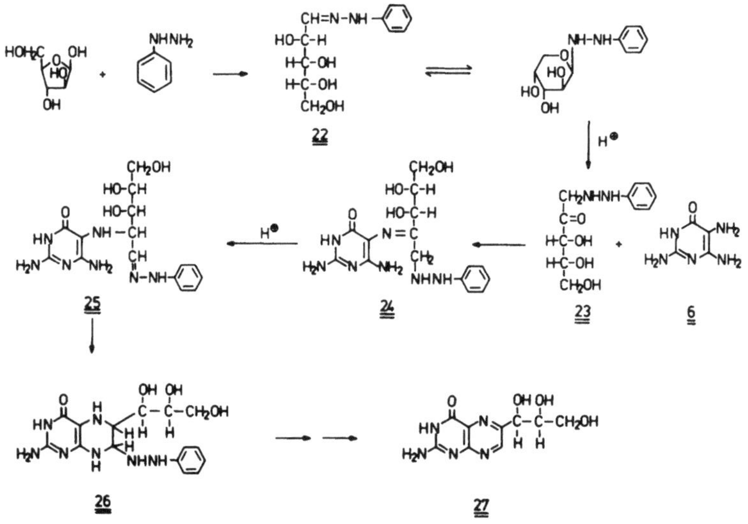

Unambiguous syntheses of pteridines originate in 1949, when Timmis discovered a new, general and unequivocal route to 6- and 7-substituted pteridines respectively in the base-catalysed condensation of a keto-methylene compound with a 6-amino-5-nitrosopyrimidine (17). The condensation reactions are not only restricted to ketones (29) and aldehydes (28) (18-20) but work also with nitriles (30)(21), esters, and acyl halides (31) possessing an activated methylene group adjacent to these functions (22-24).

TIMMIS REACTION

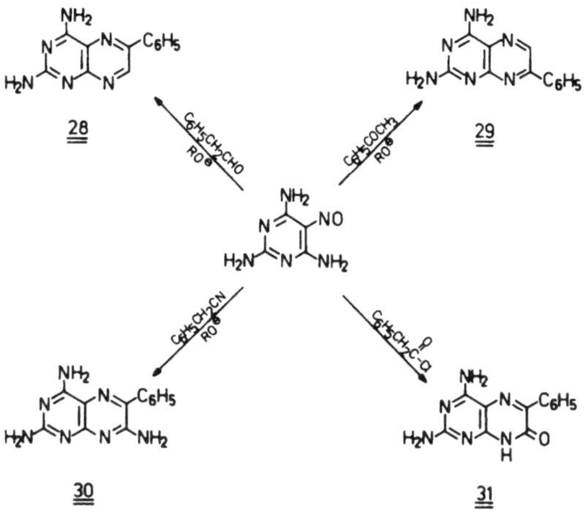

With highly activated methylene compounds no special base catalysis is necessary for the condensation reactions, since 2,6-diamino-4-isopropyloxy-5-nitrosopyrimidine (32) reacts with methyl cyanoacetate to give 2-amino-6-cyano-4-isopropyloxypteridin-7-one (33) and with malononitrile to give 2,7-diamino-6-cyano-4-isopropyloxy-pteridine (34) (25).

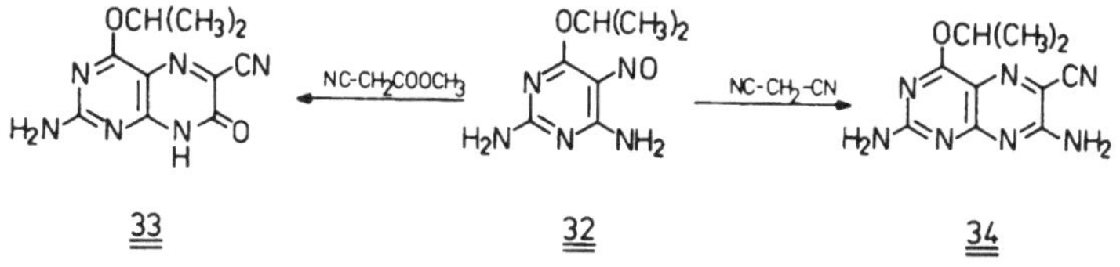

The versatility of 5-nitrosopyrimidines in pteridine syntheses was noticed by Pachter (26) during modification of the Timmis condensation between 4,6-diamino-5-nitroso-2-phenylpyrimidine (35) and benzylmethylketone. Simple condensation leads to 4-amino-7-methyl-2,6-diphenylpteridine (36), but in the presence of cyanide ion 4,7-diamino-2,6-diphenylpteridine (41) is formed. The mechanism of this reaction is still uncertain (27); it may involve cyanohydrine (37) formation, which is cleaved to the Schiff's base 38 followed by another cyanide addition to give 39. Base-catalysed cyclization will then take place easily to 4,7-diamino-2,6-diphenyl-5,6-dihydropteridine (40) which is subsequently oxidized to 41.

PACHTER REACTION

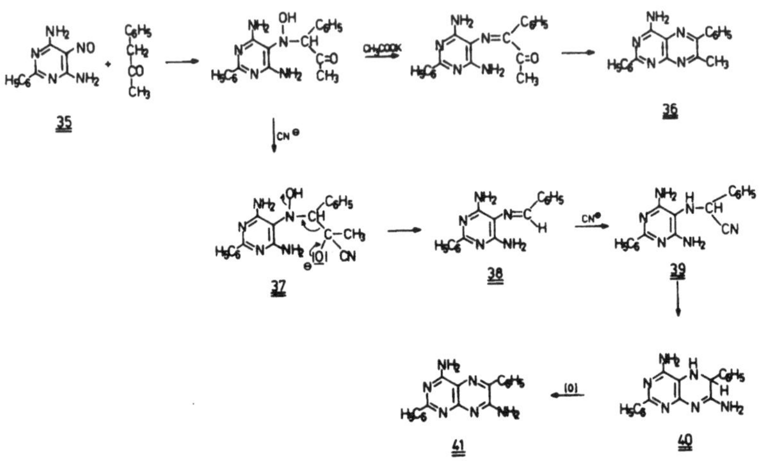

Extension of these principles to α-aryl- und α-alkyl-acetoacetonitriles omits the oxidation step and gives higher yields and forms 6-alkyl-7-aminopteridines, which cannot be obtained directly from simple aliphatic ketones. 4-Amino-5-nitrosopyrimidines also condense with benzoylacetonitrile, phenacylpyridinium bromide and acetonyl-pyridinium chloride respectively in the presence of sodium cyanide to produce 7-amino-6-pteridinyl ketones (28).
An alternative method of synthesis is the Blicke-Pachter-method (29,30) involving the formation of aminonitriles (43) from 5,6-diaminopyrimidines (42), aldehydes, and hydrogen cyanide fol-

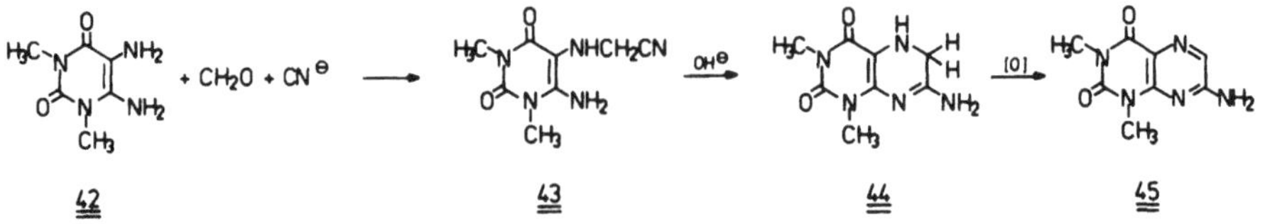

lowed by base-catalyzed cyclization with sodium methoxide and oxidation of the intermediate 5,6-dihydropteridine derivative (44) to form 7-aminopteridines (45).

The reaction works best with formaldehyde but also with homologous aldehydes and with heteroaromatic aldehydes to give 7-amino-6-alkyl- and 7-amino-6-heteroarylpteridines (29).

Another unambiguous approach to 6- and 7-substituted pteridines was found by Polonovski (30) and Boon (31) in the nucleophilic substitution of 6-chloro-5-nitropyrimidines (46) with α-aminocarbonyl compounds, which on reduction cyclize to 7,8-dihydropteridines (47). 6-Chloro-5-phenylazopyrimidines (48) react similarly with α-amino ketones or alkyl α-aminocarboxylates to yield 6-substituted 7,8-dihydropteridines (49) (32).

POLONOVSKI-BOON-REACTION

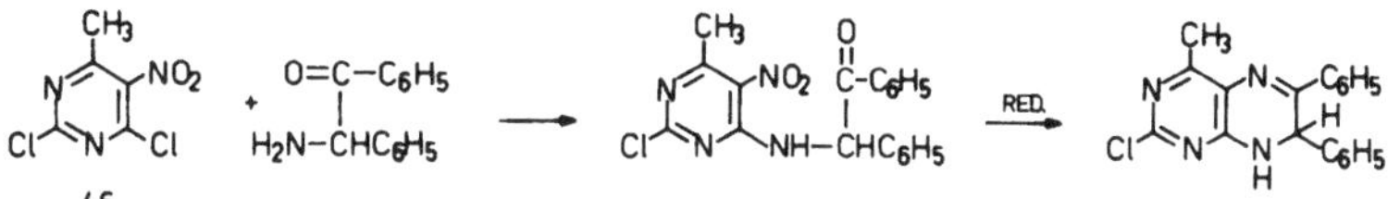

46 47

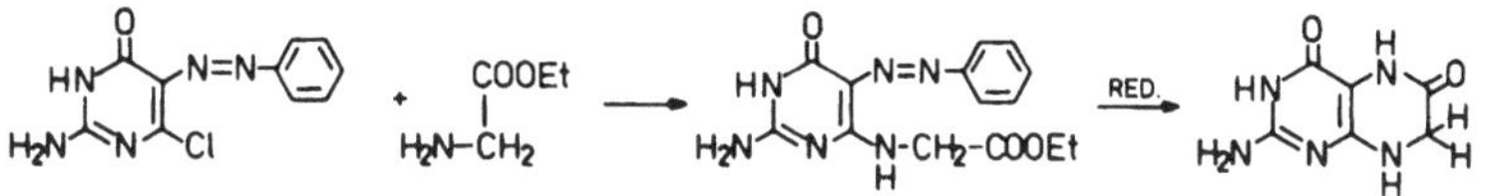

48 49

More complex molecules such as L-neopterin (33) and biopterin (54) (34) have also been synthesized by this route. 2-Amino-6-chloro-5-nitro-4-pyrimidone (50) reacts with 1-amino-1,5-dideoxy-L-erythropentulose (51) to give 52; reduction leads to 7,8-dihydrobiopterin (53) which can either be oxidized to biopterin (54) or further be reduced to its 5,6,7,8-tetrahydro derivatives (55).

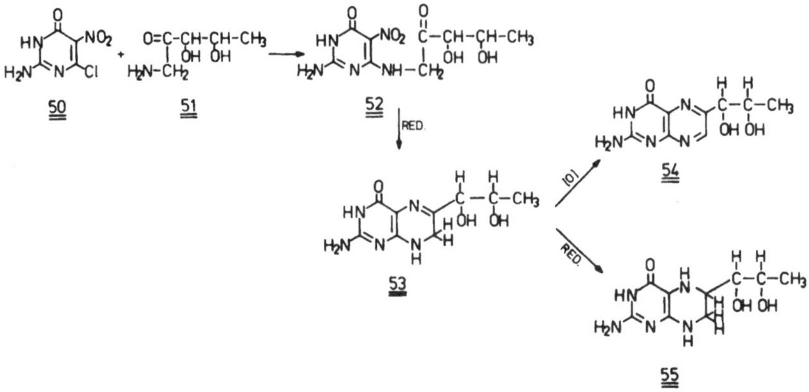

A kind of modification of the Polonovski-Boon principle can be seen in the reaction of 5,6-dihalopyrimidines with ethylendiamine derivatives. Depending on the bulkiness of the amino substituents or the adjacent carbon atom a more or less regiospecific condensation may proceed (35) as recognized from the reaction of 5-bromo-6-chloro-1,3-dimethyluracil (56) and 2-amino-n-propylamine to give 1,3,6-trimethyl-5,6,7,8-tetrahydrolumazine (57). Similar condensations with 6-amino-5-halopyrimidines (58) involving a transamidation usually give rise to the formation of the corresponding isomers (59+60) (36).

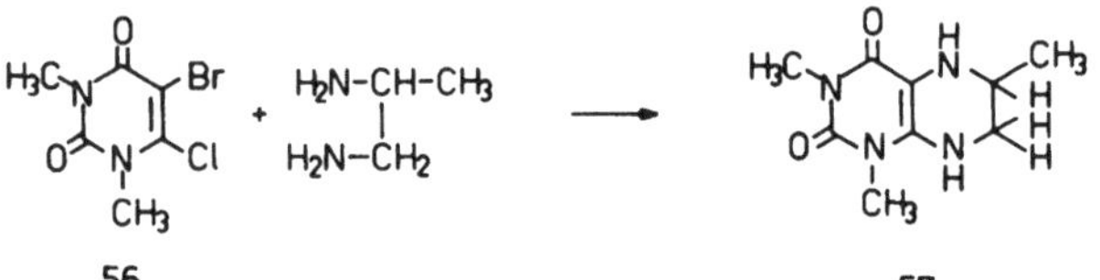

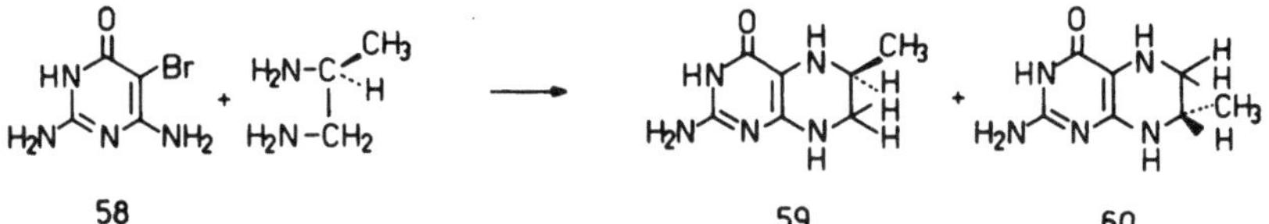

There are also some other unequivocal syntheses of pteridines as for example the electrocyclization of appropriate substituted azahexatrienes to form 6-arylpteridines (37). Treatment of 6-amino-5-benzylideneamino-1,3-dimethyluracil (61) with an excess of triethyl orthoformate in DMF affords as a plausible intermediate the 6-ethoxymethyleneamino derivative (62), which undergoes thermal cyclization by valence isomerization to 63 and subsequent aromatization by elimination of ethanol to give 1,3-dimethyl-6-phenyllumazine (64).

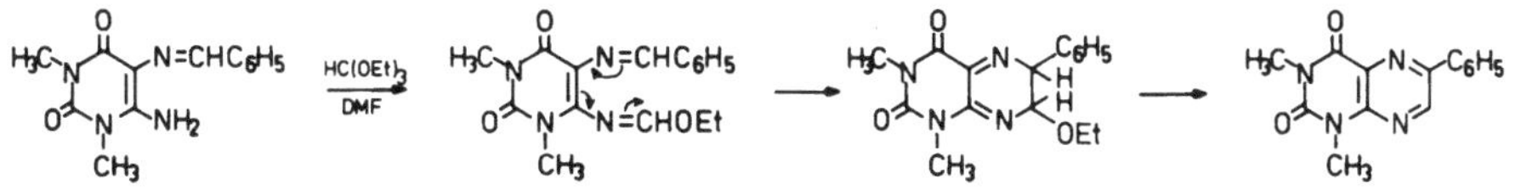

61 62 63 64

Syntheses from Pyrazines

Despite the fact that one of the first pteridine syntheses was based on an intramolecular Hofmann carboxamide degradation of pyrazine-2,3-dicarboxamide (65) by action of hypobromite and leads to lumazine (66) (38), pyrazine derivatives in general have not often been used because of availability problems.

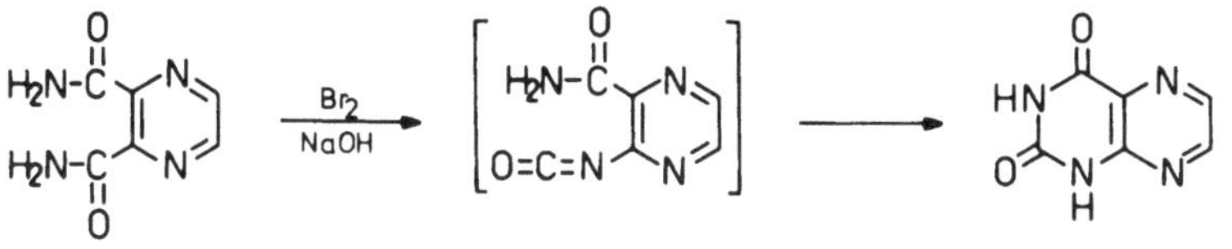

65 66

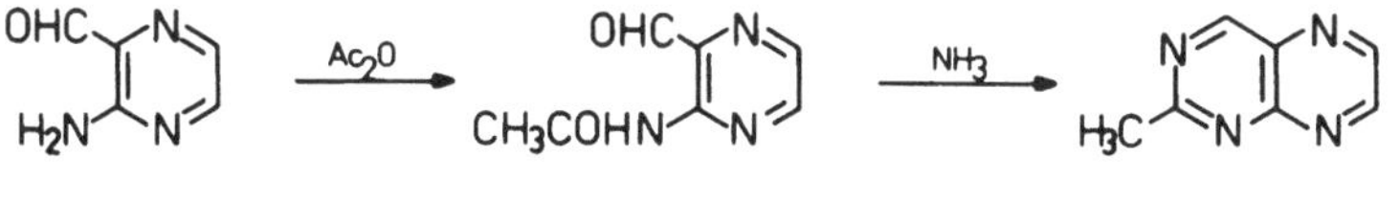

67 68 69

The reaction of alkyl 2-chloropyrazine-3-carboxylates with guanidinium salts leads to pterins (39) and cyclization of 2-aminopyrazine-3-carbohydrazide and 3-hydroxamic acid gives

3-amino-(40,41) and 3-hydroxypteridin-4-ones respectively (42). The synthesis of 3,4-dihydro and 4-unsubstituted pteridine derivatives can be achieved from 2-amino-3-aminomethylpyrazine (43) and 3-aminopyrazine-2-carbaldehyde (67) which after acetylation to 68 cyclizes to 2-methylpteridine (69) on ammonia treatment (44).

A versatile new route to pteridines was found by Taylor (45, 46) in the reaction of an oximino-aldehyde or-ketone with ethyl α-aminocyanoacetate or aminomalononitrile, to give 2-amino-3-ethoxycarbonyl-(70) or 3-cyano-pyrazine 1-oxide (75) which cyclize with guanidine to pterin-8-oxides (73) or 2,4-diaminopteridine 8-oxides (74) respectively (47,48). Reduction of these 8-oxides to the heteroaromatic analogs 71 and 72 as well as the corresponding 7,8-dihydropteridine derivatives provides a wide variety of compounds. This new approach also solves the orientation problem of the 6- and/or 7-substituents, since their structure is already determined in the α-oximino carbonyl component.

TAYLOR-SYNTHESIS

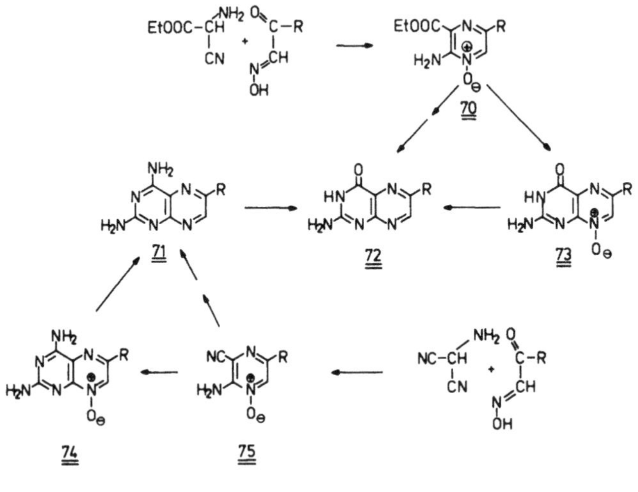

The pyrazine 1-oxide approach has also been used for the preparation of pteroic acid, folic acid, and various derivatives (49) as well as the synthesis of biopterin (79) (50). Reaction of 5-deoxy-2-keto-L-ribose oxime (76) with benzyl α-aminocyano-acetate forms in the usual manner the 2-amino-3-benzyloxycarbonyl-5-(1,2-dihydroxypropyl)-pyrazine 1-oxide (77) which is converted via cyclization by guanidine to biopterin 8-oxide (78) and reduction to biopterin (79).

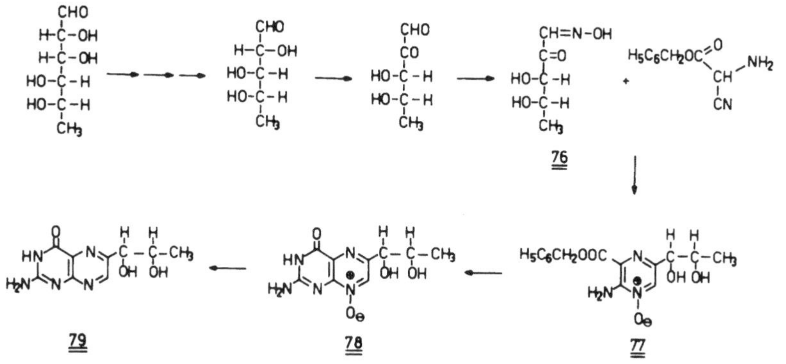

Transformations from Other Heterocyclic Ring Systems

Various condensed heterocyclic ring systems containing the masked 5,6-diaminopyrimidine or 2-aminopyrazine-3-carboxylic acid structural element may act as a potential pteridine synthon. Some acid-labile purines can be converted into pteridines through hydrolytic ring-opening of the imidazole moiety and subsequent condensation with a 1,2-dicarbonyl compound (51). More stable purine derivatives require more drastic reaction conditions and, as with uric acid, lead to a complex mixture of compounds in low yield (52). On the other hand, the imidazole ring can be labilized by quaternization to facilitate nucleophilic attack at C-8 and ring-opening. Thus, 9-methylguanine (80) was alkylated by ethyl bromoacetate to 7-ethoxycarbonylmethyl-9-methyl-guaninium bromide (81) which on treatment first with base and then with acid reacted via 82, 83, and 84 to give 8-methylisoxanthopterin (85) (53).

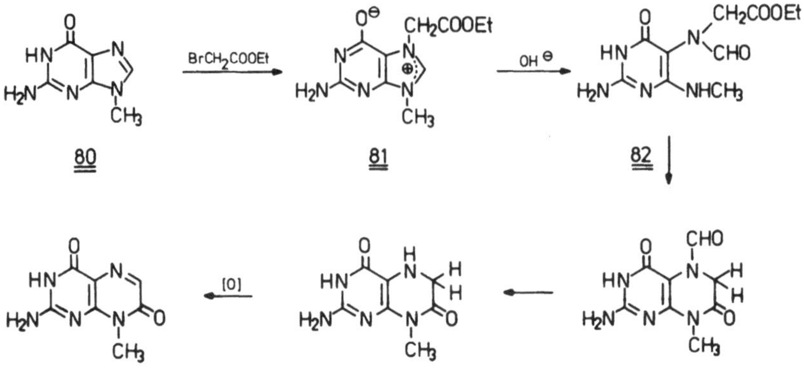

80 81 82 83 84 85

In an analogous conversion, isoxanthopterin-8-riboside resulted from guanosine (54).
7-Aminofurazano[3,4-d]pyrimidines represent latent 4,5,6-triaminopyrimidines due to the easy reductive cleavage of the N-O-N linkage. Reduction of 7-benzoylmethylamino-5-phenylfurazano[3,4-d]pyrimidine (86) followed by acid-catalized cyclization and oxidation leads to an unequivocal synthesis of 4-amino-2,6-diphenylpteridine (87) (55).

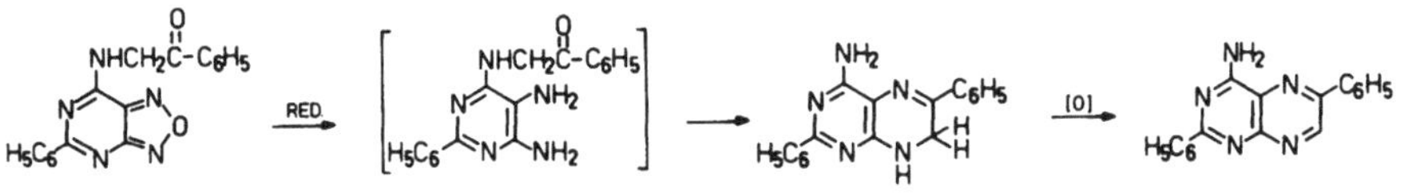

86 87

References

1. Albert, A.: Quart.Rev. 6, 197 (1952).
2. Gabriel, S., Colman, J.: Ber.Deut.Chem.Ges. 34, 1234 (1901).
3. Isay, O.: Ber.Deut.Chem.Ges. 39, 250 (1906).

4. Sachs, F., Meyerheim, G.: Ber.Deut.Chem.Ges. 41, 3957 (1908).

5. Zondler, H., Pfleiderer, W.: Liebigs Ann.Chem. 741, 64 (1970).

6. Purrmann, R.: Liebigs Ann.Chem. 548, 284 (1941).

7. Elion, G.B., Hitchings, G.H., Russell, P.B.: J.Am.Chem. Soc. 72, 78 (1950).

8. Pfleiderer, W.: Chem.Ber. 90, 2588; 2604; 2617, 2624 (1957).

9. Pfleiderer, W., Lohrmann, R.: Chem.Ber. 94, 2708 (1961).

10. Pfleiderer, W.: Chem.Ber. 92, 3190 (1959).

11. Pfleiderer, W., Rukwied, M.: ChemBer. 94, 1 (1961).

12. Albert, A., Brown, D.J.: J.Chem.Soc. 1953, 74.

13. Viscontini, M., Huwyler, S.: Helv.Chim. Acta 48, 764 (1965).

14. Sletzinger, M., Reinhold, D., Grier, J., Beachem, M., Tishler, M.: J.Am.Chem.Soc. 77, 6365 (1955).

15. Hultquist, M.E., Kuh, E., Cosulich, D.B., Fahrenbach, M.J., Northey, E.H., Seeger, D.R., Sickels, J.P., Smith, J.M., Angier, R.B., Boothe, J.H., Hutchings, B.L., Mowat, J.H., Semb, J., Stokstad, E.L.R., SubbaRow, Y., Waller, C.W.: J.Am.Chem.Soc. 70, 23 (1948).

16. Viscontini, M., Provenzale, R., Ohlgart, W., Mallevialle, J.: Helv.Chim. Acta 53, 1202 (1970).

17. Timmis, G.M.: Nature (London) 164, 139 (1949).

18. Felten, D.G.I., Timmis, G.M.: J.Chem.Soc. 1954, 2881.

19. Felten, D.G.I., Osdene, T.S., Timmis, G.M.: J.Chem.Soc. 1954, 2895.

20. Dick, G.P.G., Wood, H.S.C., Logan, W.R.: J.Chem.Soc. 1956, 2131.

21. Weinstock, J., Dunoff, R.Y., Sutton, B., Trost, B., Kirkpatrick, J., Farina, F., Straub, A.S.: J.Med.Chem. 11, 549 (1968).

22. Spickett, R.G.W., Timmis, G.M.: J.Chem.Soc. 1954, 2887.

23. Osdene, T.S., Timmis, G.M.: J.Chem.Soc. 1955, 2036; 2038.

24. Osdene, T.S.: Pteridine Chemistry, ed. Pfleiderer, W., Taylor, E.C.; Pergamon Press, Oxford, 1964, p. 65.

25. Pfleiderer, W., Lohrmann, R.: Chem.Ber. 94, 2708 (1961).

26. Pachter, I.J.: Pteridine Chemistry, ed. Pfleiderer, W., Taylor, E.C.; Pergamon Press, Oxford, 1964, p. 47.

27. Pachter, I.J., Nemeth, P.E.: J.Org.Chem. 28, 1187 (1963).

28. Pachter, I.J., Nemeth, P.E., Villaini, A.J.: J.Org.Chem. 28, 1197 (1963).

29. Weinstock, J., Graboyes, H., Jaffe, G., Pachter, I.J., Snader, K., Karash, C.B., Dunoff, R.: J.Med.Chem. 11, 560 (1968).

30. Polonovski, M., Jerome, H.: C.R. Hebd.Seances Acad.Sci. 230, 392 (1950).

31. Boon, W.R., Jones, W.G.M., Ramage, G.R., J.Chem.Soc. 1951, 96.

32. Boon, W.R., Leigh, T.: J.Chem.Soc. 1951, 1497.

33. Lagowski, J.M., Forrest, H.S., Wood, H.C.S.: Proc.Chem. Soc. 1963, 343.

34. Andrews, K.J.M., Barber, W.E., Tong, B.P.: J.Chem.Soc. (C), 1969, 928.

35. Pfleiderer, W., Deiss, H.: Chem.Ber. 104, 780 (1971).

36. Matsuura, S., Sugimoto, T.: Bull.Chem.Soc.Jpn. 54, 2543 (1981).

37. Yoneda, F., Higushi, M.: J.Chem.Soc., Perkin Trans. 1, 1977, 2529.

38. Gabriel, S., Sonn, A.: Ber.Deut.Chem.Ges. 40, 4857 (1907).

39. Dick, G.P.G., Wood, H.S.C.: J.Chem.Soc. 1955, 1379.

40. Taylor, E.C., Vogel, O., Löffler, P.K.: J.Am.Chem.Soc. 81, 2479 (1959).

41. Dalacker, F., Steiner, G.: Liebigs Ann.Chem. 660, 98 (1962).

42. Wright, W.B., Smith, J.M.: J.Am.Chem.Soc. 77, 3927 (1955).

43. Albert, A., Ohta, K.: J.Chem.Soc. (C) 1970, 1540.

44. Albert, A., Ohta, K.: J.Chem.Soc. (C) 1971, 2357.

45. Taylor, E.C., Lenard, K.: J.Am.Chem.Soc. 90, 2424 (1968).

46. Taylor, E.C.: Chemistry and Biology of Pteridines, ed. Iwai, K., Akino, M., Goto, M., Iwanami, Y.; International Academic Printing Co., Tokyo, 1970, p. 79.

47. Taylor, E.C., Perlman, K.L., Swort, I.P., Sequin-Frey, M., Jacobi, P.A.: J.Am.Chem.Soc. 95, 6407 (1973).

48. Taylor, E.C., Perlman, K.L., Kim, Y.H., Sword, I.P., Jacobi, P.E.: J.Am.Chem.Soc. 95, 6413 (1973).

49. Taylor, E.C., Henrie, R.N., Portnoy, P.C.: J.Org.Chem. 43, 736 (1978).

50. Taylor, E.C., Jacobi, P.A.: J.Am.Chem.Soc. 96, 6781 (1974).

51. Albert, A.: Biochem.J. 65, 124 (1957).
52. Pfleiderer, W.: Chem.Ber. 92, 2468 (1959).
53. Eistetter, K., Pfleiderer, W.: Chem.Ber. 106, 1389 (1973).
54. Eistetter, K., Pfleiderer, W.: Chem.Ber. 107, 575 (1974).
55. Taylor, E.C., Cocuzza, A.J.: J.Org.Chem. 44, 302 (1979).

STEREOCHEMISTRY OF REDUCED PTERINES

Max Viscontini

Organisch-Chemisches Institut der Universität
Winterthurerstrasse 190
CH-8057 Zürich

I Introduction

The new use of hydrogenated pterines in hospitals arises some stereochemical problems which I will try to explain as simply as possible. The principal difficulty results from the fact that the actual shape of hydrogenated pterines is arranged in the three-dimensional space, but that the chemist has only two-dimensional sheets of paper at disposal to describe them. Of course, it is always possible to try to draw in perspective the molecule of a compound, but this kind of representation is sometimes very complicated and should be avoided whenever possible. Chemists then use two-dimensional projections, obligatory paired with conventions allowing the mental crossing from the two-dimensional to the three-dimensional space.

II Projection of planar pterines

The achiral molecule of pterine (1) is as planar as the molecule

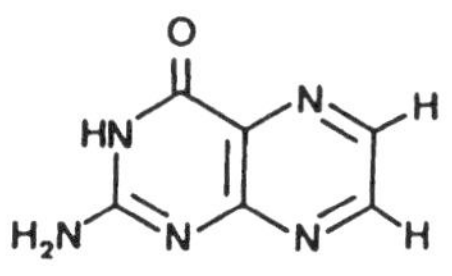

1

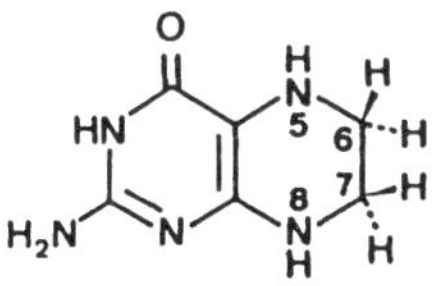

2

Biochemical and Clinical Aspects of Pteridines, Vol. 2

of benzene. Its projection on a sheet of paper does not show any difference to its spacial structure. Pterines with an achiral side chain attached at C(6) or C(7) behave like pterine 1. 6-Methyl-pterine, for example, is also planar, and the projection 3 reproduces exactly the shape of its molecule.

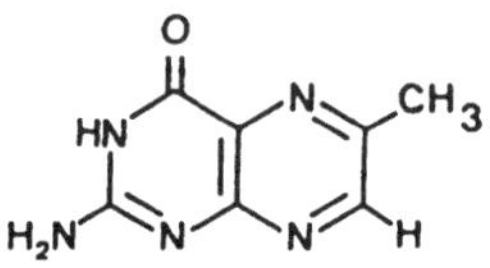

3

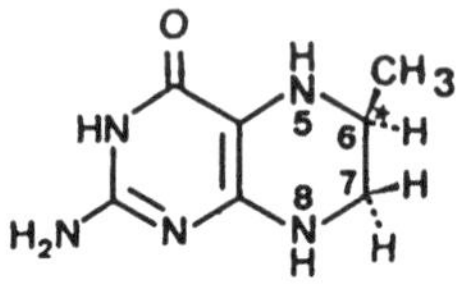

4

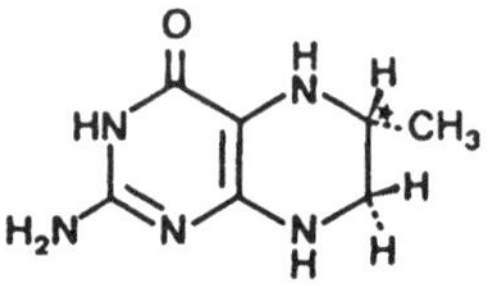

5

III Stereochemistry of reduced pterines

Planar pterines, once reduced, possess a three-dimensional structure which states the stereochemical problems I mentioned at the begin of this publication. We will treat them successively, starting with the easiest one and ending with the most difficult case of the natural products.

III 1. Stereochemistry of achiral 5,6,7,8-tetrahydro-pterine(2)

5,6,7,8-Tetrahydro-pterine (2) is like pterine (1) achiral, but no longer planar. The 4 H-atoms, which are linked to C(6) and C(7), are disposed two over the pterine-plane (straight lines) and two below it (dotted lines). However, because of the partial rotation about carbon-carbon and carbon-nitrogen single σ-bounds of the tetrahydro-pyrazine ring, the reduced pterine is no longer rigid and the 4 H-atoms can take a variety of positions in the space. Each of the resulting structure may be readily converted into another and is called conformation. Thus, in solution, the molecule 2 takes an infinity of conformations. The potential energy of these conformations is not equal and varies with the angles between the σ-bonds of the tetrahydro-pyrazine ring. Two of these conformations correspond to an energy minimum. The thermodynamics teaches us that those two must be more stable than all the others and that they represent almost 100% of all conformations. Both conformations are called conformers and are re-

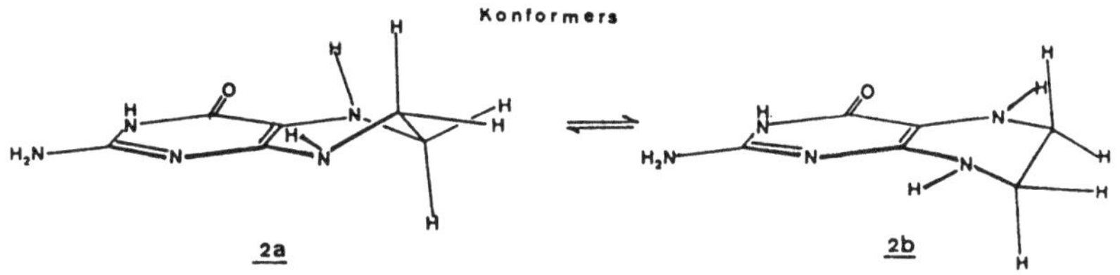

presented by the equilibrium reaction:

$$\underline{2a} \rightleftharpoons \underline{2b}$$

In this special case, the energy of the two conformers is equal. Consequently, the concentration of the two conformers in solution is the same. These conformers, in which thetetra-hydro-pyrazine ring takes a pseudo-chair conformation, cannot be represented in projection-formulas and are generally drawn in a stereoprojection (2a and 2b).

IV Stereochemistry of reduced, chiral pterines

Pterines, substituted at C(6), possess, after the reduction, a new stereofactor, because this operation introduces a three-dimensional, chiral center at C(6). Then, we have to differentiate between, firstly,pterines with a side chain which is not chiral, as, for example, 6-methyl-pterine (3) and, secondly, pterines with a C(6)-side chain which is chiral, as almost every natural, biologically active pterine (D-neopterine, L-monapterine, L-sepiapterine, L-biopterine, L-folic acid, etc).

IV 1. Stereochemistry of (RS)-6-methyl-5,6,7,8-tetrahydro-pterine

We have seen that 6-methyl-pterine is achiral and planar. It can be projected as 3. During its catalytic hydrogenation, a H_2-molecule is cis-added at each C=N double bond of the pyrazine ring. According to the symmetry of the molecule, this addition occurs on both sides of the ring (Re- and Si-addition at C(6)-atom). Consequently, two products 4 and 5 are produced, which are chiral at C(6) and enantiomer. Racemic crystals are easily obtained. To remember that they are composed of a racemic mixture, they are called, according to the sequence rule, (6RS)-6-methyl-5,6,7,8-tetrahydro-pterine (4 + 5). Each enantiomer 4 or 5 exists in solution, as we discussed for 2, in an equilibrium of two conformers:

$$\underline{4a} \rightleftharpoons \underline{4b}$$

and

$$\underline{5a} \rightleftharpoons \underline{5b}$$

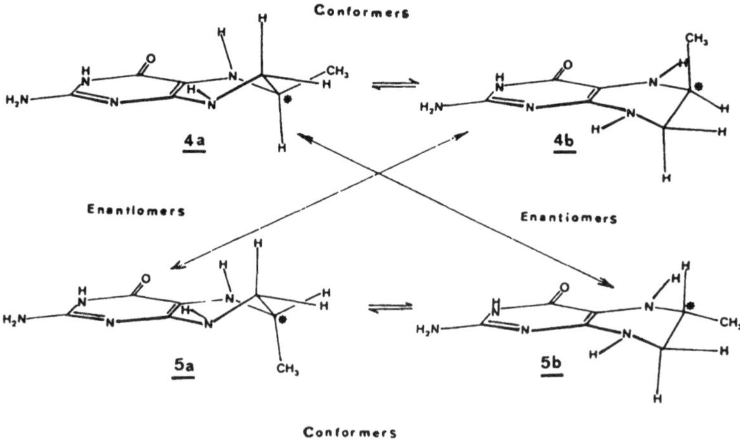

However, in this particular case,the energies of the two conformers are not equal, and the smallest one belongs to the conformer with the bigest group attached at C(6), here the methyl group, in pseudo-equatorial position. Therefore, in the equilibrium:

$$4a \rightleftharpoons 4b$$

99% ore more of 4 exists as conformer 4a, and in the equilibrium:

$$5a \rightleftharpoons 5b$$

99% or more of 5 exists as conformer 5b.Finally, it is interesting to note that the two most stable conformers 4a and 5b are enantiomers, and that the two less stable 4b and 5a are likewise enantiomers.

IV 2. Stereochemistry of reduced pterines with a C(6)-side chain which is chiral

IV 2.1. L-Biopterine

L-Biopterine (6) is also a pterine with a side chain attached at C(6), like 3, but in this case the chain itself is chiral. After catalytic reduction, we do not get two enantiomers as with 6-methyl-pterine 3, but two diastereomers 7 and 8 in different and variable concentrations. Each diastereomer is present

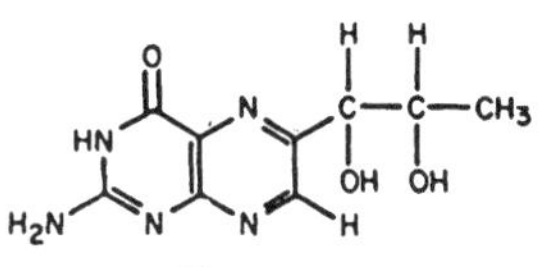

6

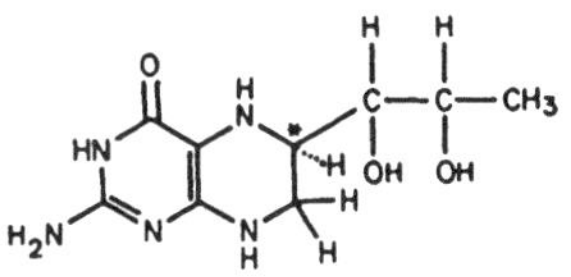

7

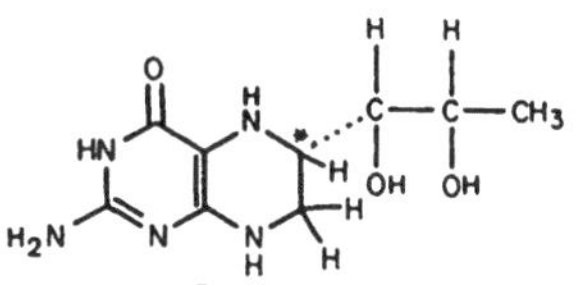

8

in solution as a mixture of two conformers. For each of them we have the following equilibrium:

$$7a \rightleftharpoons 7b$$

and

8a ⇌ 8b

7a and 8b being the most stable conformers with the chiral side chain in the pseudo-equatorial position. Both diastereomers dif-

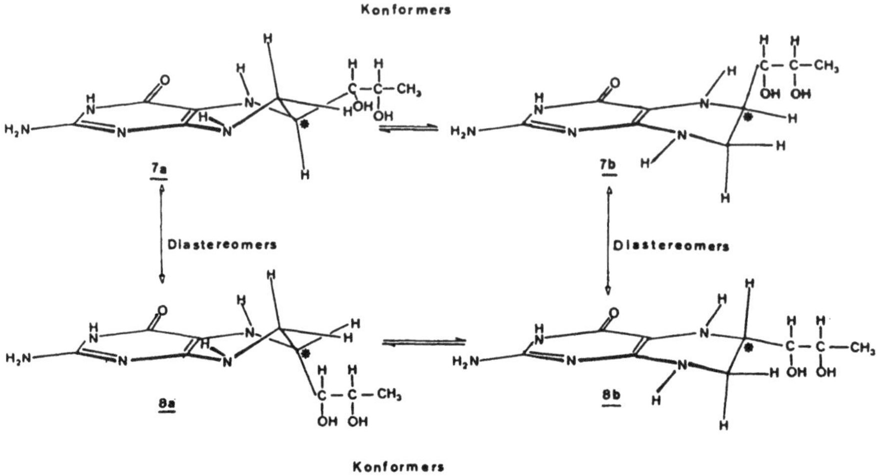

fer in physical and chemical properties. They can be distinguished by ^{1}H-, ^{13}C-NMR-spectroscopy or thin layer chromatography. It is,therefore, necessary to use a nomenclature corresponding to their different structural formulas. The chiral side chain does not undergo any configuration change during the catalytic hydrogenation, accordingly, the skeleton of both comcompounds may be futher called L-biopterine. A new configuration occurs at C(6) where a Re- or a Si-addition of H-atom takes place. According to the sequence rule, 7 is (6R)-5,6,7,8-tetrahydro-L-biopterine and 8 (6S)-5,6,7,8-tetrahydro-L-biopterine. The mixture of both diastereomers, obtained after catalytic reduction and used in the hospitals, must, therefore, be designated as (6RS)-5,6,7,8-tetrahydro-L-biopterine, without taking into account the concentration of each diastereomer in the mixture. Moreover, it is known that the proportion of the natural product reaches at least 2/3 of the total concentration. The proposed nomenclature is not univocal, first because two conventions are used simultaneously for the same product, secondly because the sequence rule itself is not univocal and assigns for the same configuration at C(6) in tetrahydropterine the symbol (6R) or (6S), according to the place of the atoms in the periodic table and in the side chain of the molecule.

IV 2.2. L-Folic acid

All these considerations are valid for the mixture (6RS) of 5,6,7,8-tetrahydro-L-folic acids (10 + 11) obtained after reduction of 7,8-dihydro-L-folic acid with $NaBH_4$. An important difference should be noted: neither the ^{13}C-NMR-spectroscopy nor the thin layer chromatography show the presence of the two

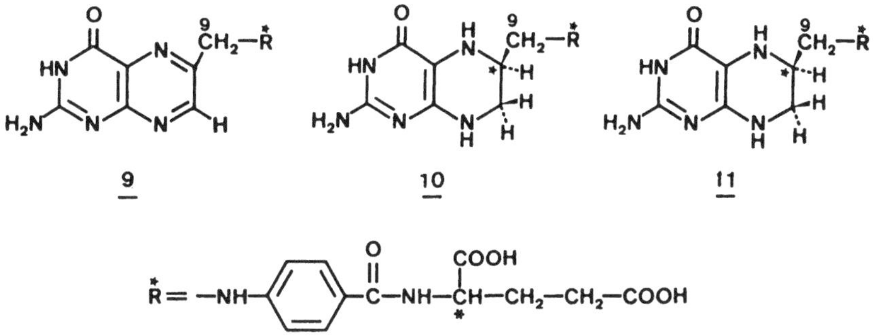

diastereomers (6R) or (6S) in the mixture [1],[2]. They could be seen by ^{13}C-NMR-spectroscopy only after treatment of 10 + 11 with formaldehyd and obtainment of (6RS)-methylene-5,6,7,8-tetrahydro-L-folic acid [3].

V Conformational analysis of the tetrahydropterines

The synthesis of tetrahydropterines is an exciting job, but the synthesis itself does not resolve all problems resulting from the stereochemistry of the substances. It is important to know afterwards the conformation of the products obtained, especially which conformer prevails in a solution. ^{1}H-NMR-spectroscopy is often employed to resolve this problem, particularly when the side chain at C(6) is achiral.

V 1. Tetrahydropterines with achiral C(6)-side chain

If we consider the racemic mixture of (6RS)-6-methyl-5,6,7,8-tetrahydro-pterine (4 + 5) in solution, we will find a large and equal concentration of both conformers 5b + 4a, with the methyl group in pseudo-equatorial position, and a small and equal concentration of the conformers 4b + 5a, with the methyl group in pseudo-axial position. In the first case, two of the three H-atoms at C(6) and C(7) represented in the Newman projection are antiperiplanar (trans-diaxial, dihedral angle = 180°); in the second case, all three H-atoms are "gauche"

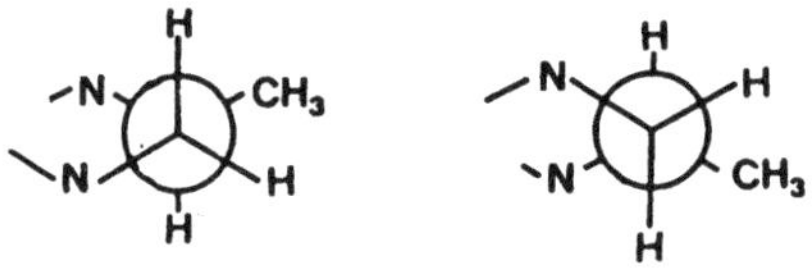

4a 5b

Newman projections of 4a and 5b

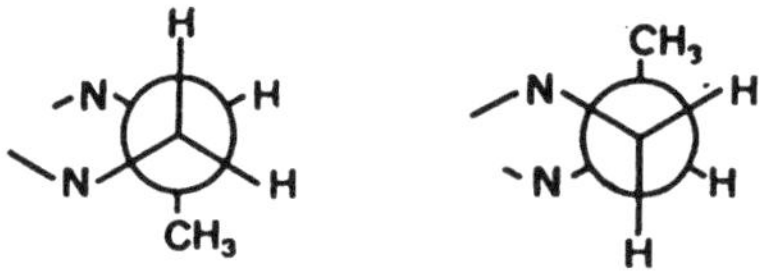

5a 4b

Newman projections of 5a and 4b

(dihedral angle = 60°). For each pair of conformers the coupling J for the dihedral angle of H-atoms, which are axial or "gauche", will be different. On the basis of Karplus calculations it is known that for a dihedral angle of 0° (syn-periplanar H-atoms, very instable conformation), J lies between 8-14Hz, diminishes to 1-3 Hz for "gauche" H-atoms (dihedral angle of 60° as for 4b and 5a), is almost zero for a dihedral angle of 90°, and reaches again 8-14 Hz for antiperiplanar, trans-diaxial H-atoms (dihedral angle = 180°, as for 4a and 5b). The application of this method allowed us to demonstrate that the mixture 4 + 5 is practically present in solution merely as 4a + 5b [4].

V 2. Tetrahydropterines with a chiral C(6)-side chain

If the side chain at C(6) is chiral and the ^1H-NMR-spectra of both diastereomers are similar as, for example, for the (6RS)-5,6,7,8-tetrahydro-L-folic acids (10 + 11), a similar argumentation can be used. Unluckily, the ^1H-NMR-spectra of this type of mixture are difficult to interpret. Our efforts in this direction with (6RS)-5,6,7,8-tetrahydro-L-biopterines (7 + 8) did not succeed. Better results were obtained with (6RS)-5,6,7,8-L-folic acids (10 + 11). Elimination of the chemical shifts of both H-atoms at C(9) through ^1H-^2H exchange allowed a good interpretation of the ^1H-NMR-spectrum and showed that the C(6)-H atom in both conformers is axial. Their corresponding structures

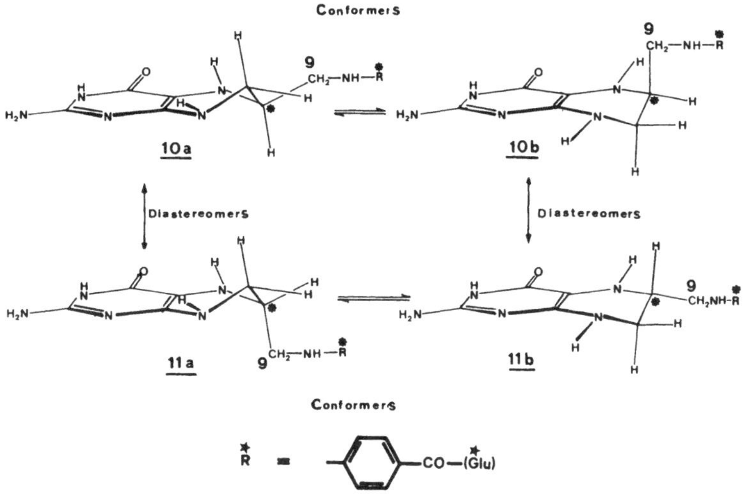

are, therefore, 10a and 11b. However, the method does not permit to decide about the size of the concentration of each diastereomer in the mixture nor about the structure of the natural product, 10a or 11b [1],[2],[3].

VI X-ray diffraction analysis

The crystal structure analysis is the best method to determine, for a racemic mixture, the conformation, and, for diastereomers, the configuration of tetrahydropterines in solid state. The most important condition required is to have good crystals at disposal. Generally, racemic mixtures of enantiomers fulfil this condition, however, mixtures of diastereomers like 7 + 8 or 10 + 11 do not. In this case, the diastereomers have to be separated one from the other. The most efficient method is the use of fractional crystallisation.

VI 1. Absolute configuration of the natural tetrahydro-L-biopterine

We did not succeed in resolving directly 7 from 8, but it is well known that a good method to separate diastereomeric alcohols is to convert them into esters, which crystallize better. Therefore, we tried the acetylation of the mixture 7 + 8. After some attempts we found that (6R)- and (6S)-pentaacetyl-5,6,7,8-tetrahydro-L-biopterines show a great difference of solubility in many organic solvents [5]. The resolution occured according to the scheme on the next page. L-Biopterine (6) was first converted into triacetate 12 and then reduced to (6RS)-1-O',2-O',2-N-triacetyl-5,6,7,8-tetrahydro-L-biopterine (13). A new acetylation led to a mixture of diastereomers (6RS)-1-O',2-O',2-N,5,8-

pentaacetyl-5,6,7,8-tetrahydro-L-biopterines (14) which could be resolved in the pure (6R)- and (6S)-diastereomers 15 and 16.

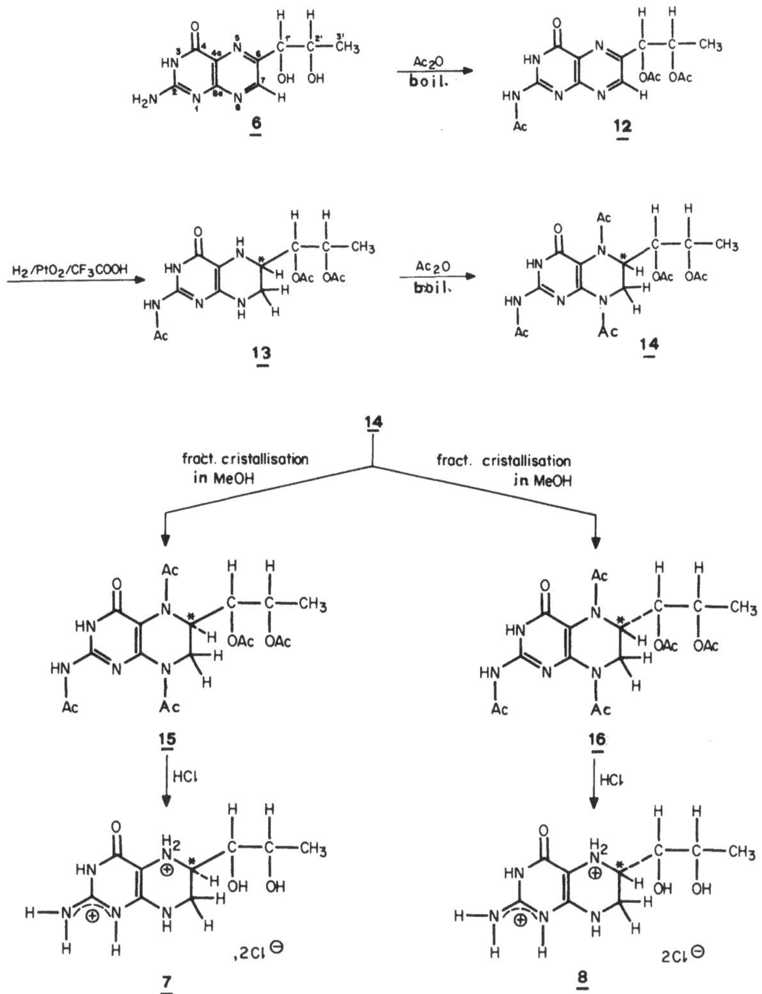

One of it gave fine crystals which could be analyzed [6]. Because X-ray analysis cannot generally distinguish, as ^{1}H-NMR-spectroscopy, between both possible configurations at C(6), the known L-configuration of the side chain was chosen for the representative structure of the molecule, and the configuration at the C(6)-atom was found to be (6R), giving structure 15 for that diastereomer. The stereoscopic representation of 15 (fig.1)

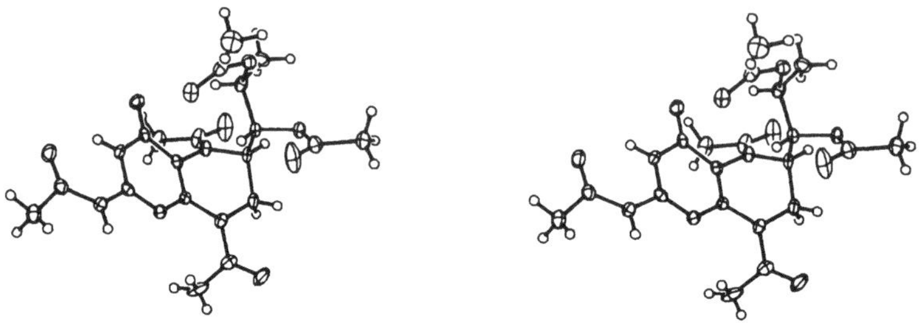

Figure 1: Stereoview of (6R)-pentaacetyl-tetrahydro-L-biopterine (15) and vibrational ellipsoids (50% probability level). The H-atoms are designed with arbitrary radius.

shows a pseudo-axial position of the side chain instead of a pseudo-equatorial,as it should be expected. This result agrees with the previous work of our laboratory, with ^{1}H-NMR-spectroscopy and X-ray diffraction analysis, in which we showed that 6-alkyl-N(5)-acyl-5,6,7,8-tetrahydro-pterines have the 6-side chain in a pseudo-axial position [7]. After HCl-hydrolysis, 15 gives the (6R)-5,6,7,8-tetrahydro-L-biopterine (7), $[\alpha]_D^{22}=-3,8^o$ (c=1,0 ; 0,1 N HCl)[5], which corresponds with the natural cofactor of the phenylalanine hydroxylase [8].

VI 2. Absolute configuration of the natural tetrahydro-L-folic acid

We could resolve any mixture of diastereomeric (6RS)-tetrahydro-L-folic acid synthetized in our laboratory, but J.C. Fontecilla-Camps et al. [9]succeeded in obtaining a small amout of pure

(-)-5-formyl-5,6,7,8-tetrahydro-L-folic acid during a large scale production of the corresponding (6RS)-mixture. They found that the product had the natural configuration at C(6) because it was twice as active as the (6RS)-mixture in supporting bacterial growth. The crystalline substance could be twice analyzed. As for the pentaacetyl-tetrahydro-L-biopterine, the X-ray results did not show any difference between the two possible C(6)-epimeric structures, but the L-configuration of glutamic acid in the side chain was chosen for the stereoview, and the analysis indicated a (6R)-configuration for C(6), which corresponds to the C(6S)-configuration for the 5,6,7,8-tetrahydro-L-folic acid part of the molecule. The authors concluded that, since all tetrahydro-L-folic acid derivatives in biological systems are interconvertible through enzymatic and chemical reactions, the configuration (6R) at C(6) of the 5-formyl-derivative is the one that would be expected for other natural compounds of tetrahydro-L-folic acid [10].

It is interesting to note that the chirality of the cofactor of phenylalanine hydroxylase and of tetrahydro-L-folic acid coenzymes is the same at the C(6)-atom, and that this chirality is formed in both cases enzymatically by a 7,8-dihydropterine-reductase, probably by the same reaction mechanism.

References

1. Frick, W., Weber, R., Viscontini, M.: Helv. Chim. Acta 57, 2658-2661 (1974).
2. Furrer, H.J. Bieri, J.H. Viscontini, M.: Helv. Chim. Acta 61, 2744-2751 (1978); Khalifa, E. Bieri, J.H., Viscontini, M.: ibid. 62, 1340-1344 (1979).
3. Kalbermatten, R., Städeli, W., Bieri, J.H., Viscontini, M.: Helv. Chim. Acta 64, 2627-2635 (1981).

4. Weber, R., Viscontini, M.: Helv. Chim. Acta 58, 1772-1780 (1975); Ganguly, A.N., Bieri, J. H., Viscontini, M.: ibid. 64, 367-372 (1981).

5. Ganguly, S.N., Viscontini, M.: Helv. Chim. Acta 65, 1090-1093 (1982).

6. Prewo, R., Bieri, J.H., Ganguly, S.N., Viscontini, M.: Helv. Chim. Acta 65, 1094-1099 (1982).

7. Weber, R., Viscontini, M.: Helv. Chim. Acta 60, 161-165 (1977); Bieri, J. H., Viscontini, M.: ibid. 60, 447-453 (1977) and unpublished results.

8. Furrer, H.J., Bieri, J.H., Viscontini, M.: Helv. Chim. Acta 61, 2577-2580 (1979); Schircks, B., Bieri, J.H., Viscontini, M.: ibid. 60,211-214 (1977); 61, 2731-2738 (1978); Bailey, S.W., Ayling, J.E.: J. Biol. Chem. 253, 1598-1605 (1978).

9. Fontecilla-Camps, J.C., Bugg, C.E., Temple, C. Jr., Rose, J.D., Montgomery, J.A., Kisliuk, R.L.: J. Amer. Chem. Soc. 101, 6114-6115 (1979).

10. Blackley, R.L.: "The Biochemistry of Folic acid and Related Pteridines" Neuberger, A., Tatum, E.L. Eds., American Elesvier, New-York 1969, pp 1-569.

QUANTUM CHEMICAL ASPECTS OF PTERINS IN COMPARISON WITH GUANINE AND DEAZAGUANINE DERIVATIVES

Dedicated to Prof. Dr. E. Schauenstein on the occasion of his 65th birthday.

G. Reibnegger, D. Fuchs, A. Hausen, O. Knosp, H. Wachter
Institute for Medical Chemistry and Biochemistry, University of Innsbruck, Fritz Pregl-Str. 3, A-6020 Innsbruck, Austria.

Introduction

There is only one biochemical function of unconjugated pteridines well established hitherto - the cofactor role of tetrahydrobiopterin in hydroxylation reactions of amino acids (1). However, there are also some studies which point to the possibility that certain unconjugated pterins might be involved in chemical interactions which could be understood by the assumption of similar binding behaviour that governs the chemical characteristics of the usual bases of nucleic acids. This idea is additionally supported by the fact that pterins have a 2-amino-4-oxo-3,4-dihydropyrimidine moiety in common with guanine.

For instance, the inhibitory effect of exogenous isoxanthopterin on the nucleic acid synthesis in developing eggs of *Oncopeltus fasciatus* has been reported (2). Furthermore, a strong enhancement of the in vitro fluorescence of isoxanthopterin when being associated with DNA has been observed in this system (2).

Pterins are potent inhibitors of the enzyme guanine-tRNA-transglycosylase which posttranscriptionally removes guanine from the wobble position of the anticodon of the Q-family of tRNAs and replaces it with either guanine, again, or with the 5-deazaguanine derivative queuine (3,4; see also H. Kersten, this volume).

Pterins seem to be involved in the regulation of growth and differentiation. A proposal has been made suggesting the formation of hydrogen bonds between pterins and normal nucleic acid bases (5,6). Thus, pterins might act directly on genetic information during transcription or translation.

These and other studies seem to justify extensive work, both theoretical and experimental, to provide intimate insight into the structural features of pterins and in particular into their relationships to structurally similar biomolecules.

Today, quantum chemistry offers the facilities of theoretically estimating molecular properties with a reasonably high degree of accuracy. This possibility is especially useful in providing information concerning properties that are not amenable easily or directly by means of experiments due to the complexity of real chemical and in particular biochemical systems.

The quantum chemical tools do not, of course, solve biochemical questions immediately, but they help to elucidate the electronic structure of the molecules of interest. Thus, they provide, at least in a semiquantitative sense, a more fundamental understanding of chemical features of molecules.

In contrast to the bases of nucleic acids, there are only a few studies available reporting on the application of quantum chemical methods for problems in pteridine chemistry. Furthermore, the studies published hitherto have been restricted to so-called "empirical" and "semiempirical" methods: In order to reduce the extensive computational demands of quantum chemical calculations, certain "empirical" parameters are introduced into the formalism that are adjusted by means of various empirically measured properties of the molecules. In contrast to the "ab initio" methods, this procedure avoids many time and storage consuming steps of the calculations. But one has to pay with the loss of essential physical exactness which causes a relatively high degree of uncertainty of interpretation.

An early example of quantum chemical computations on properties of pteridine derivatives is the application of the empirical HUECKEL-method for the estimation of spin densities of semiquinone radicals of tetrahydropterins (7). The semiempirical CNDO (Complete Neglect of Differential Overlap) scheme has been shown to yield a satisfying theoretical explanation for NMR results observed in the case of certain pterin derivatives (8).

The present contribution reports on first results of quantum chemical ab initio computations dealing with the structural relationships within a series of derivatives of pterin, guanine and deazaguanine (Table 1).

TABLE 1

MOLECULAR SPECIES STUDIED

Guanine

5-Deazaguanine
5-Methyl-5-deazaguanine

Pterin
6-Methylpterin
6-Hydroxymethylpterin
Pterin-6-aldehyde
6-Aminopterin

7,8-Dihydropterin

Methods

All quantum chemical calculations have been performed within the framework of the HARTREE-FOCK-ROOTHAAN formalism which, to date, is the most important and most frequently used "ab initio-"all electron-model (9,10). It is impossible here to give a detailed description of the mathematics involved in the formalism. However, a short outline of the basic principles of the method seems to be of some value in improving understanding and interpreting the results [Details are given in many textbooks of theoretical chemistry, e.g. (11)].

- The Hartree-Fock-Roothaan method computes an approximative wave function, which describes the behaviour of the nuclei and electrons of a molecule, according to the fundamental SCHROEDINGER equation, PAULI's exclusion principle and the BORN-OPPENHEIMER approximation. The theoretical frame is restricted to the model of independent particles: The quantum mechanical N-particle system is approximated by N one-particle systems, the mathematical solution of which is by far easier.

- The exact molecular wave function is approximated by a linear combination of "orbitals" (i.e. wave functions) of the atoms of the molecule. The coefficients of this linear combination are determined by an iterative procedure using the "variation principle": The best combination of atomic to molecular orbitals can be proven to yield the minimal total energy of the respective quantum chemical system, that can be obtained at all within this model.

- Also the atomic orbitals themselves, which are either unknown or fairly complicated functions, are approximated by a linear combination of "basis functions": These are members of a family of simpler and mathematically more conveniently treatable

functions of a suitable type. The choice of the basis influences, of course, the computational effort as well as the accuracy of the results.
From the quantum chemical point of view, in contrast to the biochemical one, pterins represent large chemical entities. Hence, the basis set has been reduced to a "minimal" size: There is at least one atomic orbital available for each electron. For computational reasons, a so-called GLO (Gaussian Lobe Orbital) basis set has been chosen, which has been shown to be well suited for similar systems (e.g. 12,13).

Although, in principle, quantum chemical methods allow a theoretical estimation of molecular geometries, this possibility is of practical value only in the case of very small molecules. For this reason, Xray diffraction data have been employed to determine the geometries of the molecules included in this study (14,15). In accordance with other authors (8), the geometry of the pyrimidine moiety has been held fixed throughout the series.

As is well known, quantum chemical computations yield, in principle the molecular wave functions of the molecules studied. This function is, however, an abstract pattern in a multidimensional "configuration space". It is only the square of this function which has a real physical meaning: It is, formally, a probabilistic distribution function defining the behaviour of the physical state which is described by the wave function. In quantum chemistry, there are two attractive ways of transforming the wave function into chemically meaningful information:

a) MULLIKEN population analysis

This approach has been developed by Mulliken already in the pioneer days of quantum chemistry (16). It is a mathematically fairly simple, but very powerful algorithm which performs the computation of the distribution of the total electronic density of a molecule among the atoms. For each atomic center, the procedure yields one single, compact number representing the partial charge of this atom within the molecule studied. Due to some arbitrariness in the definition of these MULLIKEN populations, the result of the analysis is dependent on the choice of the basis set to some extent, and, unfortunately, the populations are not physical observables in the pure sense of the word: There is no unambiguous way of dividing the total electron density of a molecule, which is indeed a physical observable, into certain portions associated with single atoms. Hence, the absolute numbers resulting from such a population analysis must not be taken too seriously. Nevertheless, the approach is very useful in explaining differences between similar molecules and in the investigation of relative trends of the charge distribution within series of structurally related compounds.

b) Electron density and electron density difference diagrams

The knowledge of the molecular wave function allows the calculation of the electron density function for each spatial position. Thus, one can compute this function for a dense grid lying within a plane which passes through the molecule in an arbitrarily selected manner. Then, by means of modern computer-assisted graphical devices, a three-dimensional (3D) plot of the electron density function along that grid can be generated. Similarly, the difference between the electron densities of two molecules can be computed and represented in just the same way. The latter possibility is obviously of particular interest for it provides a very vivid picture of the effects which are caused by certain variations in the chemical structure (17).

Contrary to population analysis, the calculation of 3D-plots of electron densities requires quite a lot of computational effort. However, this method provides information on the detailed behaviour of a physical observable. It has, therefore, a more correct and direct meaning than population numbers. Fortunately electron densities are not strongly dependent on the basis set actually chosen (17). The method is particularly suited for a detailed analysis of planar molecules. The molecular plane itself, or parallel planes of it, yield, in most cases, a maximum of chemically relevant information.

Computer and Programs

All quantum chemical calculations have been performed at the CDC cyber 74 computer of the University of Innsbruck. The quantum chemical program package used is described in detail (18). The 3D-plots have been generated by using standard CALCOMP graphical software (program "THREE-D").

Results and Discussion

a) MULLIKEN population analysis results

The results of the population analysis of the atoms which constitute the pyrimidine moiety of the molecules studied, are shown in table 2. The restriction of the analysis to this part of the molecules has been made due to the fact that just these atoms are likely to be involved in similar hydrogen bonds as are known from the chemistry of nucleic acid bases. The data are represented in a more compact manner in figure 1. The mean values of the atomic partial charges and the respective variation coefficients are given. Obviously the predominant influence of structural alterations in the remaining (not depicted) part of the molecules is experienced by the atoms combining

Table 2: Atomic partial charges of the atoms constituting the pyrimidine portion of the molecules resulting from a MULLIKEN population analysis (in atomic charge units)

	N(1)	C(2)	N(2')	H(a)	H(b)	N(3)	H(3)	C(4)	O	C'	C"
Guanine	-.689	1.044	-.927	.388	.429	-1.146	.460	.753	-.195	-.544	.594
5-Deazaguanine	-.695	1.051	-.930	.388	.427	-1.141	.452	.833	-.217	-.908	.713
5-Methyl-5-deazaguanine	-.696	1.050	-.931	.387	.427	-1.141	.451	.836	-.226	-.822	.711
Pterin	-.666	1.050	-.913	.389	.436	-1.144	.463	.721	-.164	-.418	.534
6-Methylpterin	-.668	1.049	-.915	.389	.435	-1.144	.462	.720	-.168	-.413	.534
6-Hydroxymethylpterin	-.669	1.050	-.915	.389	.435	-1.144	.462	.720	-.164	-.410	.535
Pterin-6-aldehyde	-.664	1.053	-.911	.392	.437	-1.144	.465	.721	-.158	-.412	.547
6-Aminopterin	-.668	1.038	-.923	.384	.431	-1.144	.460	.715	-.174	-.364	.483
7,8-Dihydropterin	-.725	1.072	-.916	.393	.429	-1.146	.458	.728	-.195	-.524	.743
mean	-.682	1.051	-.920	.389	.432	-1.144	.459	.750	-.185	-.535	.599
standard deviation	.021	.009	.008	.003	.004	.002	.005	.049	.025	.197	.097
variation coefficient (%)	3.0	0.9	0.8	0.7	0.9	0.2	1.0	6.6	13.4	36.8	16.2

Figure 1: Mean values and variation coefficients of the atomic partial charges

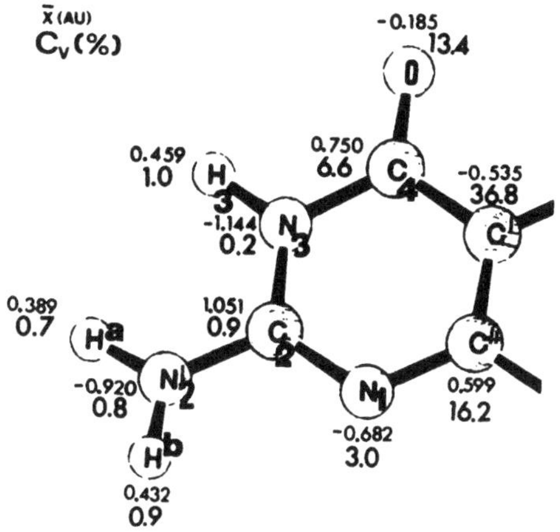

the two rings, C' and C". The atoms N(1), C(4) and O are moderately affected; the remainder of the pyrimidine moiety is scarcely influenced.

The results suggest the need for a closer inspection of the actual data of the atoms N(1), C(4), O, C' and C". For this purpose, the atomic partial charges of interest are given in figure 2. The figure has been constructed in such a way that the mean value and the standard deviation of each variable coincide. This procedure generates an optimum of comparability. The figure reveals 4 interesting features:

- Guanine is situated close to the mean with respect to all 5 variables
- Pterin and the 6-substituted derivatives form a relatively dense cluster in the range of partial charges smaller than the mean values. The largest deviations within this group are observed with 6-aminopterin
- The 5-deazaguanine and 5-methyl-5-deazaguanine species represent a cluster of their own, situated in the range of

FIGURE 2 : ATOMIC PARTIAL CHARGES OF N(1), C(4), O, C', C"
(atomic charge units)

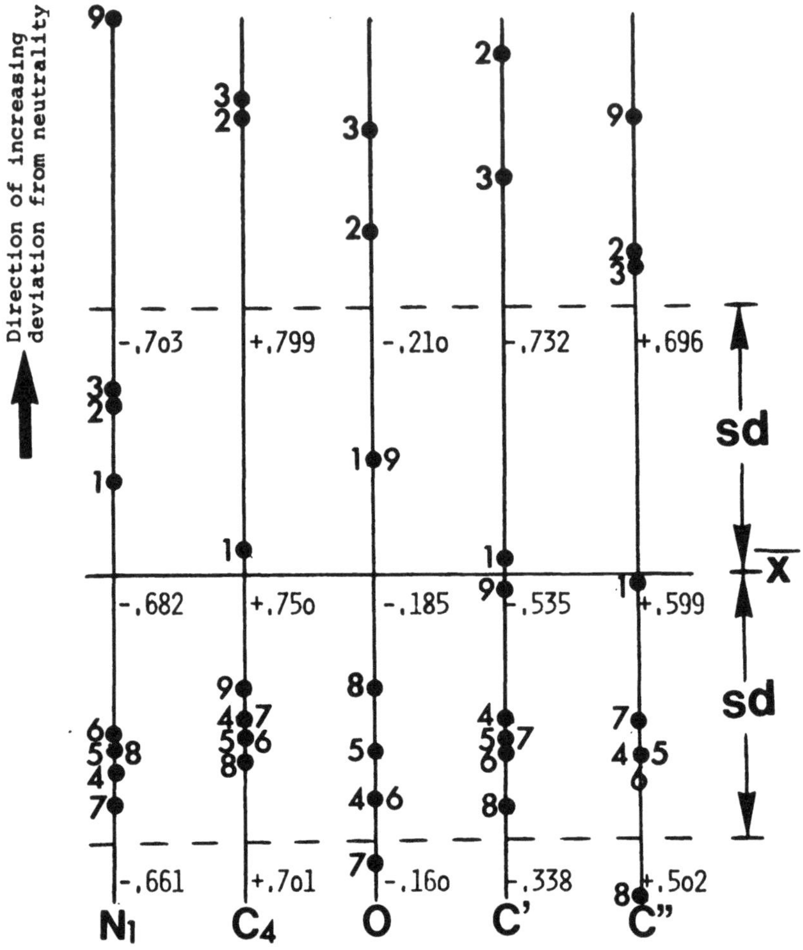

Legend: sd = standard deviation, $\bar{x}$ = mean value
1 = guanine, 2 = 5-deazaguanine, 3 = 5-methyl-5-deazaguanine, 4 = pterin, 5 = 6-methylpterin, 6 = 6-hydroxymethylpterin, 7 = pterin-6-aldehyde, 8 = 6-aminopterin, 9 = 7,8-dihydropterin

large deviations from neutrality.

- 7,8-Dihydropterin shows a more complicated behaviour with respect to the 5 variables. This indicates more fundamental differences in the electronic distribution within the pyrimidine ring, compared with the other compounds.

By employment of simple methods from general cluster analysis (19), one can represent the behaviour of the molecules with respect to the 5 partial charges in yet another form: the 5 variables (i.e. atomic partial charges) can be assumed to span a 5-dimensional (5D) space. Each molecule can be represented in this space by a point defined by the actual values of the partial charges of N(1), C(4), O, C' and C" of this molecule. Now, according to elementary assumptions of cluster analysis and pattern recognition (19), the "distance" between two points in the 5D-space can be regarded as a measure for the difference of the electron density distribution among the 5 atoms considered. Table 3 shows the distance matrix of the molecules studied in the 5D-space mentioned. Each matrix element a_{ij} indicates the "distance" between molecule i and j. Dense aggregates of molecules are indicated in the table. Obviously, the distance matrix reflects the 4 above mentioned features (Fig. 2) very well.

A quantity of practical importance is the polarity of chemical bonds. It is easily obtainable by computing the difference between the two atoms forming the bond under consideration. (As has already been mentioned, only an estimation of relative trends can be made, due to the general properties of population analysis results).

Table 4 contains the resulting bond polarities for the bonds between O-C(4), C(4)-C', C"-C', C"-N(1). In analogy to the foregoing results, the polarity of these bonds increases in the order pterin (and derivatives) guanine 5-deazaguanine.

Table 3: Distance matrix of the molecules in a 5D-space spanned by the 5 partial charges of N(1), C(4), O, C' and C" in atomic charge units x 10^{-3}.

i \ j	1	2	3	4	5	6	7	8
2	392							
3	314	87						
4	148	537	461					
5	151	542	465	6				
6	154	544	468	9	5			
7	150	539	462	16	16	15		
8	217	605	529	75	71	70	82	
9	156	401	321	243	245	245	237	311

1 = guanine, 2 = 5-deazaguanine, 3 = 5-methyl-5-deazaguanine, 4 = pterin, 5 = 6-methylpterin, 6 = 6-hydroxymethylpterin, 7 = pterin-6-aldehyde, 8 = 6-aminopterin, 9 = 7,8-dihydropterin

Again, the difference between pterin and guanine is smaller than the difference between guanine and 5-deazaguanine with respect to the polarity of the bonds studied.

Table 4: Relative polarities of the chemical bonds between C(4)-O, C(4)-C', C"-C' and C"-N(1), obtained by the difference between the respective atomic partial charges (in atomic charge units)

	C(4)-O	C(4)-C'	C"-C'	C"-N(1)
Guanine	.948	1.297	1.138	1.283
5-Deazaguanine	1.050	1.741	1.621	1.408
5-Methyl-5-deazaguanine	1.062	1.658	1.533	1.407
Pterin	.885	1.139	.952	1.200
6-Methylpterin	.888	1.133	.947	1.202
6-Hydroxymethylpterin	.884	1.130	.945	1.204
Pterin-6-aldehyde	.879	1.133	.959	1.211
6-Aminopterin	.889	1.079	.847	1.151
7,8-Dihydropterin	.923	1.252	1.267	1.468

b) Electron density and electron density difference diagrams

The results of the foregoing analysis can be visualized and extended by graphical representation of the electron density difference functions, as has been described. To illustrate the method, figure 3 shows the electron density function of pterin in a logarithmic scale. The molecular plane has been chosen for reference. Two features can be seen: sharp peaks indicate the sudden increase of the electron density in the immediate surroundings of the "heavy" atoms C, N and O. These peaks reflect the geometrical structure of pterin. They contain essentially the inner shell electrons of the atoms.

Figure 3: Electron density diagram of pterin in a logarithmic scale

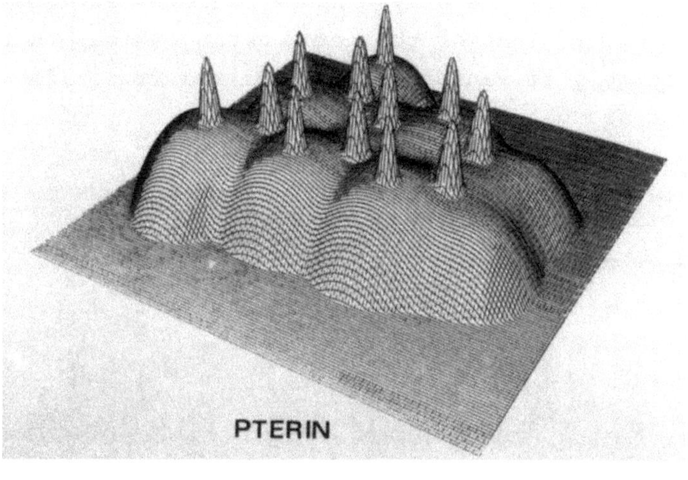

As has been explained, electron density difference plots are even better suited for the visualization of the differences between similar molecules. Contrary to electron density plots, a linear scale is favourable for the difference diagram.

Figure 4 depicts the electron density difference between pterin and 6-methylpterin. The figure clearly shows that the substi-

Figure 4: Electron density difference between pterin and 6-methylpterin (cut along the C(6)-C(7)-bond)

tution of pterin with a methyl group at position 6 has practically no influence on the electronic situation within the pyrimidine moiety: there are only minimal differences located in the inner shell regions of the heavy atoms, which can be expected to be only of vanishing chemical relevance (The diagram has been cut along the C(6)-C(7)-bond).

Figure 5: Electron density difference between guanine and pterin (pyrimidine part only)

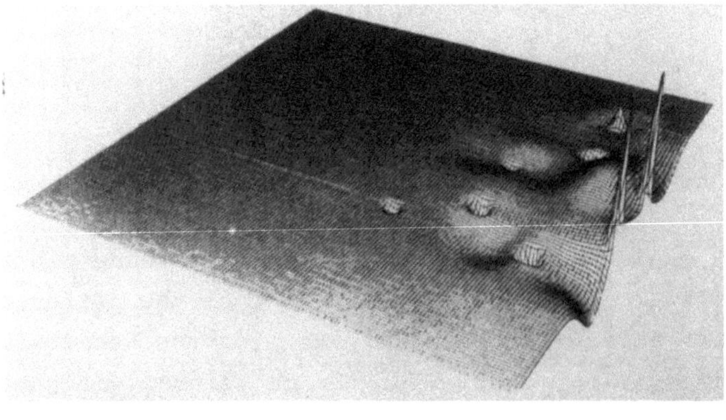

Figure 6: Electron density difference between guanine and 6-aminopterin (pyrimidine part only)

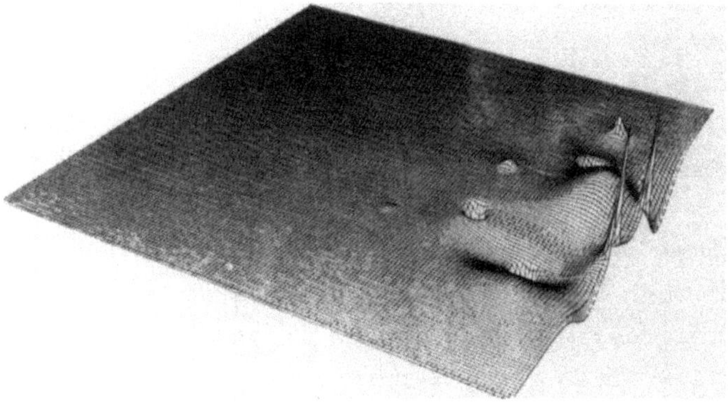

The difference between the electron density functions of guanine and pterin (fig. 5) and between guanine and 6-aminopterin (fig. 6) has practically the same functional course. There are only slight quantitative differences, located preferentially in the region of the innner shell electrons. This similarity is in good accordance to the population analysis results.

Figure 7: Electron density difference between guanine and 7,8-dihydropterin (pyrimidine part only)

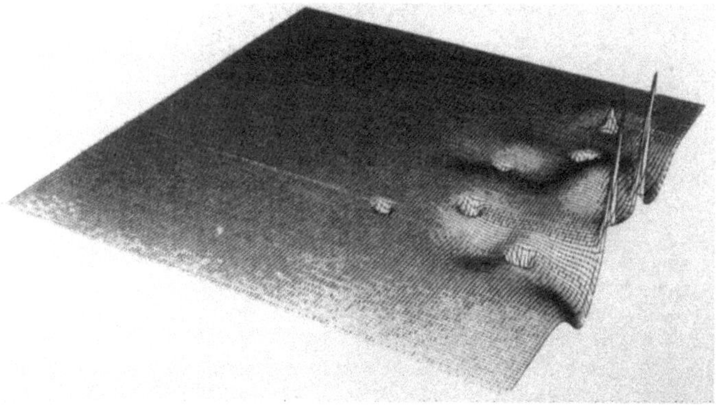

The difference between guanine and 7,8-dihydropterin (fig. 7) with respect to the electron densities exhibits quite distinct features, compared to the oxidized pterin species: the functional behaviour of the difference function is of another type, in particular in the surroundings of N(1). The inner shell as well as the valence region are affected. Again, this observation supports the population analysis findings.

The functional characteristics of the electron density difference function between guanine and 5-deazaguanine (Fig. 8) confirm very impressively the results of the population analysis: The inner as well as the outer regions of the electron shells exhibit - at nearly every location within the pyrimidine system - just the reverse behaviour compared to pterin and derivatives.

Figure 8: Electron density difference between guanine and 5-deazaguanine (pyrimidine part only).

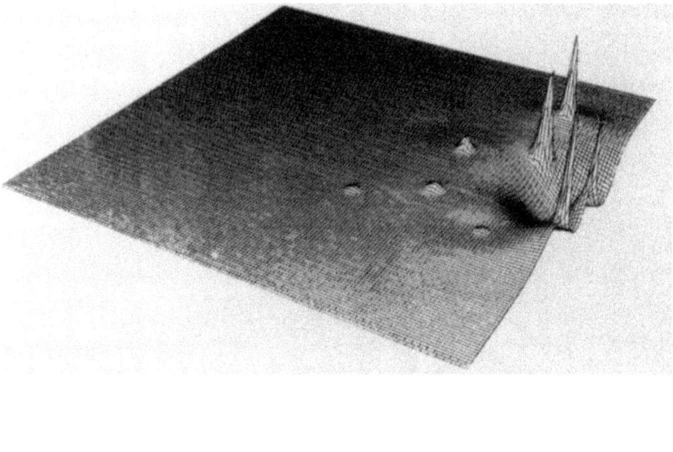

Summary

Quantum chemical calculations have been performed on pterin and some 6-substituted derivatives, 7,8-dihydropterin, guanine and two 5-deazaguanine derivatives. The framework of the all electron-Hartree-Fock-Roothaan model has been employed. The molecular wave functions have been evaluated by means of MULLIKEN population analysis as well as by computer-assisted graphical representations of the electron density differences. Only the pyrimidine moiety has been evaluated. The analysis reveals the following interesting features: Guanine is situated between the relatively dense cluster of the 6-substituted pterin derivatives (and pterin itself) and the 5-deazaguanine species with respect to the atomic partial charges and bond polarities as well as the electron density behaviour. 7,8-dihydropterin appears to possess a qualitatively different functional course of the electronic situation within the pyrimidine part. The results may be of practical value for the discussion of the inhibitory effect of pterins on the enzyme guanine-tRNA transglycosylase. An interesting extension of this study, the investigation of the intermolecular interactions of the different species studied with suited partner molecules, e.g. via hydrogen bonds, is under progress.

References

1. Kaufman, S.: Proc. Natl. Acad. Sci. USA 50, 1085 (1963).
2. Lagowski, J.M., Forrest, M.S.: Proc. Natl. Acad. Sci. USA 58, 1541 (1967).
3. Jacobson, K.B., Farkas, W.R., Katze, J.R.: Nucl. Acids Res. 9, 2351 (1981).
4. Katze, J.R., Beck, W.T., Cheng, C.S., McCloskey, J.A., in: Recent Results in Cancer Research, vol. 84, pp 147-159, Springer Verlag, Berlin-Heidelberg, 1983.
5. Wachter, H., Fuchs, D., Hausen, A., Reibnegger, G., in: Biochemical and Clinical Aspects of Pteridines, vol 1., pp 227-281, Eds.: H. Wachter, H.Ch. Curtius, W. Pfleiderer. Walter de Gruyter, Berlin-New York, 1982.
6. Wachter, H., Fuchs, D., Hausen, A., Reibnegger, G., in: Chemistry and Biology of Pteridines and Folic Acid Derivatives. Ed.: J.A. Blair, Walter de Gruyter, Berlin-New York, 1983, in press.
7. Blair,J.A., Pearson, A.J.: J. Chem. Soc. Perkin II, 80 (1974).
8. Bieri, J.H., Geiger, R.E.: Helv. Chim. Acta 58, 1201 (1975).
9. Fock, V.: Z. Phys. 61, 126 (1930).
10. Roothaan, C.C.J.: Rev. Mod. Phys. 23, 69 (1951).
11. Kutzelnigg, W.: Einführung in die theoretische Chemie, Vol. 1,2. Verlag Chemie, Weinheim-New York, 1978.
12. Rode, B.M., Sagarik, K.: Chem. Phys. Letters 88, 337 (1982).
13. Reibnegger, G.J., Rode, B.M.: Inorg. Chim. Acta, in press (1983).
14. Spencer, M.: Acta Cryst. 12, 59 (1959).
15. Shirell, C.D., Williams, D.E.: J. Chem. Soc. Perkin II, 40 (1975).
16. Mulliken, R.S.: J. Chem. Phys. 23, 1833, 1841, 2338, 2343 (1955).
17. Absar, I., van Wazer, J.R.: Angew. Chem. 90, 86 (1978).
18. Ahlrichs, R.: Theoret. Chim. Acta (Berlin) 33, 157 (1974).
19. Varmuza, K.: Pattern Recognition in Chemistry; Lecture Notes in Chemistry (21), Springer Verlag, Berlin-Heidelberg-New York, 1980.

Acknowledgement
This study was financially supported by the Austrian Research Funds "Zur Förderung der wissenschaftlichen Forschung", project P 4905.

DETERMINATION OF BIOPTERIN AND NEOPTERIN IN FLUIDS AND TISSUES BY RADIOIMMUNOASSAY

Takashi Sugimoto and Sadao Matsuura
Department of Chemistry, College of General Education, Nagoya University, Chikusa-ku, Nagoya 464, Japan

Tokio Yamaguchi, Makoto Sawada, and Toshiharu Nagatsu
Laboratory of Cell Physiology, Department of Life Chemistry, Graduate School of Nagatsuta, Tokyo Institute of Technology, Midori-ku, Yokohama 227, Japan

Introduction

The pterin requiring phenylalanine-, tyrosine-, and tryptophan-hydroxylases are key enzymes in the biosynthesis of catecholeamines (dopamine, norepinephrine, epinephrine) and indolamines (serotonin, melatonin), which are important neurotransmitters or hormones. L-erythro-Tetrahydrobiopterin is the natural cofactor for these enzymes, and the cofactor plays an important part in the kinetic regulation of the enzymes (1). Therefore, a deficiency of tetrahydrobiopterin may have physiological effects such as hyperphenylalanemia (2,3).

Besides to the cofactor activity of tetrahydrobiopterin, several clinical reports have shown recently that human urinary excretion of biopterin and neopterin (or their derivatives) reflects some pathological conditions like neoplastic diseases (4,5,6) and indicated an involvement of these pteridine compounds in cell metabolism such as proliferation and differentiation (7,8) and in immunoregulation (9,10). A good assay method for L-erythro-biopterin and D-erythro-neopterin is essential for such studies. We have recently developed a

Biochemical and Clinical Aspects of Pteridines, Vol. 2

highly sensitive and specific radioimmunoassay for these two pterins. Here we report the outline of the assay and its application to determine the biopterin and neopterin in several biological samples in connection with the neurotransmitter abnormality.

Materials and Methods

Preparation of immunogenic antigens, specific antisera, and tracers.

The immunogenic antigens (1b and 2b) were synthesized starting from $\underline{N}^2$-(5-carboxypentyl)-biopterin (1a) and -neopterin (2a) in the same way as previously described (11). Namely, the carboxylic acid was first converted to a mix anhydride by ethyl chloroformate and then condensed with bovine serum albumin (BSA). Thus obtained conjugates (1b and 2b) were found to possess 32 mol biopterin or 31 mol neopterin per mol BSA, respectively.

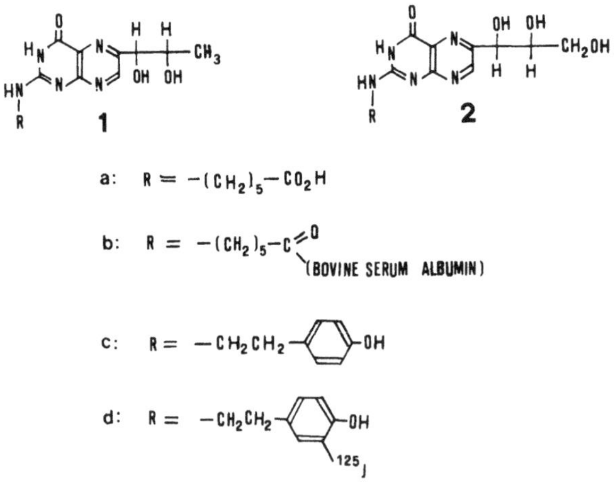

The pterin-BSA conjugate (1 mg) in complete Freund's adjuvant was intradermally injected to a New Zealand albino rabbit, which one month later received booster injections (0.4 mg) four times once a week. The antiserum was collected ten days after the final booster injection (11).

Iodination of $\underline{N}^2$-(p-hydroxyphenylethyl)-biopterin (1c) and -neopterin (2c) by the method of Hunter and Greenwood (12) afforded the tracers (1d and 2d) (13,14).

Preparation of samples from urine and tissues for the assay.
Urine and tissue homogenates were generally submitted to iodine oxidation (15) in 0.2 M HCl prior to radioimmunoassay to obtain a summed concentration of biopterin (or neopterin) and their dihydro- and tetrahydro-derivatives. The iodine oxidation was occasionally carried out in 0.2 M NaOH to obtain a summed concentration of only aromatic- and 7,8-dihydro-biopterin (15) (or neopterin).
After oxidation, urine samples were used for the assay without any additional purification, except lyophilization. Tissue homogenates (10 -200 mg/0.3 M sucrose), after oxidation, were centrifuged and the supernatants were passed through a small Dowex-50 column. The eluate from the column with 1 M ammonia was lyophilized and submitted to the assay.

Radioimmunoassay procedure
A mixture of 100 µl of sample or standard compound in a 0.02 M phosphate buffer, pH 7.5, containing 0.1 % BSA, 100 µl of 8.000-fold diluted antiserum, and 100 µl of the above buffer was incubated at 37°C for 30 min. A solution of the tracer (100 µl, 20.000 cpm) was added and the mixture was kept at 4°C for 1 h. Antigen-antibody complex was precipitated by a second antibody method (16) using anti-rabbit IgG and Dextran T-70, and the radioactivity of the precipitate was counted.

Results and Discussion

The standard curves for radioimmunoassay of L-erythro-biopterin and D-erythro-neopterin are shown in Fig. 1. The assay

Fig. 1. Standard Curves for Biopterin (BP) and Neopterin (NP) in Radioimmunoassay -

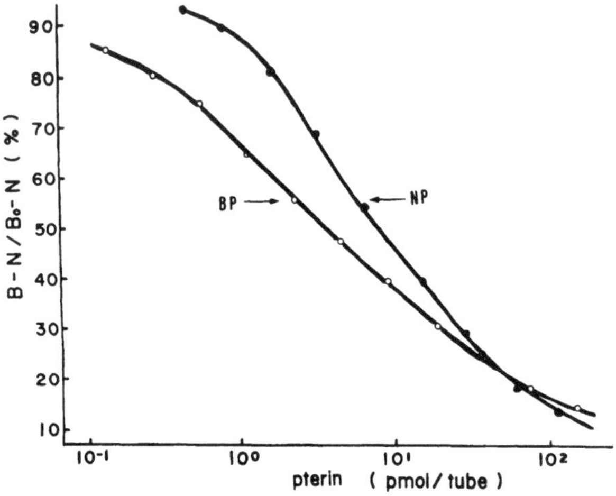

B, B_0: bound tracer in the presence (B) and absence (B_0) of pterin.

N: nonspecific binding of tracer.

T: total tracer; B_0-N/T = 30 %.

was of high sensitivity, which made possible to determine the two pterins as low as 0.5 pmol/tube. The specificity of the assay is shown in Table 1 as the percentage cross-reaction of the anti-L-erythro-biopterin antibody and anti-D-erythro-neopterin antibody with various pteridines. As seen from Table 1, both antibodies were highly specific for their hapten pteridines. We have proven the reliability of this assay by comparison with bioassay or high performance liquid chromatography (11,14).

Several examples of biopterin concentrations in human fluids and tissues are given in Table 2. It is noteworthy that the biopterin concentration in caudate nucleus of parkinsonian patients was as low as about 25 % of the control (17). The dopamine concentration in the brain region of parkinsonian

Table 1. - Cross-Reaction of Anti-L-erythro-Biopterin and Anti-D-erythro-Neopterin Antibodies with Pterins

Pterins	Cross-Reaction (%)	
	anti-L-erythro Biopterin	anti-D-erythro Neopterin
L-erythro-Biopterin	100	0.09
D-threo-Biopterin	0.5	---
L-threo-Biopterin	0.3	---
D-erythro-Biopterin	0.2	---
D-erythro-Neopterin	0.01	100
D-threo-Neopterin	---	0.2
L-threo-Neopterin	---	0.07
L-erythro-Neopterin	---	0.02
6-L-Dihydroxyethylpterin	1.1	0.02
6-Hydroxymethylpterin	0.2	0.003
6-Methylpterin	0.009	0.001
Pterin	0.03	0.009

Cross-reaction (%): the L-erythro-biopterin or D-erythro-neopterin equivalent was read off the standard curve, and this value was divided by the amount of each pterin and multiplied by 100 to give percentage cross-reaction.

patients has been found to be greatly decreased (18), for which the lowered activities of dopamine-synthesizing enzymes such as L-dopa decarboxylase (19) and tyrosine hydroxylase (only 1/20 of control) (20) have been considered to be most responsible. The present data, together with a successful result on administration of tetrahydrobiopterin to parkinsonian patients to improve the symptom (21,22) show that the dopamine abnormality in parkinsonian brain is caused, in part, also by an insufficient biosynthesis of tetrahydrobiopterin in that brain region.

Table 2. - Biopterin in Human Fluids and Brain Tissue

Tissue	Biopterin (mean ± S.E.M.)
Serum	25.5 ± 2.9 pmol/ml
Cerebrospiral fluid	10.2 ± 1.3
Caudate nucleus[a)]	
Control	1.25 ± 0.68 pmol/mg tissue
Parkinsonism	0.30 ± 0.59
Shy-Drager-Syndrome	0.11
Striato-nigral degeneration	0.66
Perioral dyskinesia	0.26, 0.78
Urine	
Control[b)]	6.53 ± 0.68 nmol/mg creatinine
Parkinsonism	5.84 ± 0.59

a) postmorten autopsy samples.

b) age-matched control for parkinsonism.

In contrast to the decreased biopterin concentration in caudate nucleus, the urinary biopterin concentration from parkinsonian patients were found to be nearly the same as that from control (Table 2). The urinary biopterin is supposed to be synthesized in kidney and excreted into urine (23), and thus the present result supports the concept of specific degeneration of dopaminergic neurons in parkinson brain. It also indicates that there is a high-blood-brain barrier against peripherally generated or administered tetrahydrobiopterin, despite of the above mentioned and other (24) administration studies of tetrahydrobiopterin.

The biopterin concentration in caudate nucleus of related diseases, i.e. Shy-Drager syndrom, Striato-nigral degeneration, and perioral dyskinesia were also found significantly lower than that of control (Table 2).

As a model study for human cerebullar abnormality, we investigated the biopterin concentration in the brain of rolling mice (25). The rolling mouse is a neurological mutant and is characterized by severe gait disturbances (26), which is originated most probably from abnormality of cerebullar neurotransmitters. Result in Table 3 shows that bio-

Table 3. - Biopterin in the Brain of Rolling and Control Mice

Tissue	Biopterin (pmol/mg tissue)	
	TRH[a] (none)	TRH (500 μg/mouse)[b]
Caudate nucleus		
control	0.35 ± 0.05	0.28 ± 0.04
rolling	0.18 ± 0.04	0.25 ± 0.04
Hypothalamus		
control	0.90 ± 0.07	0.55 ± 0.09
rolling	0.57 ± 0.09	0.88 ± 0.09

a) Thyrotropin releasing hormone.

b) One hour after an intraperitoneal injection of TRH.

pterin concentration in both caudate nucleus and hypothalamus of rolling mice were much lower than the control. The activities of tyrosine and tryptophan hydroxylases in the two regions did not differ significantly between the rolling and control mice. The effect of Thyrotropin releasing hormone, which improve ataxic gait of rolling mice (27) on the biopterin concentration and the activities of the two enzymes were complex (25), but the amplitude of the effect was not significant except the following two: biopterin concentration and tyrosine hydroxylase activity in hypothalamus increased significantly. From these results it may be concluded that ataxic gait of rolling mice is partly due to an abnormality of catecholaminergic neuron, especially noradrenergic neurons, and that biopterin may be a regulatory factor in the biosynthesis of the amines.

Table 4 shows the neopterin concentration in human urine. The total neopterin concentration obtained by radioimmunoassay agreed well with the result obtained by high performance liquid chromatography.

Alteration of the conditions of the iodine oxidation from acid to alkaline gave an interesting result. On iodine oxidation, tetrahydroneopterin behaves in the same way as tetrahydrobiopterin:tetrahydroneopterin give exclusively neopterin on iodine oxidation in acid, and alternatively 6-unsubstituted pterin in alkali. We found that the oxidation in an alkaline solution of urine samples gave a lower (about 35 % less) value of neopterin compared to the value from the oxidation in acid. It means that about 35 % of neopterin derivatives is alkaline labile, and we assume that this compound is tetrahydroneopterin.

At present, biological meanings of tetrahydroneopterin are obscure. We assume, however, that an important role of tetra-

Table 4. - Comparison of Neopterin Concentration in Human Urine Obtained by Radioimmunoassay and High Performance Liquid Chromatography

Subjects	Radioimmunoassay	High Performance Liquid Chromatography[a)]
S.M.	5.86 nmol/ml	5.25 nmol/ml
T.Y.	6.19	7.26
K.G.	1.27	1.07
C.A.	2.33	2.24
K.N.	1.01	1.37

a) On a Partsil 10-SCX column using 12.5 mM ammonium acetate, pH 3.3, as solvent. Detected by fluorometry (350/440 nm).

hydroneopterin will be found as a regulating factor in cell proliferation and differentiation.

Wachter et al. (7) have proposed preliminary hypotheses to explain the connection between proliferation and differentiation of T-lymphocytes and elevated urinary neopterin excretion from patients with various malignant diseases, viral infection, and allograft rejections.

References

1. T. Nagatsu: in "Frontiers in Catecholamine Research", Eds. E. Usdin and S. Snyder, Pergamon Press, Oxford (1979), p. 34.

2. S. Kaufman, N.A. Holzman, S. Milstein, I.J. Buther, and A. Krumholz: New Engl.J.Med. 293, 785 (1975).

3. A. Niederwieser, H.Ch. Curtius, O. Bettoni, J. Bieri, B. Schircks, M. Viscontini, and J. Schaub: Lancet, 131 (1979).

4. H. Rokos, K. Rokos, H. Frisius, and H.I. Kirstaedter: Clin.Chim. Acta 105, 275 (1980).

5. B. Stea, R.M. Halpern, B.C. Halpern, and R.A. Smith: Clin.Chim. Acta 113, 231 (1981).

6. A. Hausen, D. Fuchs, K. Grünerwald, H. Huber, and H. Wachter: Clin. Biochem. 15, 34 (1982).

7. H. Wachter, D. Fuchs, A. Hauser, and G. Reibnegger: in "Biochemical and Clinical Aspects of Pteridines", Eds. H. Wachter, H.Ch. Curtius, and W. Pfleiderer, Walter de Gruyter, Berlin (1982), p. 275.

8. I. Ziegler, H.J. Kolb, U. Bodenberger, and W. Wilmanns: Blut 44, 261 (1982).

9. H. Wachter, A. Hausen, and K. Grassmayr: Hoppe-Seyler's Z.Physiol.Chem. 360, 1957 (1979).

10. A. Hausen, D. Fuchs, K. Grünwald, H. Huber, K. König, and H. Wachter: Clin.Chim. Acta 117, 297 (1981).

11. T. Nagatsu, T. Yamaguchi, T. Kato, T. Sugimoto, S. Matsuura, K. Kobayashi, M. Akino, S. Tsushima, N. Nakazawa, and H. Ogawa: Proc. Japan Acad. 55, Ser. B, 317 (1979).

12. W.M. Hunter and F.C. Greenwood: Nature 194, 495 (1962).

13. T. Nagatsu, T. Yamaguchi, T. Kato, T. Sugimoto, S. Matsuura, M. Akino, S. Tsushima, N. Nakazawa, and H. Ogawa: Anal. Biochem. 110, 182 (1981).

14. T. Nagatsu, M. Sawada, T. Yamaguchi, T. Sufimoto, S. Matsuura, M. Akino, N. Nakazawa, and H. Ogawa: unpublished.

15. T. Fukushima and J.C. Nixon: Anal. Biochem. 102, 176 (1980).

16. C.R. Morgen and A. Lazarow: Diabetes 12, 115 (1963).

17. T. Nagatsu, T. Yamaguchi, T. Kato, T. Sugimoto, S. Matsuura, M. Akino, I. Nagatsu, R. Iizuka, and H. Narabayashi: Clin.Chim. Acta 109, 305 (1981).

18. H. Ehringer and O. Hornykiewicz: Klin.Wschr. 38, 1236 (1960).

19. K.G. Lloyd and O. Hornykiewicz: Science 170, 1212 (1970).

20. T. Nagatsu, T. Kato, I. Nagatsu, Y. Kondo, S. Inagaki, R. Iizuka, and H. Narabayashi: Adv.Neurol. 24, 283 (1979).

21. H. Narabayashi, T. Kondo, T. Nagatsu, T. Sugimoto, and S. Matsuura: Proc. Japan Acad. 58, Ser. B, 283 (1982).

22. H.Ch. Curtius, H. Muldner, and A. Niederwieser: in "Biochemical and Clinical Aspects of Pteridines", Eds. H. Wachter, H.Ch. Curtius, and W. Pfleiderer, Walter de Gruyter, Berlin (1982), p. 285.

23. D.A. Häble, H. Schiffl, G. Mayer, G. Hennings, and H. Remboldt: Pfluegers Arch. 375, 9 (1978).

24. A. Niederwieser, H.Ch. Curtius, M. Wang, and D. Leupold: Eur. J. Pediatr. 138, 110 (1982).

25. T. Yamaguchi, Y. Hirata, T. Nagatsu, S. Oda, T. Sugimoto, S. Matsuura, M. Konagaya, T. Takayanagi, and I. Sobue: Neurochen. International 4, 491 (1982).

26. S. Oda: Exp. Animals (Tokyo) 22, 281 (1973).

27. I. Sobue: Clin. Neurol. (Tokyo) 17, 791 (1977).

TETRAHYDROBIOPTERIN IS DESTROYED BY AUTOCLAVING AT pH 4.5: A COMPARISON OF CRITHIDIA AND HPLC ASSAYS

Sheldon Milstien
Laboratory of Neurochemistry
National Institute of Mental Health
Bethesda, Maryland 20205

Introduction

The measurement of tetrahydrobiopterin in human urine (1), serum (2), and cerebrospinal fluid (3), has become an important tool in the differential diagnosis of hyperphenylalaninemia as well as in monitoring replacement therapy in children with genetic defects in tetrahydrobiopterin biosynthesis (3). Recently, biopterin and other unconjugated pterins have also been reported to be useful markers of immune cell proliferation (4) and of malignancy (5).

Several different methods have been proposed for the isolation and measurement of biopterin in biological samples. The first methods used for the determination of biopterin were based on the specific growth stimulation of the organism, Crithidia fasciculata (6). In this method, samples are autoclaved at pH 4.5 in aconitate buffer to both sterilize the samples and to release "bound" biopterin (7). Usually, recoveries are measured with the addition of the fully oxidized, stable form of biopterin. It is now clear that most, if not all, of the biopterin in tissues is present in the fully reduced, tetrahydro form (8) (except in those cases where dihydropteridine reductase activity is missing). The oxidation of tetrahydrobiopterin at pH values greater than 4 does not give quantitative yields of biopterin. Instead, substantial amounts of non-Crithidia active pterins, such as deoxysepiapterin, sepiapterin, dihydroxanthopterin, and pterin are also formed (9).

Biochemical and Clinical Aspects of Pteridines, Vol. 2

In spite of these previously published results on the complexities in the oxidation of tetrahydrobiopterin-containing samples at high pH, several groups have continued to use these methods of sample preparation in conjunction with Crithidia assays for the measurement of biopterin in various disease states (10, 11). In this report, we have reexamined the effect of both autoclaving and chemical oxidation on the determination of biopterin in rat tissues. We have also compared Crithidia and high performance liquid chromatographic assays which give comparable results when proper precautions are taken during sample workup procedures.

Materials and Methods

Crithidia fasciculata (ATCC 12837) were cultured and biopterin assays carried out as described by Baker et al. (7). Biopterin was synthesized in this laboratory (12). Solutions of biopterin were standardized by absorption spectra in 0.1 N NaOH (molar extinction coefficient at 365 nm, 5700).

High performance liquid chromatographic analyses were performed as described by Fukushima and Nixon (13). The system consisted of a Wisp automatic sample injector, a Waters 660 pump, Gilson 121 fluorometer (excitation 340 nm, emission 420 nm), and a Whatman ODS-1 column. Methanol-water (5:95) was the solvent. Total biopterin (tetrahydro, dihydro, oxidized) was measured by the addition of 0.1 ml of each sample to a solution containing 0.03 ml of 1 M H_3PO_4 and 0.1 ml I_2/KI (1% I_2 dissolved in 2 % KI). After 60 minutes at room temperature, excess I_2 was reduced by the addition of 0.05 ml ascorbic acid solution (20 mg per ml). Dihydro plus oxidized biopterin were measured separately by the addition of 0.1 ml of sample to a solution containing 0.03 ml 1 N NaOH and 0.1 ml I_2KI. After 60 minutes, 0.03 ml 1 M H_3PO_4 were added and the samples treated as above. Oxidized biopterin alone was determined by direct injection of samples without any prior oxidation.

Results and Discussion

Pfleiderer (9), Fukushima, Kobayashi, Eto, and Shiota (4), as well as Fukushima and Nixon (13) have all previously shown that oxidation of tetrahydrobiopterin at pH values higher than 1 does not give quantitative yields of the fully oxidized form. As the pH of the oxidation media is increased, a larger fraction of the tetrahydrobiopterin is converted to non- Crithidia active substances. This specific loss of tetrahydrobiopterin upon oxidation in alkaline solutions, a condition which quantitatively converts dihydrobioperin to biopterin, formed the basis for the development of a method for the measurement of reduced pterins in human urine which has been applied to the diagnosis of dihydropteridine reductase deficiency (1).

In the Baker method for biopterin determinations with Crithidia (7), samples are autoclaved at pH 4.5 without any prior oxidation step to convert the labile tetrahydrobiopterin to more stable oxidized forms. In Figure 1 are presented the results of autoclaving 1 nmol of either tetrahydrobiopterin or dihydrobiopterin in 5 ml of solutions of varying pH for 20 minutes at 121 C. In the case of tetrahydrobiopterin-containing solutions, there is a large loss of total bioperin when the reactions are carried out above pH 3. In no case, even in 0.1 N HCl, is there more than 30% conversion to the fully oxidized form during the autoclaving step. Note that the arrows designate the reactions in pH 4.5 sodium aconitate buffer, the recommended solution for "oxidation" and "release" of biopterin in biological samples. After autoclaving at pH 4.5, 55% of the biopterin is destroyed, and only 18% of original amount of tetrahydrobiopterin is converted to oxidized, stable forms.

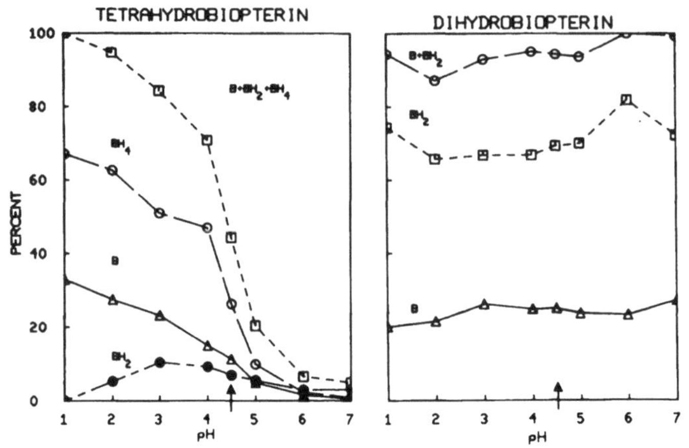

Figure 1. The effect of autoclaving on the oxidation and recovery of tetrahydrobiopterin and dihydrobiopterin. The buffers used were: 0.1 N HCl (pH 1); 0.01 N HCl (pH 2); 0.01 M sodium formate (pH 3); 0.01 M sodium acetate (pH 4, 5); 0.01 M sodium phosphate (pH 6, 7). Tubes were autoclaved for 20 minutes, cooled, and then the amounts of the various forms of biopterin determined by differential oxidation and high performance liquid chromatography.

Dihydrobiopterin is more stable at all pH values than is tetrahydrobiopterin. As is shown in Figure 1, autoclaving dihydrobiopterin in solutions ranging from pH 1 to pH 7 results in only about 20% conversion to biopterin and recovery of 80% of the original dihydrobiopterin. Therefore, dihydrobiopterin containing solutions would be expected to slowly oxidize during a Crithidia assay and give the same growth response curve as biopterin. Of course, the fully oxidized form is stable to further oxidation.

These results are in contrast to those of Pfleiderer, who suggested that the conversion of reduced biopterin forms to non-Crithidia active substances occurred during the oxidation of dihydrobioperin (9). It is clear, however, from these results on pure tetrahydro and dihydrobioperin, that autoclaving biological samples at pH 4.5 that contain tetrahydrobiopterin will lead to a substantial loss of biopterin. It would also be expected that there would be a further loss of any remaining tetrahydrobiopterin due to oxidation during the 4-day incubation period in the Crithidia assay. It is likely, in fact, that the presence of endogenous reducing substances, such as glutathione or ascorbic acid, would act to keep tetrahydrobiopterin reduced during the sterilization step. This could result in a greater loss of the tetrahydrobiopterin by autoxidation during the Crithidia assay.

		Biopterin (pmol/g or ml)	
Sample	Assay Method	Autoclave (pH 4.5)	Iodine (pH1)-Autoclave (pH 4.5)
Brain	Crithidia	122 ± 14	404 ± 30
	HPLC	203	398 ± 46
Serum	Crithidia	80.2 ± 3.8	141 ± 6.3
	HPLC	66.5	138 ± 8.4

Table 1. The effect of autoclaving and iodine oxidation on the measurement of biopterin in rat serum and brain homogenates. Samples of serum and brain homogenates (1 g per 3 ml 0.15 M KCl) from 3 rats were analyzed for biopterin content by both the Crithidia fasciculata bioassay and by HPLC. Samples were oxidized with iodine in 0.1 N HCl, treated with ascorbic acid to reduce excess iodine, adjusted to pH 4.5 with NaOH, diluted into 10 mM sodium aconitate buffer (pH 4.5), and then autoclaved for 20 minutes.

In Table 1, are presented the results of some experiments on the measurement of biopterin in rat serum and brain extracts by both the Crithidia and the high performance liquid chromatographic assays. When samples are treated by the autoclaving procedure recommended for the Crithidia assay and then measured by the Crithidia method, approximately two-thirds of the biopterin is lost, as determined by comparison to the high performance liquid chromatographic assay. However, when the samples are first oxidized with iodine, under conditions where tetrahydrobiopterin is quantitatively converted to biopterin, both the Crithidia and the high performance liquid chromatography assays give the same values for the biopterin content.

Finally, it should be noted that when serum samples from humans or rats are chromatographed of Sephadex G-25 columns to separate the low molecular weight components from proteins and high molecular weight components, essentially all of the biopterin is removed from the high molecular weight fraction. This result rules out the notion that there is a substantial amount of biopterin that is "bound" and that needs to be released in order to be measurable in the Crithidia assay.

References

1. Milstien, S., Kaufman, S., Summer, G.K.: Pediatrics 65, 806-810 (1980).

2. Leeming, R.J., Blair, J.A., Green, R.A., Raine, D.N.: Arch. Dis. Child. 51, 771-777 (1976).

3. Kaufman, S., Kapatos, G., McInnes, R.R., Schulman, J.D., Rizzo, W.B.: Pediatrics 70, 376-380 (1982).

4. Fuchs, D., Hausen, A., Huber, C., Margreiter, R., Reibnegger, G., Spielberger, M., Wachter, H.: Hoppe-Seyler's Z. Physiol. Chem. 363, 661-664 (1982).

5. Stea, B., Halpern, R.M., Halpern, B.C., Smith, R.A.: Clin. Chim. Acta 113, 231-242 (1981).

6. Dewey, V.C., Kidder, G.W.: Meth. Enzymol. 18B, 618-624 (1972).

7. Baker, H., Frank, O., Bacchi, C.J., Hutner, S.H.: Amer. J. Clin. Nutr. 27, 1247-1253 (1974).

8. Kaufman, S., Holtzman, N.A., Milstien, S., Butler, I.J., Krumholz, A.: New Engl. J. Med. 293, 785-790 (1975).

9. Pfleiderer, W.P.: In Chemistry and Biology of Pteridines, ed. by Pfleiderer, W.P., Walter de Gruyter, Berlin, pp. 941-949 (1975).

10. Leeming, R.J., Blair, J.A.: Clin. Chim. Acta 108, 103-111 (1980).

11. Ziegler, I., Fink, M., Wilmanns, W.: Blut 44, 231-240 (1980).

12. Andrews, K.J.M., Barber, W.E., Tong, B.P.: J. Chem. Soc. (C), 928-930 (1969).

13. Fukushima, T., Nixon, J.C.: Anal. Biochem. 102, 176-188 (1980).

14. Fukushima, T., Kobayashi, K.-I, Eto, I., Shiota, T.: Anal. Biochem. 89, 71-79 (1980).

ISOLATION AND IDENTIFICATION OF PTERIDINES FROM SHEEP PINEALS AND THE INFLUENCE OF SOME SYNTHETIC PTERIDINES ON THE BIOSYNTHESIS OF 5-METHOXYINDOLES IN PINEALS OF RATS

I. Ebels, H.P.J.M. Noteborn and A. de Morée

Department of Organic Chemistry, Section of Organic Chemistry of Natural Products (Head: Prof. Dr. C.A. Salemink) State University of Utrecht, Utrecht, The Netherlands

M.G.M. Balemans and J. van Benthem

Zoological Laboratory, State University of Utrecht, Utrecht, The Netherlands

Introduction

In 1973 we first observed that when an aqueous extract of sheep pineals was separated on a Sephadex G-25 or a Sephadex G-10 column, besides the wellknown fluorescence peaks of indole-like substances, another high fluorescent peak could be detected, which was not observed in a comparable extract of sheep cerebral cortex, treated in the same way (1). This high fluorescent peak showed an excitation maximum (e) at 360 nm and a fluorescence maximum (f) at about 440 nm. After ultrafiltration experiments it was found that the substance with these special fluorescence characteristics was located in an UMO5F-fraction, which contains compounds with a molecular weight (MW) < 500. Details of the isolation, purification and identification were described by Van der Have-Kirchberg et al. (2).
Some years later we could isolate from an UMO5R-fraction of the waterlayer (MW > 500 and < 1000), prepared according to an isobutanol Bensinger-extraction method (3, 4, 5) a substance which had after trimethylsilylation a mass spectrum, which was identical with that of synthetic neopterin. From experiments we

Biochemical and Clinical Aspects of Pteridines, Vol. 2

carried out till now, we have the impression, that we can detect neopterin in the isobutanol-extract when higher molecular weight material is extracted into the isobutanol (5). When that peak is absent, we could isolate neopterin from the waterlayer. Therefore, we have the impression that neopterin can be bound to higher molecular weight substances.
In this paper we will present the isolation and purification of a "biopterin-glucoside" and of pterin-6-carboxylic acid from sheep pineals. Furthermore, a survey of the influence of some synthetic pteridines on the indole-metabolism in rat pineals will be discussed.

Materials and Methods

Sheep pineals were collected in the winter months by ERSCO, San Mateo, California, U.S.A.; shipped on dry-ice and stored at -30°C.
The synthetic pteridines, which were used as reference substances, are the same as described before (2, 6).
The pineal extracts were prepared according to a modified Bensinger-extraction method and ultrafiltrated as shown in scheme 1. The Bensinger-extraction method is first described in abstract form (7).
Gelfiltration of the sheep pineal fraction UM05R was carried out on a Sephadex G-15 column (85 x 2.5 cm) equilibrated and eluted with 0.2 M acetic acid; subsequently with distilled water and later on, a linear gradient was applied from water to 0.01 M NaOH.
The excitation and fluorescence maxima of the eluate were determined with an Aminco Bowman spectrofluorimeter (4-8203) as published before (2).
For the micro determination of phosphorus the method of Chen et al. (8) was used.
Paper chromatography was carried out as described before in detail (2), using the following solvent systems; n-butanol-

Outline of the separation methods used

sheep pineal glands

↓ extraction with acetone/ammonia 1% (3:1, v/v)

- acetone/ammonia extract
  - ↓ after evaporation of the acetone extraction with chloroform/methanol (2:1, v/v)
    - chloroform/methanol extract
    - alkaline water layer
      - ↓ after evaporation of chloroform/methanol ultrafiltration through PM10
        - PM10 residue (PM10R)
        - PM10 filtrate
          - ↓ ultrafiltration through UM2
            - UM2 filtrate
              - ↓ ultrafiltration through UMO5
                - UMO5 residue (UMO5R)
                  - Sephadex G-15
                  - paper chromatography
                  - trimethylsilylation
                  - gas liquid chromatography (GC) and GC-mass spectrometry
                - UMO5 filtrate (UMO5F)
            - UM2 residue (UM2R)
- residue

SCHEME 1.

acetic acid-water, 4:1:1; ethanol-water, 1:3; n-propanol-ammonia 1%, 2:1.
All experiments were carried out under dim red light conditions (λ > 600 nm).
Gasliquid chromatography (GC) and subsequently GC-mass-spectrometry experiments (GC-MS) were carried out as described before (2).
For the gasliquid chromatography two different columns were used: a 1.5% Dexsil column (2.00 m x 2 mm) and a 3% SE-30 column (2.00 m x 4 mm).

Results

A. Isolation of pteridines from sheep pineals

Figure 1 shows the elution curve of a sheep pineal UMO5R-fraction, separated on a Sephadex G-15 (85 x 2.5 cm) column, equilibrated and eluted with 0.2 M acetic acid, subsequently with distilled water. After the elution of 1320 ml a linear gradient between 0 → 0.01 M NaOH was applied.
On this Sephadex G-15 column Dextran blue 2000 was eluted from 115-225 ml and Cl^- from 325 → 420 ml.
As can be seen in the Sephadex G-15 F6 fraction (eluted from 1260 and 1710 ml) and -F7 fraction (eluted from 1710 → 1948 ml), high peaks are present, with fluorescence characteristics like pteridines have.
The Sephadex G-15 F6 was first further studied with paper chromatography. The results of this study compared with some synthetic pteridines are presented in Table 1.
The main, bright blue fluorescent spot at 356 nm, Rf value (x 100) 61, was eluted, lyophilized and examined with GC and GC-MS after trimethylsilylation. The results of the oxidation of the main bright blue fluorescent spot (Rf 61, Table 1) according to Rembold (9) are presented in Table 2. It is clear that the fluorescent spot Rf 61 can be oxidized to a substance, which

Figure 1.

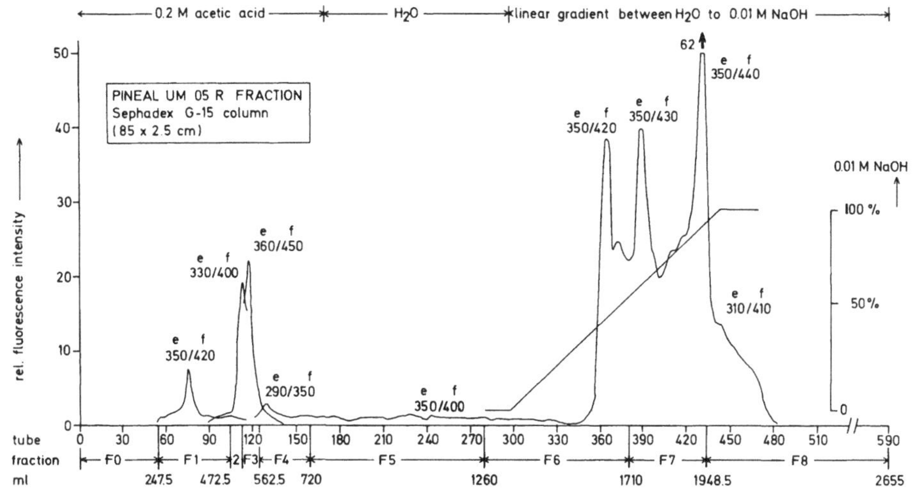

Separation of a sheep pineal UMO5R-fraction on a Sephadex G-15 column (85 x 2.5 cm) equilibrated and eluted with 0.2 N acetic acid. After 765 ml the column was eluted with distilled water and from 1320 ml a linear gradient between water to 0.01 M NaOH was applied.

e = excitation ; f = fluorescence

Table 1

Rf-values ($X10^2$) of compounds present in the sheep pineal Sephadex G-15, F6 fractions (Fig. 1) and of some synthetic pteridines, in two solvent systems

	solvents	
Compound or fractions	B-A-W	E-W
pterin-6-carboxylic acid	16	66
6-D-neopterin	15	44
6-L-erythrobiopterin	29	49
Sephadex G-15, F6 (eluted between 1260 and 1710 ml)	5,12,19,40	8,32,61*,76,86

B-A-W- = n-butanol-acetic acid water, 4:1:1

E-W = ethanol-water, 1:3

* The main spot with a pteridine-like fluorescence (light blue at 356 nm) with Rf value 61, was eluted, lyophilized and examined with GLC and GC-MS.

Table 2

Rf-values ($X10^2$) of the main spot with a pteridine-like fluorescence of the sheep pineal Sephadex G-15, F6-fraction (Rf 61 in EW) in two solvent systems after potassium-permanganate-oxidation

	solvent	
Compound or fraction	B-A-W	P-A
pterin-6-carboxylic acid	21	7
UMO5R, Sephadex G-15 F6, paper chromatography band (Rf 61) see Table 1	16,20	9
UMO5R, Sephadex G-15 F6, paper chromatography band (Rf 61), oxidized*	21	7

B-A-W = n-butanol-acetic acid water, 4:1:1

P-A = n-propanol-ammonia 1%, 2:1

* According to <u>Rembold, (1971)</u>

behaves in these two solvent systems, as synthetic-pterin-6-carboxylic acid, but it is not possible to conclude from the oxidation procedure that the pteridine-like compound (Rf 61 in EW) is transformed in "free" pterin-6-carboxylic acid. With the method of Chen et al. (8) in the main spot (Rf 61) of the Sephadex G-15 F6-fraction, a phosphorus content was calculated of 4.24 ± 0.08 μg P per 1000 μg of the pteridine-like compound. From Table 2, the obtained fluorescence spectra (Fig. 2) and the results of the phosphorus determination was concluded that a substance present in the fraction with Rf 61 was most probably a 6-substituted phosphorus-containing pteridine-derivative. In Table 3 the retention time of pineal compound Rf 61 and different synthetic pteridines are summarized, using two different GC-columns.

Although on a 1.5% Dexsil column the pineal compound Rf 61 is eluted in the same region as pterin-6-carboxylic acid, a difference in elution-volume is clearly observed when a 3% SE-30 column is used.

Figure 3 gives the mass spectrum of the, with preparative paper chromatography isolated pineal compound Rf 61 of the UMO5R, Sephadex G-15, F6 fraction, after trimethylsilylation (TMS) having a retention-time of 18.56 min. on a 1.5% Dexsil-column (see Table 3). As can be seen, the high-mass regions (m/e > 800) are not recorded due to the limitation of the used equipment.

From the bar graph presentation of the mass spectrum of Fig. 3, a fragmentation pattern which belongs to trimethylsilylated biopterin and that of hexakis-(trimethylsilyl)-α-D-glucopyranose-6-phosphate respectively described by Lloyd et al. (10) and Zinbo and Sherman (11) can be recognized. The m/e-values: 525 ($M^{\cdot +}$), 510 (M-15), 481, 409, 337, 320 are from ions involved in the fragmentation of biopterin, see also (2, 10). The m/e-values 677 (M-15), 602 (M-90), 587 (M-105), 513 (M-179), 512 (M-180), 497 (M-195) are from the spectrum of hexakis (trimethylsilyl)-α-D-glucopyranose-6-phosphate. Of particular interest in the spectrum of Figure 2, are the ions which retain

Table 3.

Retention time in minutes in gas liquid chromatography of the main fluorescent compound of the sheep pineal UMO5R, Sephadex G-15, F6 fraction and some synthetic pteridines on two different columns

Compound	Retention time (min.)
1.5% Dexsil (2.00 m x 2 mm)	
6-D-erythroneopterin	20.98
6-L-erythrobiopterin	16.44
7-L-erythrobiopterin	14.34
pterine	5.84
pterin-6-carboxylic acid	18.58
Sephadex G-15, F6 fraction paper chromatography band (Rf 61, see Table 1.)	18.56
Co chromatography pterin-6-carboxylic acid and paper chromatography band Rf 61	18.30 (shoulder), 18.60
Sephadex G-15, F6 fraction paper chromatography band Rf 61), oxidized	18.90
3% SE-30 (2.00 m x 4 mm)	
pterin-6-carboxylic acid	12.74
Sephadex G-15, F6 fraction paper chromatography band (Rf 61, see Table 1.)	19.64

Figure 3.

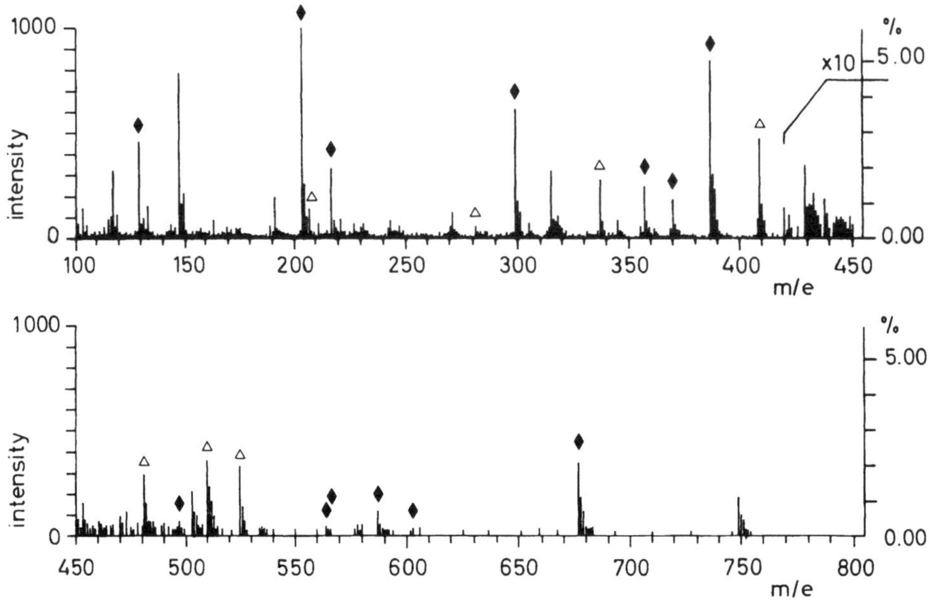

Mass spectrum of the paper chromatography band Rf 61 of an UMO5R, Sephadex G-15, F6-fraction, after trimethylsilylation, which had a retention time of 18.56 min. on a 1,5% Dexsil column (see Table 3).

△ fragmentation of TMS-biopterin

♦ fragmentation of TMS-α-D-glucopyranose-6-phosphate

Figure 2.

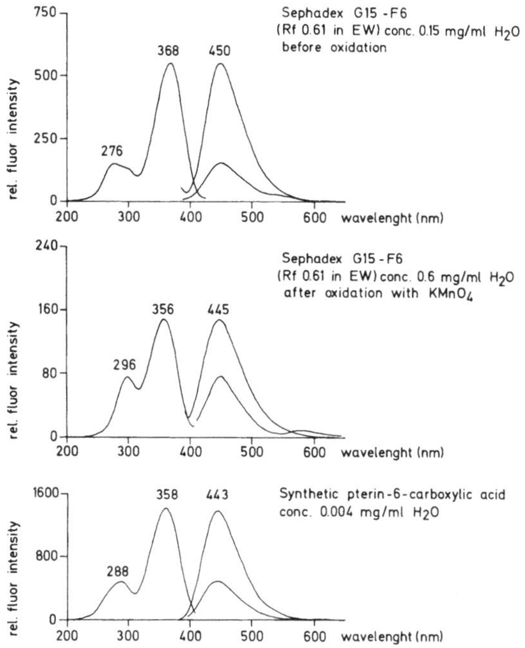

Excitation- and fluorescence spectra of UMO5R, Sephadex G-15, F6, Rf 61 before and after oxidation, and of synthetic pterin-6-carboxylic acid.

Figure 4.

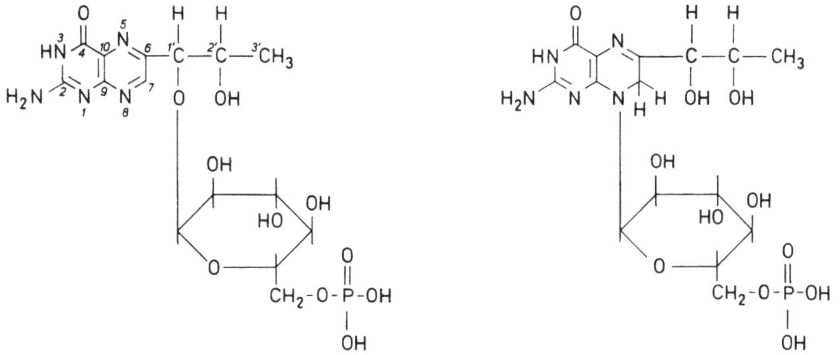

Two possibilities of how the α-D-glucopyranose-phosphate unit may be bound to biopterin.

the phosphorus atom: m/e: 387, 370, 357, 343, 315, 314, 299, 227 and 211. These ions carry a large portion of the total ion current in the spectra of trimethylsilylated-phosphorylated sugars as is shown by Zinbo and Sherman (11). Therefore we think that pineal compound Rf 61, can be a biopterin glycoside. However, we have to repeat the experiment on a mass spectrometer, which can record higher-mass-ions, in order to obtain a complete spectrum with the parent peak.
Figure 4 shows a possible interpretation of the base graph presentation of the mass spectrum in Figure 3.
The obtained data do not permit a decision as to the point of attachment of the sugar in the "biopterin-α-D-gucopyranose-phosphate", but it has been generally assumed to be at the 1'-position; MacLean et al. (12). However, Forrest et al. (13) proposed an attachment of the sugar at position 2' (see Fig. 4a). Distinct from the by these authors supposed structure of the biopterin-glucoside (Fig. 4a) we suggest that the phosphorylated sugar may be attached at the N-atom in the 8-position (Fig. 4b) and that the obtained ions in the fragmentation pattern (Fig. 3) are those resulting from hydrolysis of the carbohydrate derivative under mild acid conditions, during the preparation of the trimethylsilyl compound of pineal substance Rf 61. It is known from the literature that weak acid conditions causes the selective hydrolytic removal of purine bases, without affecting the pyrimidine-deoxyribose or the phosphodiester bonds of the backbone, by gentle acid hydrolysis of DNA at pH 3.0 (14). Till now the hydrolysis of the phosphate-sugar attached at position 1' or 2' of the pteridine-side-chain is carried out with stronger acid: 2N HCl or 2N H_2SO_4 at 90°C, (13, 15, 16). Synthetic reference compounds are needed to study at which position the α-D-glucopyranose-phosphate group is really attached, using enzymes and different hydrolysis conditions to compare the isolated compound Rf 61 and synthetic components.
From the UMO5R Sephadex G-15, F7 fraction, eluted between 1710 ml and 1948 ml from the column, a compound identical with pterin-6-carboxylic acid could be isolated and identified with

the same methods as described above (5). Since aeration or mild oxidizing conditions can increase the yield of degradation products of compounds like biopterin-glucoside, even if the sample is stored in the freezer and in darkness, it is possible that this pterin-6-carboxylic acid is formed during our isolation and separation procedures.
It is clear that acid compounds as "biopterin-glucoside" and pterin-6-carboxylic acid can only be eluted from a Sephadex G-15 column when a basic solvent is used. These results are comparable with those we have obtained previously with the acidic indoles, 5-methoxy indole acetic acid and 5-hydroxy indole acetic acid, which were retained on a Sephadex G-10 column, using water as eluent. These two indoles were eluted by gradient elution of water and NaOH (17).

B. The influence of some synthetic pteridines on the indole-metabolism in pineals of rats

Cremer-Bartels and Hollwich (18) were the first authors who indicated an influence of pteridines (triaminophenylpteridine or triamterene) on hydroxyindole-O-methyltransferase (HIOMT), the enzyme responsible for all O-methylating processes in indole-metabolism in the pineal and the retina (19). These authors postulated that naturally occuring pteridines might be involved in the regulation of HIOMT in dependence of light.
Cremer-Bartels (20) showed the effect of two synthetic pteridines: triamterene and xanthopterin on HIOMT of retina and pineal gland. Several HIOMT enzymes were detected which differed in isoelectric point and concentrations in dependence on light adaptation. The blocking effects of synthetic pteridines varied depending on light adaptation and species.
Cremer-Bartels and Ebels (21) have shown that pterin-6-aldehyde, one of the photolytic products of folic acid can decrease the activity of bovine pineal HIOMT. In all these experiments only the HIOMT-activity at one point in a 24 h period was tested.

In the experiments summarized in the following part, the influence of pteridines on the capacity of methylation in pineals is determined every 2 or every 4 hours during a 24 h period. Balemans et al. (19, 6, 22) studied the influence of three synthetic pterins (2-amino, 4-oxo-pteridines) on the diurnal fluctuations of HIOMT-activity in the isolated pineal gland, of 42-day old male Wistar rats. The method used to determine the HIOMT-activity of the pineal permits the separate determination of the synthesis of 5-methoxyindoles (23). The results of these experiments showed that reduced neopterin stimulated the methylation of the substances: 5-hydroxytryptophan (5-HTP) 5-hydroxytryptamin (5-HT) and 5-hydroxyindole-3-acetic acid (5-HIAA), during the night; HIOMT-action on the combination N-acetylserotonin with 5-hydroxytryptophol (N-Ac-5-HT/5-HTL) was shifted to a later moment in the dark period.
Pterin-6-aldehyde stimulated HIOMT-action on 5-HT during the day-time; HIOMT action on the substrates 5-HTP, 5-HIAA and N-Ac-5-HT/5-HTL was shifted to an earlier period.
Isoxanthopterin did not exert any influence on diurnal variation in the HIOMT-activities studied in this period of the year. From later experiments, carried out in other seasons, could be concluded that the circadian rhythmicity of HIOMT-activity could only be influenced by these pteridines, when the maximal HIOMT-activity was measured at 24.00 o'clock.
The effects of the above mentioned pteridines on HIOMT-activity in the pineal of 42-day old rats resembled those which could be observed when 28-day old rats were exposed to red and green light.
In experiments performed in December and January with 28-day old rats it could be demonstrated that red light causes a shift of the synthesis of melatonin/5-methoxytryptophol (mel/5-MTL) towards an earlier period. Exposure to green light had an opposite effect and caused a shift of the synthesis of mel/5-MTL to a later period. When the mean value of mel/5-MTL was calculated over a 24 hour period, it could be observed that red light inhibited mel/5-MTL synthesis whereas green light

stimulated this synthesis. In identical experiments, carried out in the same time of the year, incubation with reduced neopterin inhibited mel/5-MTL synthesis in the pineal of rats exposed to white, red or green light (25, 26). Pterin-6-aldehyde, however, stimulated this synthesis in white and green light exposed animals, whereas in red light pterin-6-aldehyde intensified the inhibitory effect of this wavelength. The hypothesis that a relationship may exist between light of different wavelengths, pteridines, indole metabolism and gonadal growth will be tested in the near future in a seasonal breeder: the hamster, an animal which shows a pineal mediated (27) gonadal regression when kept under short photo-period.

Conclusions

- With rather simple and mild methods, different pteridines could be isolated and identified from sheep pineals. Some of the pteridines are most probably bound to other compounds.
- Synthetic pteridines can influence the biosynthesis of methoxyindoles, depending on the time of the year.
- There may exist a relationship between light of different wavelengths, pteridines, indole metabolism and gonadal growth, which has to be studied more in detail in the near future in a seasonal breeder, the hamster, an animal which shows a pineal mediated gonadal regression when kept under short photo-period.

References

1. Zurburg, W., Ebels, I.: J. Neural Transm. 35, 117-124 (1974).

2. Van der Have-Kirchberg, M.M.L., Morée, A. de, Laar, J.F. van, Gerwig, G.J., Versluis, C., Ebels, I., Hus-Citharel, A., L'Héritier, A., Roseau, S., Zurburg, W., Moszkowska, A.: J. Neural Transm. 40, 205-220 (1977).

3. Ebels, I., Benson, B., Bria, C.F., Richardson, D., Larsen, B.R., Hruby, V.J.: J. Neural Transm. 45, 43-61 (1979).

4. Ebels, I.: Pteridines in the Pineal Organ. in: The Pineal Organ, pp. 281-302, Eds. A. Oksche, P. Pévet, Elsevier/ North-Holland, Amsterdam-New York-Berlin 1981.

5. Ebels I., Balemans, M.G.M., Benthem, J. van, Noteborn, H.P.J.M., Morée, A. de,: Pteridines in the Pineal and Effects of these Substances on the Indole Metabolism of this Organ. in: The Pineal Gland and its Endocrine Role, Eds. J. Axelrod, G.P. Velo, F. Fraschini, Plenum Press, London (in press).

6. Balemans, M.G.M., Benthem, J. van, Legerstee, W.C., Morée, A. de, Noteborn,H.P.J.M., Ebels, I.: Reprod. Nutr. Develop. 20 (4A), 1051-1060 (1980).

7. Bensinger, R., Vaughan, M., Klein, D.C.: Fed. Proc. 32, 252 abs. neuroendocrines IV (1973).

8. Chen, P.S. jr., Toribara, T.Y., Warner, H.: Anal. Chem. 28, 1756-1758 (1956).

9. Rembold, H.: Isolation of Pteridines by Ion-Exchange Chromatography. in: Methods of Enzymology, XVIII B, pp. 652-670, Eds. D.B. McCormick, L.D. Wright, Academic Press, New York-London 1971.

10. Lloyd, T., Markey, S., Weiner, N.: Anal. Biochem. 42, 108-112 (1971).

11. Zinbo, M., Sherman, W.R.: J. Am. Chem. Soc. 92, 2105-2113 (1970).

12. MacLean, F.I., Forrest, H.S., Meyers, J.: Arch. Biochem. Biophys. 114, 404-413 (1966).

13. Forrest, H.S., Baalen, C. van, Meyers, J.: Arch. Biochem. Biophys. 78, 95-99 (1958).

14. Lehninger, A.L.: Biochemistry, The Molecular Basis of Cell Structure and Function, pp. 322, Worth Publishers Inc, New York, second edition, 1975.

15. Goto, M., Kobayashi, K., Sato, H., Korte, F.: Ann. Chem. 689, 221-228 (1965).

16. Kamerling, J.P.: Structure Determination of Oligosaccharides, an Investigation of Pertrimethylsilyl Derivatives by Mass Spectrometry and PMR-Spectrometry, Thesis, Utrecht 1972.

17. Zurburg, W., Ebels I.: J. Neural Transm. 36, 59-69 (1975).

18. Cremer-Bartels, G., Hollwich, F.: J. Neural Transm. Suppl. 13, 360 (1978).

19. Balemans, M.G.M.: Progr. Brain Res. 52, 221-229 (1979).

20. Cremer-Bartels, G.: Progr. Brain Res. 52, 231-239 (1979).

21. Cremer-Bartels, G., Ebels, I.: Proc. Natl. Acad. Sci. U.S.A. 77, 2415-2418 (1980).

22. Balemans, M.G.M.: Indole Metabolism in the Pineal Gland, the Harderian Gland and the Retina in Mammals. in: The Pineal Organ, pp. 261-279, Eds. A. Oksche, P. Pévet, Elsevier/North Holland, Amsterdam-New York-Berlin 1981.

23. Balemans, M.G.M., Noordegraaf, E.M., Bary, F.A.M., Berlo, F.M. van: Experientia 34, 887-888 (1978).

24. Balemans, M.G.M., Ebels, I., Hendriks, H.G., Berlo, M.F. van: J. Neural Transm. 53, 293-303 (1982).

25. Balemans, M.G.M., Ebels, I., Hendriks, H.G., Berlo, M.F. van, Morée, A. de: J. Neural Transm. 56 - (1983).

26. Balemans, M.G.M., Ebels, I., Hendriks, H.G., Berlo, M.F. van, Morée, A. de: submitted for publication

27. Reiter, R.J.: The Pineal and Seasonal Reproduction. in: The Pineal Gland and its Endocrine Role. Eds. J. Axelrod G.P. Velo, F. Fraschini, Plenum Press, London (in press).

PTERIDINES IN BLOOD CELLS AND PLASMA

Methods for the quantification of free and conjugated (protein-bound) pteridines

Hans-Jörg Zeitler and Berta Andondonskaja-Renz

Institut für Hämatologie der Gesellschaft für Strahlen- und Umweltforschung,
Abteilung für Klinische Hämatologie
D-8000 München 2, Germany

Michael Fink and Wolfgang Wilmanns

III. Medizinische Klinik der Universität, Klinikum Großhadern
D-8000 München 70, Germany

Introduction

Pteridines are important cell metabolites in higher mammals (1) and some of them seem to have key positions. The studies of different biological mate= rials and the isolation of some pterins (2, 3) from the proliferation media of transformed cells led to the result that the metabolism of different pteridines and of folic acid in tumor cells must be altered: increased levels of 6-hydroxymethylpterin (3, 6), biopterin/tetrahydrobiopterin (2, 7) and of neopterin, as well as changes of the level of pterin, xan= thopterin and sepiapterin (8, 9) in some cancer patients were found. Recently, the importance of the determination of the biopterin level during autologous bone marrow transplantation in Beagle dogs (10) and during leukemia (11), resp., and of neopterin in hematological neoplasias (12) was reported.

The analyses in the different materials (e.g., liver, brain, urine, plasma, corpusculate blood fraction) for the investigated pteridines were achieved with methods of high-performance liquid chromatography (HPLC) (12, 13) and for Crithidia active pterins with the Crithidia bioassay (10, 11, 14, 15),

Biochemical and Clinical Aspects of Pteridines, Vol. 2

respectively. Above all the ion-exchange HPLC of STEA et al. (16) allowed a quantification of a great many pteridines in biological samples (urine, plasma). But in order to determine free and conjugated (protein-bound) pteridines the pteridine releasing procedure (17) for biological materials had to be optimized (18). The parameters which influence (19) pteridine release were investigated by us in detail (18) and are discussed here together with our HPLC results of normal and pathological (AML, acute myelocytic leukemia; ALL, acute lymphocytic leukemia; CLL, chronic lymphocytic leukemia; bronchial cancer) blood samples.

Using reverse-phase high-performance liquid chromatography (rp HPLC) on 5µ C_{18}-ODS (mobile phase, K_2HPO_4 buffer; pH = 7.1 - 7.8) up to 10 pteridines (pterin-6-carboxylic acid, xanthopterin, neopterin, monapterin, isoxanthopterin, lumazine, biopterin, 6-hydroxymethylpterin, pterin, 6-methylpterin) could be separated in the extracts of blood samples and were determined fluorometrically (femtomolar range). In addition, three sepiapterins (sepiapterin, desoxysepiapterin, 3'-hydroxysepiapterin) were separated with aqueous methanol as eluent (uv and fluorometrical detection, respectively.

The described methods are suited to compile pteridine patterns for cells and biological fluids, resp., because the determination of only one or two pteridines is not sufficient for an understanding and a biochemical interpretation of alterations in pteridine metabolism of pathological cells.

Materials and Methods

All glassware for pteridine containing material was siliconized with silicon solution (SERVA, Heidelberg, GFR). The chemicals were of HPLC and of p.a. quality, respectively. The pteridines were from Dr.B.Schircks (Wettswil a.A., Switzerland). The buffers were filtered through a 0.45 µ filter (type HA; MILLIPORE, Neu-Isenburg, GFR) and degassed by sonication or by bubbling with helium gas. All reverse-phase HPLC analyses (BECKMAN 112 Solvent Delivery Module; SPECTRA-PHYSICS integrator 4270; KONTRON spectrofluorometer SFM 23 for the fluorometry: exc/em = 360/460 nm, blue fluorescence, and 425/530 nm, yellow fluorescence, resp.) were carried out with selfpacked columns (250 x 4.6 mm; Spherisorb 5µ C_{18}-ODS; SHANDON HPLC packing pump; packing pressure, 550 bar and 8,000 psi, resp.;GYNKOTEK, München, GFR; temperature, 20°C; flow rate, 1 ml/min). The optimum pH value of the mobile phase (3.5 mM K_2HPO_4; pH, 7.10 - 7.80) has to be determined individually for each column by a test chromatogram with a pteridine standard solution.

Blood fractionation. The anticoagulated venous blood from human blood units or from patients was separated into plasma and the blood cells (erythrocytes, granulocytes, lymphocytes, buffy coat) by normal and by FICOLL centrifugation, respectively. The detailed procedure is described in (18).

Pteridine releasing procedure at 120°C (17). The autoclave was heated to 100°C and the samples which had been diluted with the eight-fold amount of a 0.07 M KH_2PO_4 buffer (pH, 4.45) were homogenized by sonication for some minutes and then autoclaved for 5 min at 120°C. Subsequently, the pressure in the autoclave was lowered to 0 bar within 10 minutes. The samples were centrifuged for 20 minutes at 70,000 x g and the supernatants frozen in liquid nitrogen and lyophilized. The residue of each sample was extracted with a minimum amount of HPLC buffer (mobile phase) by sonication. After centrifugation the clear supernatant was directly applied on the HPLC column.

Pteridine releasing procedure at 100°C (18). The cleavage of conjugated pteridines at this temperature is well reproducible since the pressure reduction, essential to the 120°C procedure, can be omitted here. According to this method the samples were put in the pre-heated (100°C) autoclave and cleaved for a definite time. The further experimental procedure was to be carried out as above.

Oxidation of reduced pteridines. For the oxidation of the unstable (13), reduced forms of some pteridines in tissues and fluids some oxidation methods (16, 19, 20) were described from which we adopted the oxidation procedure with iodine at acidic pH. The samples (plasma, cell fractions) were diluted with the eight-fold volume of bidistilled water and sonicated for some minutes. The oxidation was accomplished by addition to the acidified (pH = 1, with 1 N HCl) samples of a solution of 0.5% J_2/1% KJ in 0.1 N HCl

until iodine persisted. The magnetically stirred suspensions were incubated for at least 45 min because shorter incubation times in accordance with other laboratories (16) were not sufficient for a complete oxidation. The excess iodine was reduced by addition of a 1% solution of ascorbic acid until the starch test was negative. As there is no protein precipitation at pH = 1 the pH was brought to about 5 with 1 N KOH. Thereupon, the precipitated proteins were centrifuged down for 20 min at 70,000 x g ($-4^{o}C$) and the clear supernatant lyophilized. The residue was extracted with the minimum amount of HPLC buffer (mobile phase) and after centrifugation the supernatant was directly applied for HPLC analysis.
To optimize the analysis for free and conjugated pteridines in cells and protein containing fluids we completed the oxidation procedure by a pteridine releasing for 20 min at $100^{o}C$ as described above.

Results and Discussion

1) From the literature we know that there is a correlation between the pteridine levels and some diseases. According to our first analyses this correlation could also be suggested for different hematological and tumor diseases. Because the Crithidia bioassay and the cofactor assay (21) are relevant only for biopterin, biopterin-like and Crithidia active substances, resp., with the different kinds of HPLC we have suitable and sensitive systems for the analysis of pteridines in biological materials. Most of the reported methods, however, enable separation of only one or a few unconjugated pteridines in biological samples (mostly urine). Although some HPLC systems have been described recently for the analysis of different pterins (13, 16, 22) these methods were not suited for our pteridine screening because lumazine cannot be separated from biopterin. Moreover, ion-exchange techniques are difficult to perform on the small extraction volumes from lymphocytes and buffy coat. For our studies we prefer the reverse-phase HPLC on 5µ C_{18}-ODS with a 3.5 mM potassium phosphate buffer (pH, 7.1 - 7.8) as mobile phase with which we separate (see Figure 1) up to 10 pteridines. The sensitivity was about 7.5 - 20 pg for most pteridines at a signal/noiseratio of about 3. The yellow fluorescing pteridines (3'-hydroxysepiapterin, sepiapterin, des-

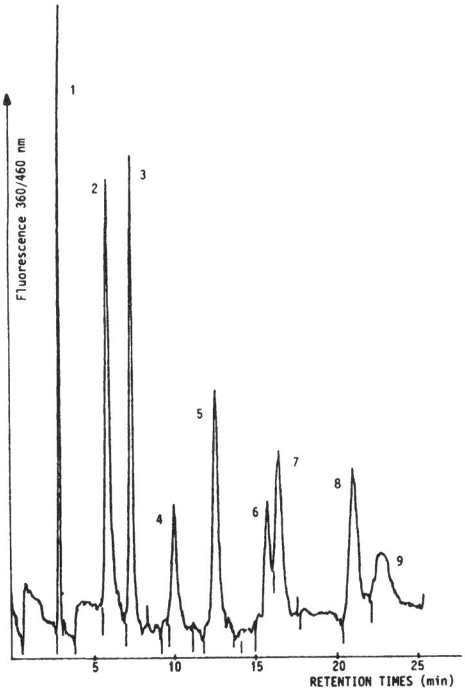

Fig.1 Reverse-phase HPLC separation on 5µ C_{18}-ODS of a standard solution of 10 pteridines (isocratic elution). Abbreviations: 1, pterin-6-carboxylic acid; 2, xanthopterin; 3, neopterin; 4, monapterin; 5, isoxanthopterin; 6, lumazine; 7, biopterin; 8, 6-hydroxymethylpterin; 9, pterin; 10, (not shown; retention time, approx. 35 min) 6-methylpterin.

oxysepiapterin) were determined on 5µ C_{18}-ODS with 15% methanol-85% water (isocratic elution) as eluent.

2) In tissues and protein containing fluids we have to differentiate between the free and the protein-bound (conjugated) pteridines. For this reason, an effective releasing procedure for the determination of the conjugated part was needed. A simple deproteinization (e.g., precipitations with alcohols) appeared to be unsuitable for plasma and blood cells because under these conditions the complete cleavage of the pteridine-protein-binding could not be ensured nor could the possibility of

a re-adsorption of released pteridines on the protein precipitation be excluded. An effective procedure was that of BAKER et al. (17) with which the release of the conjugated pteridines is achieved by an irre= versible heat denaturation (autoclaving at 120°C).
We have studied different parameters of this procedure and prefer the well reproducible and more gentle modification (18) at 100°C (without excess pressure). However, according to FUKUSHIMA et al. (19) the reduced forms of biopterin are air-oxidized and converted to biopterin, pterin, xanthopterin, sepiapterin and deoxysepiapterin in different buffers and at elevated temperatures. Under the described conditions of pteridine release at 100°C we found a relatively constant conversion of tetrahydro= biopterin to xanthopterin (1 - 2%), biopterin (approx. 26 %), pterin (approx. 23 %), sepiapterin (approx. 0.8%) and deoxysepiapterin (5 - 6%). 6-Hydroxymethylpterin was found in only trace amounts. At 120°C the conver= sion of tetrahydrobiopterin to the above mentioned pteridines was increa= sed by about 35 - 50 %; but when tetrahydrobiopterin and other reduced pteridines were incubated at 100°C with human whole blood (blood unit) the decomposition was lowered to some extent. The reason for this variation is not known and requires further investigation. Similar and characteristic results were obtained for other reduced forms of pteridines: dihydrobio= pterin was converted mainly to xanthopterin, biopterin and 6-hydroxymethyl= pterin. Pterin-6-carboxylic acid and pterin were found only in small amounts. Tetrahydroneopterin and dihydroneopterin gave mainly xanthopterin, neopterin, 3'-hydroxysepiapterin and pterin (only from tetrahydroneopterin). Pterin and 6-hydroxymethylpterin were produced from dihydroneopterin only in small amounts. The most important fact was that isoxanthopterin and lumazine were n o t formed from any reduced form of biopterin or neo= pterin, respectively.
The recovery experiments for some pteridines (neopterin, monapterin, luma= zine, biopterin, pterin) were made with plasma and erythrocytes under the conditions of the iodine oxidation and of the releasing procedure, respec= tively. The results are given in Table 1 and Table 2, respectively.
The influence of the temperature and of the heating time on the pteri= dine release from blood samples can be seen in Table 3 and it should be clear that the 100°C modification is more suitable than the autoclaving procedure at 120°C.

Table 1.

Recovery of pteridines from erythrocytes.
To each of 14 samples of erythrocytes (20 ml) were given a solution of 4 pteridines (neopterin, 2.44 µg; monapterin, 2.49 µg; lumazine, 4.97 µg; biopterin, 2.35 µg). 6 Samples were oxidized with iodine and 8 samples were heated for 20 min to 100°C (releasing procedure). The lyophilized supernatants of the centrifuged samples (20 min at 70,000 x g) were dis=solved each with 20 ml of HPLC buffer and analyzed three times by rp HPLC (see "Materials and Methods").- The table gives the overall recovery of the investigated pteridines (mean value of all samples in per cent) and the individual recovery from each erythrocyte sample (mean value of 3 analyses in per cent).

	Neopterin	Monapterin	Lumazine	Biopterin
OVERALL RECOVERY				
Iodine oxidation	62.9 ± 7.6	58.9 ± 6.0	53.4 ± 5.1	61.9 ± 6.7
Releasing procedure	88.5 ± 11.4	86.9 ± 8.9	86.4 ± 11.9	90.2 ± 8.6
INDIVIDUAL RECOVERY				
Iodine oxidation				
Sample 1	66.0 ± 3.1	62.7 ± 2.0	53.1 ± 2.3	63.0 ± 0.8
2	68.0 ± 1.1	63.5 ± 1.1	56.1 ± 4.0	65.1 ± 2.7
3	62.7 ± 4.1	58.6 ± 2.0	53.9 ± 2.4	62.1 ± 1.7
4	58.6 ± 0.8	53.8 ± 0.7	54.1 ± 1.2	58.3 ± 0.6
5	59.8 ± 0.5	57.4 ± 1.1	50.7 ± 1.0	60.9 ± 2.1
6	62.3 ± 2.0	57.8 ± 2.5	52.3 ± 0.7	62.1 ± 1.3
Releasing procedure				
Sample 1	82.9 ± 3.4	82.3 ± 4.0	91.6 ± 0.5	88.1 ± 7.2
2	89.1 ± 1.1	86.7 ± 0.7	81.1 ± 1.4	90.2 ± 3.7
3	78.5 ± 1.2	78.6 ± 0.6	74.9 ± 0.8	91.9 ± 0.5
4	91.3 ± 6.6	87.3 ± 1.6	80.7 ± 0.9	85.1 ± 2.3
5	91.1 ± 1.6	89.7 ± 2.0	95.6 ± 3.9	91.1 ± 4.0
6	98.6 ± 1.3	92.2 ± 1.3	86.4 ± 5.5	90.2 ± 0.7
7	90.2 ± 0.6	90.9 ± 1.1	89.9 ± 6.4	95.3 ± 3.5
8	87.9 ± 1.6	88.3 ± 3.3	84.4 ± 1.5	86.8 ± 1.8

Table 2.

Recovery of pteridines from plasma.
To each of 20 samples of plasma (20 ml) were given a solution of 4 pteri= dines (neopterin, 2.44 µg; monapterin, 2.49 µg; lumazine, 4.97 µg; bio= pterin, 2,35 µg). 10 Samples were oxidized with iodine and 10 samples were heated for 20 min to 100°C (releasing procedure). For details see Table 1.- The table gives the overall recovery of the investigated pteri= dines (mean value of all samples in per cent) and the individual recovery from each plasma (mean value of 3 analyses in per cent).-
The mean recovery of a mixture of biopterin (2.57 µg) and pterin (3.04 µg) from 20 ml of a plasma after the oxidation step was found to be 95.9 % (± 3.2 %) and 82.5 % (± 3.1 %), respectively.

	Neopterin	Monapterin	Lumazine	Biopterin
OVERALL RECOVERY				
Iodine oxidation	96.1 ± 8.0	90.0 ± 7.1	78.1 ± 5.1	93.3 ± 7.9
Releasing procedure	98.5 ± 5.7	96.4 ± 5.0	93.8 ± 6.0	96.4 ± 5.3
INDIVIDUAL RECOVERY				
Iodine oxidation				
Sample 1	92.0 ± 3.2	88.6 ± 2.0	76.3 ± 2.3	87.8 ± 1.4
2	96.4 ± 2.1	91.2 ± 3.8	77.3 ± 1.7	87.0 ± 2.0
3	98.6 ± 1.7	94.8 ± 3.0	79.9 ± 1.1	86.9 ± 2.1
4	94.5 ± 0.9	94.1 ± 2.1	78.9 ± 2.2	94.2 ± 1.5
5	99.2 ± 2.3	94.5 ± 1.7	78.5 ± 1.7	98.3 ± 3.5
6	96.3 ± 2.8	96.5 ± 2.6	75.6 ± 1.4	98.8 ± 0.8
7	95.3 ± 0.9	94.9 ± 3.7	79.2 ± 1.6	91.9 ± 5.1
8	95.3 ± 4.2	84.8 ± 2.0	79.3 ± 3.5	99.7 ± 0.8
9	98.2 ± 1.4	95.7 ± 2.5	76.4 ± 1.0	94.7 ± 3.6
10	94.3 ± 2.2	97.7 ± 1.6	78.3 ± 3.6	93.8 ± 4.0
Releasing procedure				
Sample 1	100.0 ± 0.0	92.2 ± 1.4	95.0 ± 3.9	95.8 ± 0.8
2	99.8 ± 3.5	94.4 ± 1.0	95.4 ± 1.4	98.3 ± 0.9
3	100.4 ± 2.8	97.6 ± 0.2	83.3 ± 0.8	94.5 ± 1.7
4	99.6 ± 1.1	95.6 ± 0.3	93.6 ± 4.0	83.6 ± 1.8
5	96.3 ± 5.1	94.4 ± 2.7	90.3 ± 2.6	94.0 ± 0.8
6	95.1 ± 1.7	96.4 ± 3.4	95.8 ± 0.9	97.0 ± 3.4
7	97.1 ± 2.7	97.2 ± 3.3	91.0 ± 3.2	98.7 ± 2.9
8	101.2 ± 0.9	92.8 ± 1.4	98.0 ± 3.8	98.3 ± 2.8
9	93.8 ± 3.5	94.7 ± 5.1	92.9 ± 5.3	93.6 ± 1.3
10	99.2 ± 2.6	97.2 ± 3.3	95.0 ± 4.7	95.3 ± 2.6

Table 3

Influence of the temperature and of the heating time on pteridine releasing.

Details of the releasing procedure and of the HPLC system are described in "Methods and Materials".

Protein concentrations: 48.4 mg/ml plasma (CLL) and 65.1 mg/ml lymphocytes (CLL), respectively.

	releasing min	°C	pterin-6-carboxylic acid	xanthopterin	neopterin	monapterin	isoxanthopterin	lumazine	biopterin	6-hydroxymethylpterin	pterin
		pg/ml									
Whole blood (blood unit)	5	120	2,527	---	2,041	656	535	---	8,092	---	13,559
	20	100	1,312	3,013	2,552	1,361	851	6,926	10,911	---	13,049
plasma (CLL)	5	120	2,368	464	5,703	1,491	98	---	399	458	1,171
	15	100	458	798	4,232	523	33	---	190	124	543
	25	100	700	536	4,807	1,145	59	---	320	203	517
lymphocytes (CLL)	5	120	---	116	85,078	20,930	446	2,597	911	5,833	53,682
	15	100	---	---	83,527	17,442	678	3,314	717	1,434	49,030
	25	100	---	---	82,945	19,574	368	2,926	1,143	2,965	55,620

The only sensitive pteridines are xanthopterin and isoxanthopterin and, for that reason, the optimum cleavage time should be reached after 20 minutes.Because there were no detectable amounts of sepiapterins in the human whole blood, unlike in Beagle dogs (18), and the levels of biopterin, neopterin and pterin remained relatively constant, normal human blood should not have highly elevated levels of reduced forms of biopterin and/or neopterin. However, the drastic increase of 6-hydroxymethylpterin in lymphocytes of a CLL-patient with increasing temperature could be caused to some extent by an elevated level of dihydrobiopterin as mentioned above.

3) To examine the influence of oxidation of reduced pteridines on the pteridine analyses subsequent to the releasing procedure at 100°C we adopted the iodine method (16, 19, 20). Due to our phosphate containing HPLC buffer we compared the iodine oxidation of the reduced forms of biopterin at pH = 1 in an aqueous solution of hydrochloric acid with the oxidation in an phosphoric acid solution at different incubation times (30 min, 45 min, 60 min). The best results were achieved in the HCl solution with an incubation time of at least 45 min, the more so as the oxidation in phosphoric acid led to pterin as a by-product.

The influence of the oxidation procedure on the pteridine release and a comparison of the analyses with the one following the releasing procedure are shown in Table 4. In general, in the plasma and blood cells of normal human blood (blood units) the yield of some pteridines (neopterin, monapterin, isoxanthopterin, biopterin) were improved but not necessarily as a result of the oxidation of reduced pteridines. This could be deduced from the constant pterin values and the increasing plasma levels of 6-hydroxymethylpterin. The decreasing 6-hydroxymethylpterin levels in the erythrocytes could possibly correlate with the increasing neopterin (conversion of dihydroneopterin).

A disadvantage of the oxidation procedure is the insufficient protein precipitation. As even at pH 5 the amounts of the precipitated proteins were substantially less than in the releasing procedure at elevated temperatures the lyophilized residues of the oxidized samples (especial-

Table 4

Influence of the iodine oxidation on the pteridine releasing in plasma and in blood cells.
(The pteridine concentrations in the buffy coat refer to pg/ml blood)

Sample	method / pg/ml	pterin-6-carboxylic acid	xanthopterin	neopterin	monapterin	isoxanthopterin	lumazine	biopterin	6-hydroxymethyl-pterin	pterin
plasma (blood unit)	releasing	3,086	656	1,944	680	705	---	1,555	851	4,204
	oxidation	4,374	510	3,572	1,142	948	---	2,090	1,580	4,350
	oxidation + releasing	4,229	3,402	2,989	1,920	1,045	---	4,034	3,086	4,253
erythrocytes (blood unit)	releasing	4,374	1,312	5,006	729	729	875	2,454	705	16,354
	oxidation	3,645	1,750	6,707	3,378	1,215	---	2,309	389	16,475
	oxidation + releasing	5,176	1,507	12,879	2,916	1,312	---	2,880	---	17,010
buffy coat (blood unit)	oxidation	26	3	41	7	---	---	---	---	30
	oxidation + releasing	21	3	34	7	---	---	---	---	31
lymphocytes (CLL)	releasing	18,837	---	80,059	18,837	---	43,326	2,355	3,673	32,695
	oxidation	42,384	---	279,264	23,547	---	41,913	3,720	235	63,906
	oxidation + releasing	61,221	---	262,310	21,663	---	7,535	3,391	---	60,374

ly plasma) were hard to extract with small buffer volumes and led to dis= proportionate high dilutions. To achieve the oxidation of reduced pteri= dines and optimal pteridine release we combined both methods and obtained some interesting results (see Table 4): some pteridines in normal human blood were unaffected (e.g., pterin !) but some of them were considerably improved (xanthopterin, monapterin, isoxanthopterin, biopterin and 6-hydr= oxymethylpterin in plasma; neopterin in erythrocytes). But possibly, the releasing procedure could be substituted by another suited method because their is no difference between the oxidation and the releasing procedure.

Altered pteridine levels and dramatic metabolic alterations were clearly shown from the analyses of CLL-lymphocytes (see Figure 2).

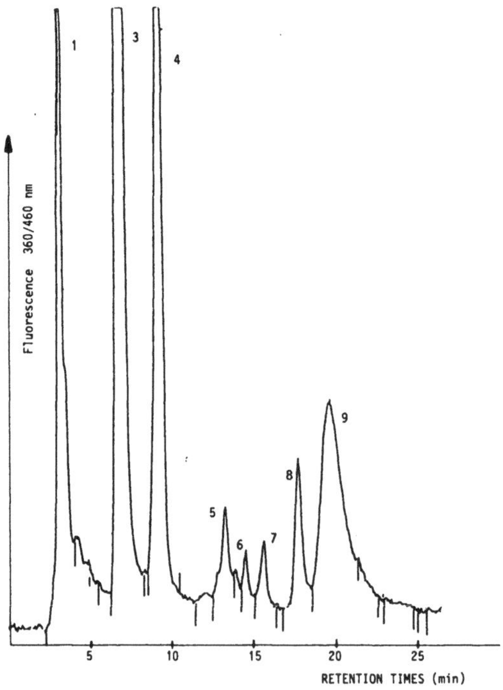

Fig. 2 Rp HPLC on 5μ ODS of human lymphocytes (chronic lymphocytic leukemia) following pteridine release for 20 min at 100 °C. Peak identification, see Figure 1.

The very high amounts of lumazine and pterin could not be attributed to a conversion (releasing procedure) of tetrahydrobiopterin because lumazine is no conversion product and the pterin values were increased by oxidation (xanthopterin was not detectable). In contrast to other samples, in CLL-lymphocytes we found 3'-hydroxysepiapterin (but no sepiapterin and desoxysepiapterin) following the releasing procedure which disappeared completely after iodine oxidation. Together with the drastic increase of the neopterin after oxidation a strongly disturbed pteridine metabolism in the investigated CLL-lymphocytes with an overproduction of tetrahydroneopterin is suggested. An additional contribution of dihydrobiopterin could be assumed from the increasing biopterin and the decreasing 6-hydroxymethylpterin, respectively.

4) The described methods were also applied for the quantification of pteridines in human plasma and blood cells of 3 kinds of leukemia and of bronchial cancer. As can be read from Table 5 the pteridine levels in the plasma and in the different cell fractions are quite different for the investigated dideases and the analyses indicate a correlation between the desease state and the pteridine content.

Another study on human CLL-lymphocytes following leukophoresis or after incubation in plasma for 5 days at +4^{0}C, resp., revealed that these cells seem to have the most intensive (most altered ?) pteridine metabolism of the investigated types of leukemia. Differences in the pteridine levels of the lymphocytes during the incubation were evident (Table 6) for neopterin (↓), lumazine (↑ ; lacking in the plasma), biopterin (⬇) and pterin (↑), whereas in the plasma the more drastic alterations were observed.

Table 5 Pteridine levels in plasma and blood cells of different leukemias and of bronchial cancer, resp. (releasing without oxidation). Concentrations in parentheses refer to pg/mg protein.

Sample	desease	pterin-6-carboxylic acid	xantho-pterin	neo-pterin	mona-pterin	iso-xantho-pterin	lumazine	bio-pterin	6-hydroxy-methyl-pterin	pterin
plasma (pg/ml)	AML	579 (15)	618 (16)	17,686 (465)	943 (25)	176 (5)	---	1,495 (40)	306 (8)	832 (22)
	ALL	---	943 (37)	4,668 (186)	1,671 (67)	1,008 (40)	---	2,159 (86)	---	---
	CLL	1,021 (27)	780 (21)	7,217 (190)	904 (24)	429 (11)	---	618 (16)	---	956 (25)
	bronchial cancer	235 (6)	200 (5)	1,038 (28)	260 (7)	112 (3)	---	532 (14)	---	---
erythrocytes (pg/ml)	AML	658 (5)	---	2,556 (21)	551 (5)	1,118 (9)	---	---	107 (1)	---
	ALL	31 (0.3)	429 (5)	1,867 (20)	796 (9)	123 (1)	---	474 (5)	276 (3)	1,133 (12)
	CLL	---	---	1,516 (12)	857 (7)	---	---	689 (6)	---	---
	bronchial cancer	520 (4)	880 (7)	4,880 (40)	1,420 (12)	---	2,010 (16)	1,450 (12)	5,170 (42)	14,160 (116)
lymphocytes (pg/ml blood)	AML	---	---	---	---	---	---	17	---	---
	ALL	---	---	---	30	---	---	28	---	---
	CLL	6,797 (486)	922 (66)	26,710 (1,911)	3,620 (259)	132 (9)	2,770 (198)	554 (40)	327 (23)	8,860 (634)
buffy coat (pg/ml blood)	bronchial cancer	---	110 (57)	640 (330)	89 (40)	2 (1)	---	10 (5)	294 (151)	439 (226)

Table 6

Alterations in the pteridine levels of incubated human leukophoresis samples (CLL).
The plasma and the lymphocytes were analysed (rp HPLC)
(a) directly after a leukophoresis of a CLL patient (approx. 10^9 lymphocytes per ml and
(b) after 5 days of incubation of the lymphocytes in plasma at $+4^{o}C$.

pg/ml	plasma (a)	plasma (b)	lymphocytes (a)	lymphocytes (b)
pterin-6-carboxylic acid	700	2,557	---	3,062
xanthopterin	536	98	---	---
neopterin	4,807	13,080	82,945	59,690
monapterin	1,145	3,793	19,574	21,124
isoxanthopterin	59	59	368	407
lumazine	---	---	2,926	4,535
biopterin	320	307	1,143	368
6-hydroxymethylpterin	203	1,432	2,965	2,597
pterin	517	5,847	55,620	62,791

References

1. Kaufman, S.: Proc. Nat. Acad. Sci. USA 50, 1985-1092 (1963).

2. Albrecht, A.M., Biedler, J.L., Baker, H., Frank, O. and Hutner, S.H.: Res. Commun. Chem. Pathol. Pharmacol. 19, 377-380 (1978):

3. Stea, B., Backlund, P.S., Berkey, P.B., Cho, A.K., Halpern, B.C., Halpern, R.M. and Smith, R.A.: Cancer Res. 38, 2378-2384 (1978).

4. Barford, P.A. and Blair, J.A.: in Chemistry and Biology of Pteridines (W.Pfleiderer, ed.) pp. 413-427, Walter de Gruyter, Berlin (1975).

5. deleted

6. Trehan, S., Rao, K.N., Shetty, P.A. and Noronha, J.M.: Cancer 50, 114-117 (1982)

7. Kokolis, N. and Ziegler, I.: Cancer Biochem. Biophys. 2, 79-85 (1977).

8. Wachter, H., Hausen, A. and Grassmayer, K.: Ber. ÖGKC 2, 90-91 (1979).

9. Rokos, H., Rokos, K., Frisius, H. and Kirstaedter, H.J.: Clin. Chim. Acta 105, 275-286 (1980).

10. Ziegler, I., Fink, M. and Wilmanns, W.: Blut 44, 231-240 (1982).

11. Ziegler, I., Kolb, H.J., Bodenberger, U. and Wilmanns, W.: Blut 44, 261-270 (1982).

12. Hausen, A., Fuchs, D., König, K. and Wachter, H.: J. Chromatogr. 227, 61-70 (1982).

13. Fukushima, T. and Nixon, J.C.: Anal. Biochem. 102, 176-188 (1980).

14. Leeming, J.R. and Blair, J.A.: Clin. Chim. Acta 108, 103-111 (1980).

15. Ziegler, I. and Fink, M.: Biochem. Medicine 27; 401-404 (1982).

16. Stea, B., Halpern, R.M., Halpern, B.C. and Smith, R.A.: J. Chromatogr. 188, 363-375 (1980).

17. Baker, H., Frank, O., Shapiro, A. and Hutner, S.H.: in Methods of Enzymology (McCormick, D.B. and Wright, L.D., eds.), Academic Press New York, Vol. 66, pp. 490-500 (1980).

18. Andondonskaja-Renz, B. and Zeitler, H.J.: Anal. Biochem. (in press) 132 (1983).

19. Fukushima, T. and Nixon, J.C.: in Chemistry and Biology of Pteridines (Kisliuk, R.L. and Brown, G.M., eds.) Elsevier, North-Holland, pp. 31-34 (1979).

20. Fukushima, T., Kobayashi, K.I., Eto, I. and Shiota, T.: Anal. Biochem. 89, 71-79 (1978).

21. Kaufman, S., Berlow, S., Summer, G.K., Milstien, S., Schulman, J.D., Orloff, S., Spielberg, S. and Pueschel, S.: N. Engl. J. Med. 299, 673-679 (1978).

22. Stea, B., Halpern, R.M. and Smith, R.A.: J. Chromatogr. 168 385-393 (1979).

2

BIOCHEMISTRY AND METABOLISM OF PTERIDINES

PTERIDINE CATABOLISM

Heinz Rembold

Max-Planck-Institute for Biochemistry,
D-8033 Martinsried, Germany

Catabolism of pterin pools

The mammalian pterin cofactor pools are balanced by different factors (1). Besides biosynthesis, which may be regulated through hormonal or feedback control of the circulating blood pterins, and by compartmentation in different organs (2), a constant pterin cofactor level is maintained in the higher organisms by excretion and by catabolism. Reduced pterins are transported by high affinity binding proteins (3) and are balanced in the kidney primarily. Most of the pterins, once filtered through the renal tubules, are not reabsorbed but are excreted with the urine (4,5). The kidney therefore seems to play a central role for the pterin cofactor balancing in the mammal. Little is known, however, about the origin of such pteridines which have a completely dehydrogenated nucleus, like biopterin, neopterin, xanthopterin etc., and which are common in the urine. Are these compounds really catabolites with no function or are there other properties of the pteridine ring besides the role as a redox catalyst in oxidases? A dihydropterin oxidase has been described which uses a

Biochemical and Clinical Aspects of Pteridines, Vol. 2

variety of 7,8-dihydropterins (dihydropterin, 6-lactoyl-7,8-dihydropterin, 7,8-dihydrobiopterin) as substrates and converts them to the fully oxidized pterins (6). Such an enzyme could be of importance for the synthesis of isoxanthopterin from dihydropterin which is the central catabolite in pterin cofactor breakdown (1). This aspect will be discussed later on.

At least in mammals, the pterin cofactors seem to be excreted in different forms, as can be seen after injection of the labelled pterins (Table 1). After i.p. injection of tetrahydrobiopterin, only 15 percent of the total excreted label is present as biopterin, whereas almost half of it has been degraded to xanthopterin and its deamination product, 6-hydroxylumazine. The almost identical catabolism is shown by tetrahydroneopterin. According to these results, two groups of enzymes seem to dominate pterin catabolism, xanthine oxidase which catalyzes the synthesis of xanthopterin, isoxanthopterin, leucopterin, and of their corresponding lumazine derivatives, and the pterin deaminases which control another irreversible step in pterin breakdown, i.e. hydrolytic deamination of the pterin to its corresponding lumazine.

Oxidation reactions

Three enzymes are capable of pteridine oxidation, xanthine oxidase, aldehyde oxidase, and xanthine dehydrogenase. Milk

Tab. 1

Metabolites identified in rat urine after intraperitoneal injection of ^{14}C-labeled reduced pterins. Yields are expressed as percentage of total excreted radioactivity.

metabolites (%)	[2-^{14}C]tetrahydrobiopterin (0.9 mg)	[2-^{14}C]tetrahydroneopterin (0.9 mg)	[$^{14}C_5$]tetrahydroneopterin (3.6 mg)	[$^{14}C_3$]tetrahydro-6-hydroxymethylpterin (3.6 mg)	[$^{14}C_3$]7.8-dihydro-6-hydroxymethylpterin (2.6 mg)
biopterin	15	0	0	0	0
neopterin	0	9.6	10	0	0
xanthopterin +	18.6	8.9	3.5	19	25.3
7.8-dihydroxanthopterin	0	1.9	0.7	0	0
6-hydroxylumazine	27	30.8	37	19.2	3.2
6-hydroxymethylpterin	0	0	0	20	40

xanthine oxidase was the first enzyme which was studied for its activity against pteridine substrates and most of our present knowledge on pteridine catabolism has been collected from studies with this enyzme. It shall be discussed in more detail therefore.

Xanthine oxidase (xanthine : oxygen oxidoreductase EC 1.2.3.2) is a metalloflavoprotein which contains FAD, molybdenum, and iron in the ratio of 1:1:4. Labile sulfide and cysteinyl residues may be involved in the active site of the enzyme. Johnson et al. (7,8) isolated a molybdenum cofactor from chicken liver sulfite oxidase which seems to be common for all the molybdenum-containing enzymes. The authors proposed the following stucture (I) which could be degraded and excreted as urothione (II) with the urine:

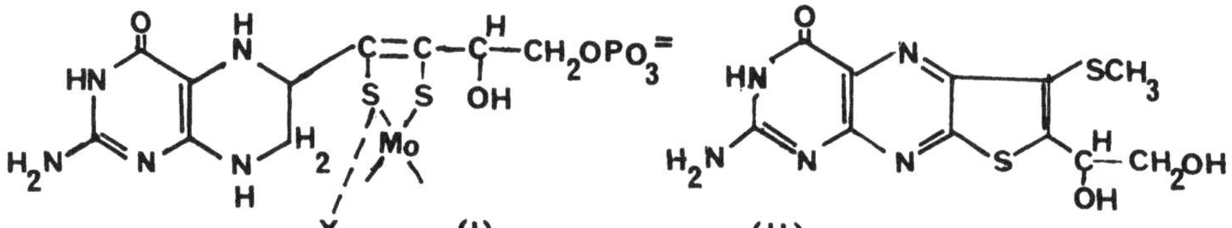

(I) (II)

Xanthine oxidase catalyzes oxidation of xanthine, hypoxanthine and other purines, aldehydes, NADPH, and several pteridines. In the order of one hundred different compounds are known to serve as substrates and electron acceptors. For our present context three pterins (and their respective lumazines) are of interest. Dihydropterin (III) is converted to 7,8-dihydroxanthopterin IV (9) which as such is no further substrate, whereas xanthopterin (V) is converted to leucopterin (VI) by xanthine oxidase. Vice versa, pterin (VIII) is not hydroxylated in position 6 but only in position 7 which finally yields isoxanthopterin (VIII):

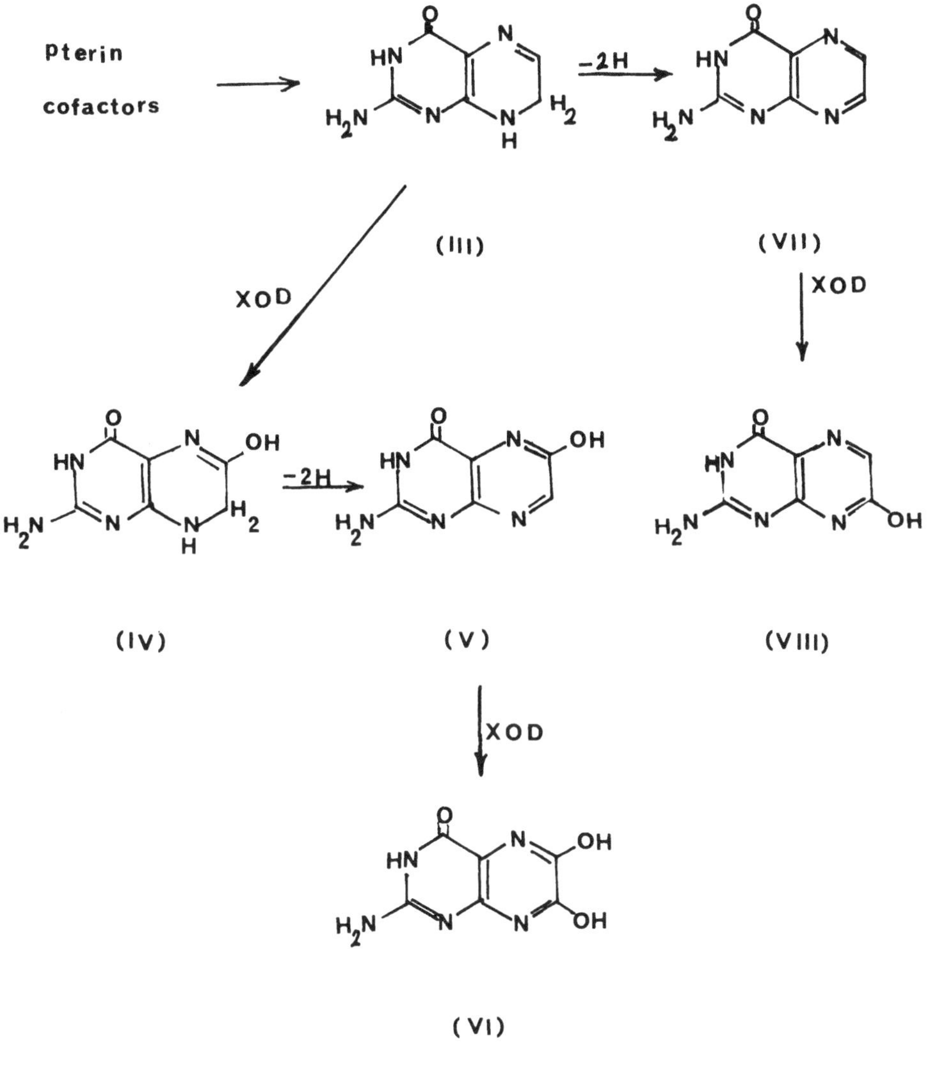

It is easy to understand from this scheme, that one of the main pterins excreted by man is xanthopterin (Table 2). Its amount is almost identical with that of biopterin and exceeds that of neopterin and of isoxanthopterin by a factor of 2-3.

Tab. 2
Urinary excretion of main pterins by man. The pteridine leads are expressed in µmol pteridine per mol creatinine.

biopterin	300-1100	xanthopterin	200-700
neopterin	100- 200	isoxanthopterin	120

Pterin-6-carboxaldehyde is a strong competitive inhibitor of xanthine oxidase with an inhibition constant of about 10^{-9}M (10-12). Xanthopterin is another effective inhibitor with K_i 1.6×10^{-6}M (13). However, no decrease in xanthine oxidase activity was found in such animals which had been treated with these two compounds.

Aldehyde oxidase (aldehyde : oxygen oxidoreductase EC 1.2.3.1) is another molybdenum - containing enzyme with many characteristics in common with xanthine oxidase. Both these enzymes are found in mammalian liver (14) which can be distinguished by inhibition analysis from each other. Xanthine oxidase (but not aldehyde oxidase) is inhibited more than 95 % by 1×10^{-4}M allopurinol, whereas aldehyde oxidase (but not xanthine oxidase) is more than 90 %

inhibited by 1x10^{-5}M menadione (2-methyl-2,4-naphthoquinome, 15).

Xanthine dehydrogenase (xanthine : NAD^{+} oxidoreductase EC 1.2.1.37) is the third enzyme which uses pteridines as substrates. One basic difference from xanthine oxidase is the inability of the dehydrogenase to use molecular oxygen as the immediate electron acceptor. The enzyme has been purified from different sources, mainly from insects, and its properties are similar to xanthine oxidase. It is inhibited by its reaction products, leucopterin and uric acid, and by dihydroxanthopterin.

Deamination reactions

Two pterin aminohydrolases have been purified and characterized: pterin deaminase (EC 3.5.4.11) and sepiapterin deaminase (EC 3.5.4.24). Some characteristics of various pteridine deaminases are collated in Table 3.

Pterin deaminase (2-amino-4-hydroxypteridine aminohydrolase EC 3.5.4.11) has been first reported in the bacterium, Alcaligenes metalcaligenes (16). When grown on 6-carboxypterin as the sole source of carbon and nitrogen, this compound is hydrolyzed to the corresponding 6-carboxylumazine. Pteroylglutamate is deaminated to the 2-deamino products. Another bacterial pterin deaminase was purified from Bacillus megaterium (17). This enyzme is relatively

Tab. 3
Characteristics of different pteridine deaminases

characteristic	Pterin deaminase EC 3.5.4.11 A.metalcaligenes	 rat	sepiapterin deaminase EC 3.5.4.24 B.mori	bacterial pterin deaminase Bacillus megaterium	isoxanthopterin deaminase B.mori	 P.americana
pH optimum	6.3-6.7	6.5	8.0	7.3	6.6	6.6
stability	unstable at pH 7, 0^oC; stable at pH 9, 0^oC	unstable at pH 6.5, 4^oC; activity decreases: 50% in 5 days	stable	unstable at pH 7, 4^oC	unstable at pH 7, 0^oC; activity decreases: 50% in 5 days	unstable at pH 7, 0^oC
K_m		$3x10^{-5}$mol/l	$5.9x10^{-4}$mol/l	$1.3x10^{-3}$mol/l	$1.2x10^{-5}$mol/l	
substrates	pterin and all 6-substituted derivatives	tetrahydropterin, pterin, isoxanthopterin	sepiapterin and isosepiapterin	pterin, 6-carboxypterin, biopterin, xanthopterin, 6-hydroxymethylpterin, sepiapterin, isosepiapterin, folate, 6.7-dimethylpterin, 7-methylpterin		
not deaminated	xanthopterin, 7-methylpterin, 7-carboxypterin, 6.7-dimethylpterin	xanthopterin, biopterin, neopterin, 6-carboxypterin	xanthopterin, pterin, 6-methylpterin, isoxanthopterin	isoxanthopterin, leucopterin, 6-methyl-isoxanthopterin, 7-carboxypterin	xanthopterin, leucopterin, pterin, 6-carboxypterin, 7.8-dihydropterin	
inhibitors	p-chloromercuribenzoate, NaF	KCN, lumazine, azaguanine	pterin, biopterin, xanthopterin	p-chloromercuribenzoate, 8-azaguanine	KCN, KF, xanthopterin, guanine	KCN, KF, xanthopterin, 8-azaguanine

unspecific, however. Presence of pterin deaminase in the rat can be followed from the high excretion of 6-hydroxglumazine after injection (Table 1) or incubation with rat liver homogenate of the reduced pterin cofactors (18,19). The rat liver pterin deaminase responsible for lumazine formation has been purified and separated from guanase (20). In contrast to the bacterial enzymes, rat liver pterin deaminase has a very high substrate specificity (Table 3). The natural substrate seems to be tetrahydro- or dihydropterin.

Sepiapterin deaminase (sepiapterin animohydrolase EC 3.5.4.24) has been isolated by Tsusue from the silkworm, Bombyx mori (21). It specifically catalyzes the hydrolysis of sepiapterin to 2-hydroxy-(2-deamino)-sepiapterin (xanthopterin B_2). Isoxanthopterin deaminase was discovered by Gyure (22) in a crude extract from Bombyx mori. This seems to be another very specific aminohydrolase. It is not known, however, whether these two enzymes are also present in mammals.

Catabolism of pterins in insects

Insects sometimes contain large amounts of pterins and lumazines which either may be used as pigments or may have still unknown biological functions. For example, isoxanthoperin and other pterins are stored in the fat body of Bombyx mori during the spinning stage. In the pupal stage, the fat body is degraded and stored 7-hydroxylumazine is released to the meconium and to the developing eggs. Isoxanthopterin deaminase is present in the developing adult wings and integuments and 7-hydroxylumazine is produced in these tissues (23). Such transport phenomena indicate a biochemical function for such simple pterin derivatives like xanthopterin or isoxanthopterin under certain conditions of development.

As a result of studies with labelled precursors, Descimon (24) distinguishes three branches in the metabolic conversion of pierid pteridines:

(1) The isoxanthopterin branch includes dihydropterin, pterin, and isoxanthopterin.

(2) The xanthopterin branch includes dihydropterin, 7,8-dihydroxanthopterin, xanthopterin, leucopterin, and erythropterin.

(3) The sepiapterin branch is on the normal biosynthetic route to the pterin cofactors.

Whereas these studies were done with Colias croceus, Harmsen (25) followed the pteridine pattern in Pieris brassicae. These authors propose two alternative pathways: the Diptera pathway without C-6 oxidation, and the Lepidoptera pathway with oxidation at C-6. The relative importance of these two pathways is influenced by the oxygen partial pressure in which the pupae are grown. However, whether such schemes are useful for the interpretation of pterin pathways is still open for further investigations. Table 4 gives an example from an extensive analysis of *Apis mellifera* pteridines during two developmental stages (26). Obviously synthesis and storage of the 6-hydroxyalkyl pterins (Diptera pathway) is unchanged in both stages, whereas the degradation products (Lepidoptera pathway) is for several pteridines completely different between the pupal and the adult stage.

Tab. 4

Pteridines isolated by ion exchange chromatography from *A. mellifera*. The concentrations are calculated on ppm basis (µg/g fresh weight).

pteridine	pupa	adult
L-*erythro*-biopterin	1.2	1.2
D-*erythro*-neopterin	0.09	0.03
7-hydroxylumazine	1.5	3.2
isoxanthopterin	0.05	1.8
pterin	0.12	0
6-carboxypterin	0.08	0.08
6-carboxy-isoxanthopterin	0	0.03

Catabolism of pterins in mammals

The catabolic pathway of pterin cofactors has been discussed in the preceeding volume already (1). In principal, pterin pools are balanced in two ways: the reduced cofactor is either excreted in the urine without reabsorption by the kidney (4,5) or it undergoes irreversible breakdown which is initiated by an irreversible oxidative elimination of the side chain (13,27). The latter pathway has been studied with rat liver homogenate and has proved to be of general validity. The catabolism of tetrahydroneopterin is initiated by cleavage of the cofactor, in presence of oxygen, to 7,8-dihydroxanthopterin and glyceraldehyde. It is till an open question, whether this irreversible step is controlled enzymatically, but it also requires the participation of xanthine oxidase (9). In a similar way, rat liver homogenate degrades tetrahydrobiopterin to pterin, 7,8-dihydroxanthopterin (totaling 30 %), lumazine, 6- and 7-hydroxylumazine (totaling 40 %). The main breakdown product is 6-hydroxylumazine.

The in vivo studies, using the labelled tetrahydrobiopterin, indicate a primary incorporation into liver and kidney as the main cofactor pools (1,2). The metabolites identified in rat urine after intraperitoneal injection of various labelled reduced pterins are collated in Table 1. As in the in vitro experiments, the two main catabolites are found besides the reoxidized pterin. In all cases xanthopterin and its deamination product 6-hydroxylumazine, are the predominant metabolites. It is an interesting question, whether a xanthopterin deaminase is involved in this pterin cofactor metabolism. From these data it becomes clear, that the in vivo metabolism follows the scheme proposed on the basis of in vitro experiments. The same situation is found if human urine is analyzed for pterins (Table 2).

References

(1) Rembold,H.: In "Biochemical and Clinical Aspects of Pteridines" (Wachter,H., Curtius,H.Ch., Pfleiderer,W., eds.), pp. 51-63, W. de Gruyter, Berlin 1982.

(2) Hennings,G., Rembold,H.: Internat. J. Vit. Nutr. Res. 52, 36-43 (1982).

(3) Rembold,H., Buff,K., Hennings,G.: Clinica Chimica Acta 76, 329-338 (1977).

(4) Häberle,D.A., Schiffl,H., Mayer,G., Hennings,G., Rembold,H.: Pfügers Arch. 375, 9-16 (1978).

(5) Häberle,D.A., Schiffl.H., Mayer,G., Davis,J.M., Hennings,G., Rembold,H.: Naunyn-Schmiedeberg's Arch. Pharmacol. 306, 287-293 (1979).

(6) Unnasch,T.R., Brown,G.M.: J. Biol. Chem. 257, 14211-14216 (1982).

(7) Johnson,J.L, Rajagopalan,K.V.: Proc. Natl. Acad. Sci. USA 79, 6856-6860 (1982).

(8) Rajagopalan,K.V., Johnson,J.L., Hainline,B.E.: Federation Proceedings 41, 2608-2612 (1982).

(9) Rembold,H. Gutensohn,W.: Biochem. Biophys. Res. Commun. 31, 837-841 (1968).

(10) Lowry,O.H., Bessey,O.H., Crawford,E.J.: J. Biol. Chem. 180, 399-410 (1949).

(11) Kalckar,H.M., Kjeldgaard,N.O., Klenow,H.: J. Biol. Chem. 174, 771-772 (1948).

(12) Lowry,O.H., Bessey,O.H., Crawford,E.J.: J. Biol. Chem. 180, 389-398 (1949).

(13) Rembold,H., Metzger,H., Gutensohn,W.: Biochim. Biophys. Acta 230, 117-126 (1971).

(14) Rajagopalan,K.V., Fridovich,I., Handler,P.: J. Biol. Chem. 237, 922-928 (1962).

(15) Hodnett,C.N., McCormack,J.J., Sabean,J.A.: J. Pharm. Sci., 65, 1150-1154 (1976).

(16) Levenberg,B., Hayaishi,O.: J. Biol. Chem. 234, 955-961 (1959).

(17) Takikawa,S., Kitayama-Yokokawa,C., Tsusue,M.: J. Biochem. (Tokyo) 85, 785-790 (1979).

(18) Rembold,H.: Korean Biochem. J.. 3, 1-14 (1970).

(19) Rembold,H., Metzger,H., Sudershan,P., Gutensohn, W., Biochim. Biophys. Acta 184, 386-396 (1969).

(20) Rembold,H., Simmersbach,F.: Biochim. Biophys. Acta 184, 589-596 (1969).

(21) Tsusue,M., Mazda,T.: Experientia 33, 854-855 (1977).

(22) Gyure,W.L.: Insect Biochem. 4, 113-121 (1974).

(23) Gyure,W.L.: Insect Biochem. 4, 303-312 (1974).

(24) Descimon,H.: In "Chemistry and Biology of Pteridines" (Pfleiderer,W., ed.), pp. 805-840, W. de Gruyter, Berlin 1975.

(25) Harmsen,R.: J. Exp. Biol. 45, 1-13 (1966).

(26) Rembold,H., Buschmann,L.: Liebigs Ann. Chem. 662, 72-82 (1963).

(27) Rembold,H.: In "Chemistry and Biology of Pteridines" (Iwai,K., Goto,M., Akino,M, Iwanani,Y., eds.) pp. 163-178, Internat. Academic Printing Company, Tokyo, 1970.

STUDIES ON THE BIOSYNTHESIS OF TETRAHYDROBIOPTERIN IN BOVINE ADRENAL MEDULLA PREPARATIONS

Gary K. Smith and Charles A. Nichol
Department of Medicinal Biochemistry, The Wellcome Research Laboratories, Research Triangle Park, N.C. 27709

Introduction

H_4biopterin[1] is synthesized from GTP. However, the identity of the intermediates and the properties of the biosynthetic enzymes on the pathway have not been completely elucidated. Evidence has been presented indicating that H_2neopterin-PPP, an unidentified compound "X" and sepiapterin are intermediates in the biosynthesis of biopterin in various organisms and in several mammalian tissues (1-9). Sepiapterin has in turn been shown to be converted to H_2biopterin by the enzyme sepiapterin reductase (10,11), and the last step of the de novo pathway in mammalian tissues has been proposed to be the reduction of H_2biopterin by DHFR (12-14). Based on these reports, one representation of the overall pathway as proposed recently (7) is shown in Scheme I.

Dihydrofolate reductase has a central role in maintaining the folate cofactor pools essential for C-1 metabolism, and this enzyme is a prime target for inhibition in a variety of neoplasms including those of the brain. If

[1]Abbreviations used: H_4biopterin, tetrahydrobiopterin; H_2biopterin, 7,8-dihydrobiopterin; H_2neopterin-PPP, 7,8-dihydroneopterin triphosphate; GTP, guanosine triphosphate; NADPH, nicotinamide adenine dinucleotide phosphate, reduced form; MTX, methotrexate; TCA, trichloroacetic acid; DHFR, dihydrofolate reductase; DTT, dithiothreitol.

Biochemical and Clinical Aspects of Pteridines, Vol. 2

SCHEME I

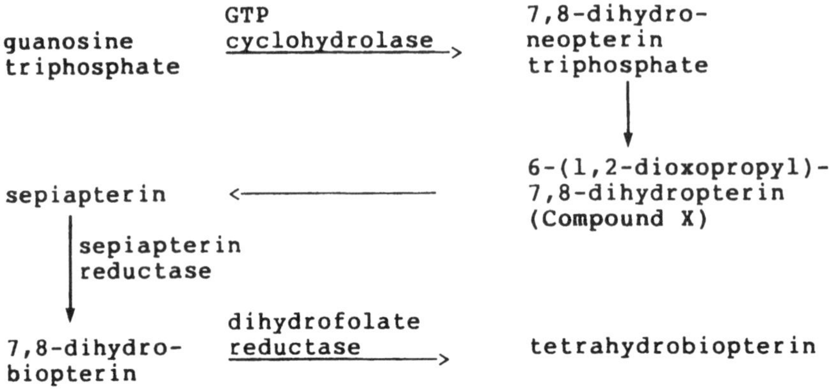

DHFR is involved in the biosynthesis of H_4biopterin, then DHFR inhibitors used in cancer chemotherapy might inhibit catecholamine and indoleamine synthesis, both of which utilize H_4biopterin-requiring hydroxylases for the initial step in their biosynthesis. However, daily administration of MTX (10 mg/kg, I.P. plus leucovorin to prevent toxicity) to rats for 4 days does not affect endogenous levels of H_4biopterin in brain, pineal, or adrenal medulla and cortex (15). Similarly, treatment of a variety of cells in culture with up to 0.1 mM MTX does not inhibit H_4biopterin biosynthesis. Moreover a cell line (DUKX-B11) which does not contain DHFR also synthesizes H_4biopterin (16,17). Thus, these studies cast doubts on the role of DHFR in the de novo biosynthesis of H_4biopterin.

In contrast, the synthesis of H_4biopterin from sepiapterin in the above systems does appear to require DHFR. Thus, the cells (DUKX-B11) which lack DHFR do not produce H_4biopterin from sepiapterin, and in cells and tissues which can synthesize H_4biopterin from sepiapterin, the conversion is profoundly inhibited by MTX. The synthesis from exogenous sepiapterin thus behaves differently from the endogenous synthesis from GTP and suggests the presence of two pathways for the synthesis of H_4biopterin: 1) *de novo* synthesis independent of DHFR; 2) synthesis via sepiapterin dependent upon DHFR (16,17).

In Vitro Studies

Our procedure for the determination of total biopterin and H_4biopterin synthesis in cell-free fractions is based on differential oxidation by iodine in acid and base followed by HPLC separation of oxidized pterins (18-20).

The bovine adrenal medulla, like that of the rat, is rich in the enzymes for the biosynthesis of H_4biopterin (21). The desalted homogenate supernatant of this tissue will produce 45 pmol of biopterin/mg protein/hr from 2 mM GTP in the presence of 2 mM NADPH, 1 mM $MgCl_2$, 5 mM DTT, and 50 mM Tris-HCl, pH 7.5. In the presence or absence of 0.1 mM MTX, 90% of this biopterin is found as H_4biopterin (21). Similarly, when 10 µM H_2neopterin-PPP (plus 5 mM sodium pyrophosphate to inhibit phosphatases) replaces GTP, 82-84 pmol of biopterin/mg protein/hr is produced, of which $\geq$85% is H_4biopterin. As with GTP as substrate, MTX has no effect on H_4biopterin synthesis. Thus, these *in vitro* results support the *in vivo* observations that a DHFR inhibitor does not affect the *de novo* synthesis of H_4biopterin.

However, when 50 µM sepiapterin replaces GTP, synthesis of H_4biopterin drops from 395 pmol/mg protein/hr in the absence of MTX to 3 pmol/mg protein/hr in the presence of 0.1 mM MTX. Since total biopterin synthesis under these

Table 1

Purification of H_2neopterin-PPP ——> H_4biopterin Activity*

Preparation	[MTX] (mM)	Activity: pmol biopterin / mg protein (hr)	Recovery: % H_4biopterin	% of homogenate
Homogenate	0	84	89	100
	0.1	82	88	
40-60% $(NH_4)_2SO_4$ pellet	0	269	91	48
	0.1	283	86	
AcA-34 Gel Filtration	0	2600	85	75
	0.1	2600	85	

*Assays were carried out at 37°C at pH 7.4 in 0.05 M Tris-HCl with 7 µM H_2neopterin-PPP, 2 mM NADPH, 5 mM DTT, 1 mM $MgCl_2$ and 5 mM pyrophosphate. Homogenate and $(NH_4)_2SO_4$ fractions were desalted using Sephadex G-25 prior to assay. H_2neopterin-PPP was synthesized with *E. coli* GTP cyclohydrolase and purified on DEAE-Sephacel prior to use.

conditions is approximately 2200 pmol/mg protein/hr in the presence or absence of MTX, inhibition of sepiapterin reductase is not responsible for the decreased synthesis of H_4biopterin. Similarly, if H_2biopterin is used in place of GTP, H_4biopterin synthesis is 804 pmol/mg protein/hr in the absence of MTX and 14 pmol/mg protein/hr in the presence of 0.1 mM MTX (21). Thus, with both sepiapterin and H_2biopterin, MTX causes the rate of synthesis to drop from 10-20X above to well below the rate from GTP or H_2neopterin-PPP. Consistent with the *in vivo* results then, the synthesis of H_4biopterin from either sepiapterin or H_2biopterin in the cell free preparations is greatly inhibited by MTX and is thus different from the *de novo* synthesis from GTP.

Further, according to kinetic analysis, a competitive inhibitor, such as MTX is with DHFR, should produce greater inhibition at lower substrate concentrations (22). Thus, if DHFR reduction of H_2biopterin occurs in the H_4biopterin biosynthesis from GTP and H_2neopterin-PPP, the percent inhibition of H_4biopterin synthesis from 10 μM H_2neopterin-PPP should be greater than that with 50 μM H_2biopterin. The total lack of inhibition with 10 μM H_2neopterin-PPP as substrate compared to nearly complete inhibition with 50 μM H_2biopterin as substrate again indicates that DHFR-catalyzed reduction is not a step common to both pathways.

Partial purification of the enzyme system for H_4biopterin synthesis. Partial purification of the enzymes required for the conversion of H_2neopterin-PPP to H_4biopterin is shown in Table 1. The biosynthetic system is purified approximately 30-fold by these procedures, and the conversion of H_2neopterin-PPP to H_4biopterin remains insensitive to MTX throughout the purification. The AcA-34 gel filtration elution profile indicates a MW in the range of 60,000 to 70,000 daltons for the enzymes. GTP cyclohydrolase is removed by this treatment since it fractionates into the 30-40% ammonium sulfate pellet and elutes near the AcA-34 void volume. The enzyme fraction purified through the AcA-34 step does contain sepiapterin reductase (20,000 pmol/mg protein/hr). However, neither exogenously added H_2biopterin nor the H_2biopterin produced by sepiapterin reductase in this enzyme fraction is reduced to H_4biopterin by this preparation. Thus, the enzyme required for the reduction of H_2biopterin, presumably DHFR, is removed by this purification procedure. Since the biopterin produced from H_2neopterin-PPP by this preparation is 85% H_4biopterin, the data provide further evidence that sepiapterin and H_2biopterin are converted to H_4biopterin by a pathway different from that used by GTP and H_2neopterin-PPP.

In this purified fraction, both NADPH and Mg^{++} are required for activity. In the absence of NADPH, no biopterin is synthesized; however, pterin is found following either acid or base oxidation. In the presence of EDTA, neither biopterin nor pterin is produced, either in the presence or absence of NADPH. This Mg^{++}-dependent production of pterin in the absence of NADPH may indicate the synthesis of a labile intermediate which decomposes to pterin upon oxidation. The biosynthesis of pterins in human liver and kidney, rat liver, chicken liver, and *Drosophila* also requires Mg^{++} and NADPH, and a labile compound is also produced in the absence of NADPH (4,6,9).

Proposed Pathway for *De Novo* H_4biopterin Biosynthesis.

Based upon the results discussed above, we propose the following general scheme for *de novo* biosynthesis of H_4biopterin.

Scheme II

pterin ↑ I_2 (from X)

GTP —(GTP cyclohydrolase)→ H_2neopterin-PPP —(Mg^{++})→ X —(NADPH)→ H_4biopterin

The identity of the intermediates between H_2neopterin-PPP and H_4biopterin on the *de novo* pathway is unknown; however, as stated above, the evidence indicates that neither sepiapterin nor H_2biopterin is among these intermediates. Rather, the synthesis of H_4biopterin from these compounds proceeds using DHFR as shown in Scheme III.

Scheme III

$$\text{sepiapterin} \xrightarrow{\text{sepiapterin reductase}} H_2\text{biopterin} \xrightarrow{\text{DHFR}} H_4\text{biopterin.}$$

Since DHFR inhibitors do not inhibit the *de novo* production of H_4biopterin in most tissues, the biological relevance of this second pathway (Scheme III) is unclear. However, this pathway could have a role in the recycling of H_2biopterin or sepiapterin in tissues and organs capable of the H_4biopterin biosynthesis.

Several groups involved in elucidating the *de novo* biosynthesis of H_4biopterin have detected sepiapterin in reaction mixtures which have GTP or H_2neopterin-PPP as starting materials (2-7). This has led to the proposal that sepiapterin is an intermediate in H_4biopterin biosynthesis (Scheme I). However, since we find that H_4biopterin is formed from sepiapterin both *in vivo* and *in vitro* by a pathway other than the *de novo* pathway, we suggest that sepiapterin, rather than being produced enzymatically as an intermediate in the pathway, is produced by decomposition of the true intermediates.

One compound which might give rise to sepiapterin upon oxidative decomposition is 6-lactyl-H_4pterin (tetrahydrosepiapterin). This oxidation of 6-lactyl-H_4pterin to sepiapterin has previously been proposed to be involved in the conversion of H_4- and H_2biopterin to sepiapterin (23). The suggestion that 6-lactyl-H_4pterin may be an intermediate on the pathway infers that some of the intermediates are H_4pterins. While the data indicate that it is very unlikely that sepiapterin and H_2biopterin are intermediates, our evidence is entirely consistent with the possibility that H_4pterins, including tetrahydrosepiapterin, are intermediates. By analogy with the conversion of H_2biopterin to tetrahydrosepiapterin proposed by Pfleiderer (23), the reduction of the pterin moiety to the tetrahydro level might proceed through oxidation of a side chain carbon-oxygen bond from a hydroxyl to a carbonyl.

Such an intramolecular redox reaction would circumvent any involvement of DHFR for production of the tetrahydropterin, and thus MTX would have no effect on this biosynthesis. Further, the relative ease of oxidation of tetrahydropterins at neutral pH (12) makes decomposition of such intermediates to dihydropterins during isolation very plausible. Studies are currently under way to determine whether such a mechanism involving the conversion of H_2neopterin-PPP to tetrahydropterin intermediates truly occurs in the de novo pathway of H_4biopterin biosynthesis.

References

1. Fukushima, T., Shiota, T.: J. Biol. Chem. 249, 4445-4451 (1974).
2. Fan, C.L., Krivi, G.G., Brown, G.M.: Biochem. Biophys. Res. Commun. 67, 1047, (1975).
3. Eto, I., Fukushima, K., Shiota, T.: J. Biol. Chem. 251, 6505 (1976).
4. Krivi, G.G., Brown, G.M.: Biochem. Genet. 17, 371-390 (1979).
5. Otsuka, H., Sugiura, K., Goto, M.: Biochim. Biophys. Acta 629, 69-76 (1980).

6. Tanaka, K., Akino, M., Hagi, Y., Doi, M., Shiota, T.: J. Biol. Chem. 256, 2963-2971 (1981).

7. Kapatos, G., Katoh, S., Kaufman, S.: J. Neurochem. 39, 1152-1162 (1982).

8. Dorsett, D., Flanagan, J.M., Jacobson, K.B.: Biochem. 21, 3892-3899 (1982).

9. Curtius, H.C., Hauserman, M., Niederwieser, H., Chisla, S.: Biochemical and Clinical Aspects of Pteridines 1, 27-50 (1982) Walter de Gruyter and Co., Berlin.

10. Matsubara, M., Katoh, S., Akino, M., Kaufman, S.: Biochim. Biophys. Acta 122, 202-212 (1966).

11. Matsubara, M.N.: Arch. Biochem. Biophys. 126, 426-435 (1968).

12. Kaufman, S.: J. Biol. Chem. 242, 3934-3943 (1967).

13. Pollack, R., Kaufman, S.: J. Neurochem. 30, 253-256 (1978).

14. Spector, R., Fosburg, M., Levy, P., Abelson, H.T.: J. Neurochem. 30, 899-901 (1978).

15. Duch, D.S., Lee, C.L., Edelstein, M.P., Nichol, C.A.: Molecular Pharmacol. In Press (1983).

16. Nichol, C.A., Lee, C.L., Edelstein, M.P., Chao, J.Y., Duch, D.S.: Proc. Nat. Acad. Sci. 80, 1546-1550 (1983).

17. Nichol, C.A., Viveros, O.H., Duch, D.S., Abou-Donia, M.M., Smith, G.K.: Seventh International Symposium on Pteridines and Folic Acid Derivatives: Chemical, Biological and Clinical Aspects, Walter de Gruyter, Berlin (1983).

18. Fukushima, T., Nixon, J.C.: Chemistry and Biology of Pteridines, Eds. R.L. Kisluik and Brown, G.M., 31-34, Elsevier/North Holland, New York (1979).

19. Fukushima, T., Nixon, J.C.: Anal. Biochem. 102, 176-188 (1980).

20. Woolf, J.H., Nichol, C.A., Duch, D.S.: J. Chromatog., (1983).

21. Nichol, C.A., Smith, G.K., Duch, D.S.: Seventh International Symposium on Pteridines and Folic Acid Derivatives: Chemical, Biological and Clinical Aspects, Walter de Gruyter, Berlin (1983).

22. White, J.C.: J. Biol. Chem. 254, 10889-10895 (1979).

23. Pfleiderer, W.: Chemistry and Biology of Pteridines: Ed. W. Pfleiderer, 941-949 (1975).

DIHYDROFOLATE REDUCTASE CATALYZES THE REDUCTION OF 7,8-DIHYDROBIOPTERIN IN LIVER AND BRAIN

Sheldon Milstien and Seymour Kaufman
Laboratory of Neurochemistry
National Institute of Mental Helath
Bethesda, Maryland 20205

INTRODUCTION

The tetrahydrobiopterin biosynthetic pathway has not yet been completely elucidated. There is general agreement that the first, and probably rate-limiting, step is the conversion of guanosine triphosphate to dihydroneopterin triphosphate. This reaction is catalyzed by the enzyme, GTP-cyclohydrolase (1). The nature of the intermediates between neopterin triphosphate and tetrahydrobiopterin is still under investigation. Some evidence has recently been presented that indicates that sepiapterin is an intermediate on the pathway, presumably being converted to 7,8-dihydrobiopterin by sepiapterin reductase (2). Dihydrobiopterin could then be reduced to 5,6,7,8-tetrahydrobiopterin by the action of dihydrofolate reductase (3).

Recently, several hyperphenylalanimemic children with peripheral defects in tetrahydrobiopterin biosynthesis have been discovered who have no central pterin defect and thus no neurotransmitter defects (4,5). This finding suggested the possibility that there could be two pathways for biopterin biosynthesis. In 1976, Stone showed that injection of rats with trimethoprim, a dihydrofolate reductase inhibitor, had no effect on liver tetrahydrobiopterin levels and concluded that the biosynthesis of tetrahydrobiopterin was unlikely to involve reduction of 7,8-dihydrobiopterin (6). Also, Nichol _et al._ have recently demonstrated that tetrahydrobiopterin biosynthesis in both rat brain and bovine adrenal medulla appears to proceed normally in the presence of methotrexate, a potent dihydrofolate reductase inhibitor (7).

One hypothesis that could explain the observation of a difference between central and peripheral biopterin biosynthesis as well as a pathway that is insensitive to folate reductase inhibitors would be the occurrence of inhibitor-insensitive reductases, distinct from dihydrofolate reductase, which could catalyze the reduction of 7,8-Dihydrobiopterin.

In this report we compare the 7,8-Dihydrobiopterin reductase activity in rat liver and brain and show by methotrexate titrations and by DEAE-cellulose chromatography of extracts that there is only a single methotrexate-sensitive reductase present in both tissues which appears to be identical to dihydrofolate reductase.

MATERIALS AND METHODS

7,8-Dihydrobiopterin (3) and dihydrofolic acid (8) were prepared in this laboratory. Solutions of dihydrobiopterin were prepared in 50 mM phosphate buffer (pH 7) to promote the oxidation of a small amount of contaminating tetrahydrobiopterin. 7,8-Dihydrobiopterin activity was measured by the use of the phenlalanine hydroxylase-catalyzed tetrahydrobiopterin-dependent conversion of phenylalanine to tyrosine (9). Dihydrofolate reductase activity was determind as described (10). Quinonoid dihydropterin reductase activity was measured as previously described (11).

Rat livers and brains were homogenized in 3 volumes of 10 mM Tris-HCl (pH 7.4) and centrifuged at 100,000 x g for one hour.

Biochemical and Clinical Aspects of Pteridines, Vol. 2

RESULTS AND DISCUSSION

Kaufman first demonstrated that dihydrofolate reductase could catalyze the reduction of 7,8-Dihydrobiopterin to 5,6,7,8-tetrahydrobioperin and suggested that this enzyme might also play a role in the biosynthesis of tetrahydrobiopterin (3). During the purification of sheep liver quinonoid dihydropteridine reductase (11), the enzyme which maintains tetrahydrobiopterin in its fully reduced state, two forms of dihydrofolate reductase activity were found on DEAE-cellulose chromatography. One form coeluted with the quinonoid reductase activity and the other was eluted at a lower salt concentration. Further purification separated most of the dihydrofolate reductase activity but even the most highly purified quinonoid reductase preparations (95% pure) still have a small, but detectable, activity towards 7,8-Dihydrobiopterin. thus, it seemed possible that quinonoid dihydropterin reductase itself might have the intrinsic capacity to catalyze the reduction of 7,8-Dihydrobiopterin in analogy to the ability of dihydrofolate reductase to catalyze the reduction of folic acid (8). However, as shown in Figure 1, rabbit anti-sheep liver quinonoid dihydropteridine reductase antiserum (12), which we had previously demonstrated does not inhibit dihydrofolate reductase (12), has no effect on the 7,8-Dihydrobiopterin reductase activity in a purified quinonoid reductase preparation. It is unlikely, therefore, that quinonoid dihydropteridine reductase from liver catalyzes the reduction of 7,8-dihydrobiopterin.

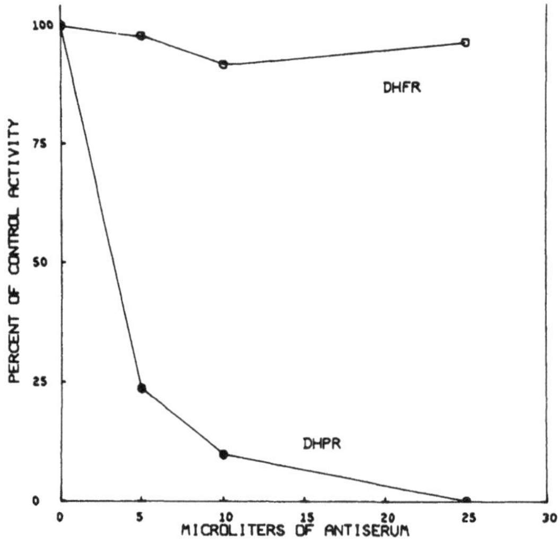

Figure 1. Immunotitration of sheep liver quinonoid dihydropteridine reductase activity. Samples (30 ul) were incubated at 4° for 17 hours. After centrifugation, aliquots were removed and both quinonoid reductase and dihydrobiopterin reductase activities were measured. Each sample contained 10 ug "pure" sheep liver dihydropteridine redutase. The total serum volume was maintained at 25 ul by the addition of preimmune serum.

Rat brain and liver extracts were chromatographed on DEAE-cellulose columns under conditions which led to a high recovery of applied dihydrobioperin reductase activity. In this way, the presence of minor forms of reductase activity distinct from dihydrofolate reductase activity might be detected. However, as shown in Figure 2, in both liver and brain extracts, essentially all of the dihydrobiopterin reductase activity was eluted as a single peak. The elution volume of the dihydrobiopterin reductase activity corresponded to the elution volume of dihydrofolate reductase activity. As discussed above, on DEAE-cellulose chromatography, sheep liver extracts give two peaks of dihydrofolate reductase activity which both have 7,8-Dihydrobiopterin reductase activity (data not shown). This result suggests that if there is a specific reductase for dihydrobioperin in brain or liver, it has a similar elution profile on DEAE-cellulose as does dihydrofolate reductase. Note that in both brain and liver extracts, quinonoid dihydropteridine reductase is eluted at the same ionic strength as the other reductase activities.

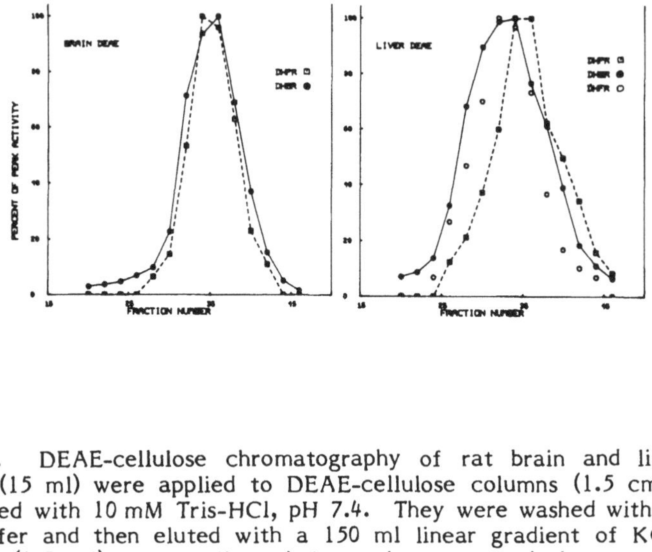

Figure 2. DEAE-cellulose chromatography of rat brain and liver extracts. Extracts (15 ml) were applied to DEAE-cellulose columns (1.5 cm by 8.5 cm) equilibrated with 10 mM Tris-HCl, pH 7.4. They were washed with 15 ml of the same buffer and then eluted with a 150 ml linear gradient of KCl to 0.5 M. Fractions (1.5 ml) were collected from the start and the enzyme activities determined as indicated. Note that all the activities are normalized to the peak fraction. Quinonoid dihydropteridine reductase was eluted at 0.1 M KCl. For clarity, only one part of the gradient is shown, since except for a few percent of the activity eluting at the void volume, no other peaks were detected. Total recoveries were 75-90%. The peak activities were: brain quinonoid reductase, 275 nmol/min/ml; liver quinonoid reductase, 2774 nmol/min/ml; liver dihydrofolate reductase, 30 nmol/min/ml. The 7,8-Dihydrobiopterin reductase activites were expressed in relative amounts of tyrosine formed in a coupled assay with phenylalanine hydroxylase.

In order to determine whether there are distinct 7,8-Dihydrobiopterin reductases which do not separate on DEAE-cellulose, methotrexate titrations of the activities were performed. These results are shown in Figure 3. Liver dihydrofolate reductase activity as well as both liver and brain dihydrobiopterin reductase activity show the same sensitivity to inhibition by methotrexate. In fact, the total dihydrobiopterin reductase activity approaches zero at the higher methotrexate concentrations used in this study. This suggests that there is no residual methotrexate-insensitive dihydrobiopterin reductase activity in either brain or liver. Because of the very low dihydrofolate reductase activity in adult rat brain (13), it was not possible with the normal assay to measure methotrexate sensitivity. However, a comparison of the rate of reduction of both dihydrobiopterin and dihydrofolic acid by pure bovine liver dihydrofolate reductase and by rat liver extracts, showed that the activity of brain dihydrobiopterin reductase was consistent with the small amount of dihydrofolate reductase activity present in brain.

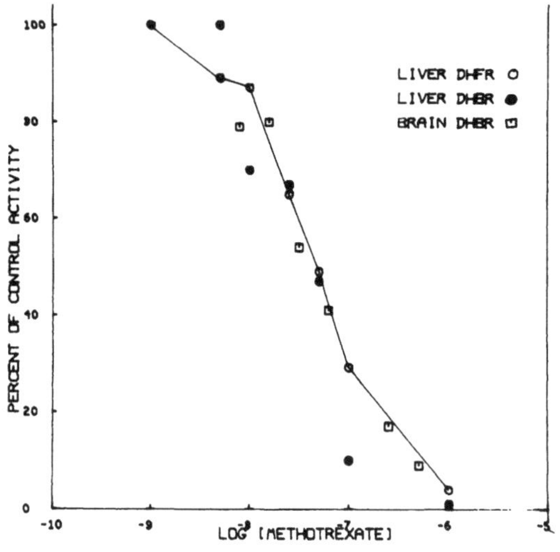

Figure 3. Methotrexate titrations of rat liver dihydrofolate reductase activity and liver and brain dihydrobiopterin reductase activities. Samples (0.1 ml) containing brain or liver extracts were incubated for 10 minutes at 37° at the indicated methotrexate concentrations. The methotrexate was then removed by gel filtration on Sephadex columns and the enzyme activities determined.

In conclusion, there is at present no evidence for a methotrexate-insensitive dihydrobiopterin reductase in either brain or in liver. If, in fact, 7,8-dihydrobiopterin is an intermediate in the biosynthesis of tetrahydrobiopterin as proposed by several groups (14), then it is likely that dihydrofolate reductase would catalyze the final reaction. It still is possible that the lack of methotrexate-sensitivity could be due to the presence of the reductase in an enzyme complex that is not accessible to methotrexate.

REFERENCES

1. Yim, J.J., Brown, G.M.: J. Biol. Chem., 251, 5087-5094 (1976).

2. Kapatos, G., Katoh, S., Kaufman, S.: J. Neurochem., 39, 1152-1162 (1982).

3. Kaufman, S.: J. Biol. Chem., 242, 3934-3943 (1967).

4. Hreidarsson, S., Valee, D., Holtzman, N., Coyle, J., Singer, H., Kapatos, G., Kaufman, S.: Soc. Ped. Res. (Abstract), (1982).

5. Berlow, S., Hoganson, G., Kaufman, S., Milstien, S.: unpublished.

6. Stone, K.J.: Biochem. J., 157, 105-109 (1976).

7. Nichol, C.A., Lee, L.L., Edelstein, M.P., Chao, J.Y., Duchs, D.S.: Proc. Natl. Acad. Sci. U.S.A., 80, 1546-1550 (1983).

8. Futterman, S.: Meth. Enzymol., 6, 801-802 (1963).

9. Milstien, S., Kaufman, S., Summer, G.K.: Pediatrics, 65, 806-810 (1980).

10. Mathews, C.K., Scrimgeour, K.G., Huennekens, F.M.: Meth. Enzymol., 6, 364-368 (1963).

11. Craine, J.E., Hall, E.S., Kaufman, S.: J. Biol. Chem., 247, 6082-6091 (1972).

12. Milstien, S., Kaufman, S.: Biochem. Biophys. Res. Comm., 66, 475-481 (1975).

13. Pollock, R.J., Kaufman, S.: J. Neurochem., 30, 253-256 (1978).

14. Tanaka, K., Akino, M., Hagi, Y., Doi, M., Shiota, T.: J. Biol. Chem., 256, 2963-2972 (1981).

INHIBITION OF DIHYDROPTERIDINE REDUCTASE BY PTERIDINES, MONOAMINES AND METALLIC COMPOUNDS.

Jean-Louis Dhondt, Zina Bellahsene

Faculté libre de Médecine, Laboratoire de Biochimie
56 rue du port, 59046 LILLE-Cedex, France

Introduction

Tetrahydrobiopterin (BH_4) is the natural cofactor of Phenylalanine-hydroxylase, Tyrosine-hydroxylase and Tryptophan-hydroxylase. During the hydroxylation process, BH_4 is oxidized to quinonoid-dihydrobiopterin which is then reduced to BH_4 by Dihydropteridine-reductase (DHPR, EC 1.6.99.10). In absence of DHPR, quinonoid-dihydrobiopterin is unstable and rapidly rearranges to 7,8 dihydrobiopterin.

The rate of the BH_4 synthesis from GTP appears very slowly in human. It is therefore possible that the rate-limiting factor for the hydroxylases is the concentration of BH_4; consequently DHPR may be an important salvage pathway to maintain efficient BH_4 concentrations in cells. The essential role of DHPR has been clearly demonstrated by the observation of children congenitally deficient in this enzyme.

Many endogenous and exogenous agents have been reported to inhibit DHPR (1,2,3). In the present paper, we reinvestigated the effect of three classes of compounds: pteridines, aromatic aminoacids derivatives and metallic compounds.

Biochemical and Clinical Aspects of Pteridines, Vol. 2

Material and Methods

Enzyme sources: DHPR was prepared from rat tissues by homogenization in Tris-HCl buffer, pH 7.0. After centrifugation at 30.000 g for 45 min at +4°C, supernatant fractions were stored frozen at -20°C. More highly purified preparations of liver DHPR were obtained by the affinity chromatography method of Cotton _et al_(4).

Assay of enzyme activity: DHPR was measured by two methods. Method 1 was derived from that of Nielsen (5) which uses H_2O_2 and peroxidase to generate the quinonoid-dihydropterin and Method 2 (6) in which 2,6 dichlorophenol-indophenol is used. Both methods use 0.05 mM 6,7-dimethyl-tetrahydropterin ($DMPH_4$) as substrate and 0.05 mM of NADH. Enzyme was monitored by measuring the decrease in absorbance at 340 nm and corrected from blank values obtained by omiting enzyme or $DMPH_4$.

Protein was assayed by the Lowry method.

Chemicals: 6,7 $DMPH_4$ was obtained from Calbiochem, other pteridines from Schircks (Wettswill, Switzerland), aminoacids and their derivatives from Sigma, cis-platinium from Roger-Bellon.

Results and Comments

(1) Pteridine effectors (Table I): Among the unconjugated pteridines, sepiapterin exhibited the highest inhibitory effect, a 50% inhibition was observed at a concentration of 0.06 mM. DHPR was more inhibited by conjugated pteridines, confirming the previous demonstration of the affinity of the rat enzyme for methotrexate (7,8).

Table I: Inhibition of DHPR by pteridines (% of inhibition).

pterins	DHB	biopterin	neopterin	sepiapterin
0.4 mM	78%	75%	94%	95%
0.1 mM	8%	28%	35%	79%
folates	THF	DHF	folic acid	
0.15 mM	77%	51%	76%	
0.06 mM	31%	36%	46%	

DHB: dihydrobiopterin; THF: tetrahydrofolic acid; DHF: dihydrofolic acid.

The significance of these results regarding eventual regulatory properties of pteridines on DHPR activity remains unclear. Katoh (9) recently reported similar results of inhibition of sepiapterin-reductase by unconjugated pteridines, sepiapterin-reductase being an enzyme probably also involved in the biosynthesis of BH_4 (10). As far as the precise metabolic pathway of the biosynthesis of BH_4 is completely established, pteridines appear to participate in the regulation mechanism of the in situ BH_4 supply at, at least, 3 steps: GTP-cyclohydrolase (11), sepiapterin-reductase (9) and DHPR.

(2) Derivatives of aromatic aminoacids: End-product feedback regulation of Tyrosine-hydroxylase by catecholamines is considered as a major mechanism for the control of synthesis of dopamine. Kinetic studies have shown that catecholamines inhibit adrenal tyrosine-hydrolxylase by competing with the pterin cofactor (12). More recently, an inhibition of DHPR by dopamine has been reported by Purdy et al (1). Consequently, we extended the observation to other compounds derived from aromatic aminoacids.

Table II: Inhibition of DHPR by aromatic derivatives (% of inhibition).

enzyme source		liver	kidney	brain
Dopa	($3.10^{-3}M$)	0%	26%	37%
Dopamine	($6.10^{-3}M$)	64%	54%	40%
Norepinephrine	($3.10^{-3}M$)	36%	66%	46%
Epinephrine	($3.10^{-3}M$)	50%	77%	88%
Phenylpyruvate	($10^{-3}M$)	39%	55%	25%
Serotonin	($3.10^{-3}M$)	48%	-	50%
	($6.10^{-4}M$)	23%	-	37%

1. phenylalanine derivatives: Phenylpyruvic acid inhibited liver and kidney DHPR and at lesser extent brain DHPR (Table II). Similar results were obtained with Phenyllactic acid (results not shown), whereas phenylalanine up to 1 mM did not have any effect.
2. tyrosine derivatives: The inhibition of DHPR by dopamine previously reported by Purdy et al (1) was presently confirmed (Table II). Dopa was also found to inhibit DHPR but only in kidney and brain. Epinephrine and Norepinephrine were also potent inhibitors. In contrast, tyrosine up to 1 mM did not have any effect. In regard to the inhibition of sepiapterin-reductase by the same compounds (9), DHPR appeared more sensitive than sepiapterin-reductase.
3. tryptophan derivatives: As well as with phenylalanine and tyrosine, no inhibition was noted with tryptophan. In contrast, serotonin was found to be a potent inhibitor, semi-purified DHPR was inhibited by 70% with $3.10^{-3}M$ of serotonin.

The regulatory significance of these inhibitions remains unclear. Purdy (1) suggested that DHPR may have a role in control of the dopamine biosynthesis. On the contrary, Turner et al (13) did not support this suggestion on the basis of a distribution of DHPR which did not parallel distribution of tyrosine-hydroxylase nor that of tryptophan-hydroxylase in rat brain.

(3) Metallic compounds: DHPR is also sensitive to metallic compounds, Leeming (14) and Purdy (2) reported the inhibition of DHPR by aluminium and lead. Our attention was first focused on aluminium inhibition following reported evidences that aluminium toxicity is the cause of the dialysis encephalopathic syndrome (15).

The activity of a semi-purified enzyme showed a 46% reduction in presence of 0.1 mM of aluminium and 20% with 0.01 mM. Administration of $Al_2(SO_4)_3$ (3% in drinking water) for 10 days to young hamsters (Table III) induced a significant reduction of DHPR activities and of the biopterin content of organs.

Table III: *In vivo* effect of aluminium cation in young hamsters (mean values)

	tissue	liver	kidney	brain
DHPR (µmol/min x mg protein)				
	controls	0.313	0.446	0.113
	+Al	0.219	0.320	0.069
biopterin (pmol/mg protein)				
	controls	30.2	-	9.5
	+Al	22.2	-	6.5

+Al: 3% $Al_2(SO_4)_3$ in drinking water for 10 days.

Abnormal concentrations of pterins in serum of chronic uraemic patients have been previously reported (16), abnormalities found to be more important in patients receiving aluminium gel (17). Ultrafiltrates of sera obtained from these latter patients inhibited semi-purified rat enzyme (unpublished results).

Many reports point out the existence of disturbances of monoamine metabolism in chronic uraemia, suggesting that chronic uraemia leads to a minimal brain dysfunction status. In addition, an impaired metabolism of pterins in chronic uraemia is ascertained. Therefore, any additional agent, such as aluminium, could alter a critical cerebral metabolism by inhibiting the salvage of BH_4.

Cis-platinium (cis-diamminedichloroplatinium) is an anticancer drug whose activity is related to chemical interaction with DNA; no other biological target was yet known (18). The present results show that DHPR and dihydrofolate-reductase (DHFR) are irreversibly inhibited by cis-platinium (Table IV).

Table IV: Inhibition of DHPR by cis-platinium (% of inhibition).

		DHPR	DHFR
50 mg/l* DDP		28%	70%
25 mg/l DDP		10%	57%
5 min preincubation	second addition		
Enz + DDP **	none	60%	85%
Enz + KCl	DDP	17%	0%
Enz + BSA	DDP	3%	0%
Enz + DDP	BSA	50%	62%

* final concentration, DDP: cis-platinium (** 100 mg/l final volume); BSA: bovine serumalbumin (1 g/l); KCl: 50 mM.

The protective effect of KCl is explained by the kinetic of chloride ion displacement by water (Figure), the compound [2] being the active form.

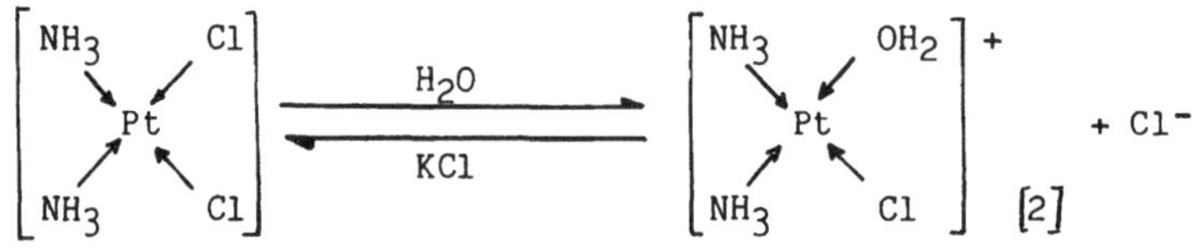

[1]

Cis-platinium is also known to bind strongly to albumin, that explained the protective effect of albumin addition prior to that of the enzyme.
To check in vivo effect of cis-platinium on reductases, cis-platinium was intraperitoneally injected to rats (10 mg/kg), DHPR and DHFR were measured in tissues 24 hours after the injection. Reductases activities were significantly reduced in liver and kidney, whereas brain DHPR was not affected, cis-platinium does not easily cross the blood brain barrier. These new side-effects of cis-platinium therapy in cancerous patients are under investigation.

Conclusion:

DHPR which assures the salvage of intracellular BH_4 is sensitive to many endogenous compounds, some of them (monoamines, pteridines) could act in the regulation of BH_4 metabolism. In addition, several environmental agents also appear to inhibit DHPR and therefore to limit the BH_4 availability.

References:

1. Purdy,S.E., Blair,J.A., Barford,J.: Biochem.J. 195,769-771 (1981)
2. Purdy,S.E., Blair,J.A., Leeming,R.J., Hilburn,M.E.: Intern.J.Enviromental Studies 17,141-145 (1981)
3. Leeming,R.J., Purdy,S.E., Blair,J.A.: Arch.Franç.Ped. in press (1983)
4. Cotton,R.G.H, Jennings,J.: Eur.J.Biochem. 83, 319-324 (1978)
5. Nielsen,K.H., Simonsen,V., Lind,K.E.: Eur.J.Biochem. 9, 497-502 (1969)

6. Cheema,S., Saldin,S.J., Knapp,A., Hofman,T., Scrimgeour,K.G.: Can.J.Biochem. 51, 1229-1239 (1973)

7. Craine,J.E., Hall,E.S., Kaufman,S.: J.Biol.Chem. 247, 6082-6091 (1972)

8. Webber,S., Deits,T.L., Snyder,W.R., Whiteley,J.M.: Anal.Biochem. 84, 491-503 (1978)

9. Katoh,S., Sueoka,T., Yamada,S.: Biochem.Biophys.Res.Commun. 105, 75-81 (1982)

10. Häusermann,M., Ghisla,S., Nierderwieser,A., Curtius,H-Ch.: FEBS Letters 131, 275-278 (1981)

11. Dhondt,J.L., Bellahsene,Z., Farriaux,J.P., Dautrevaux,M.:

12. Ikeda,M., Fahien,L.A., Udenfriend,S.: J.Biol.Chem. 241, 4452-4456 (1966)

13. Turner,A.J., Ponzio,F., Algeri,S.: Brain Res. 70, 553-558 (1974)

14. Leeming,R.J., Blair, J.A.: Lancet 1, 556 (1979)

15. Alfrey,A.C., LeGendre,G.R., Kaehny,W.D.: N.Engl.J.Med. 294, 184-188 (1976)

16. Dhondt,J.L., Vanhille,Ph.: in Biochemical and Clinical Aspects of Pteridines (Wachter,H., Curtius,H-Ch., Pfleiderer,W. Eds) WdeGruyter Publ. Berlin, pp 329-335 (1982)

17. Dhondt,J.L., Bellahsene,Z., Vanhille,Ph., Noël,C.: Lancet 2, 491 (1982)

18. Lippard,S.J.: Science 218, 1075-1082 (1982)

THERAPEUTIC POTENTIAL OF 6,6-DISUBSTITUTED PTERIDINES

June E. Ayling and Steven W. Bailey

Department of Pharmacology, College of Medicine
University of South Alabama, Mobile, Alabama 36688, USA

Tyrosine hydroxylase and tryptophan hydroxylase, the rate limiting enzymes in the biosynthesis of norepinephrine (1) and serotonin (2), respectively, each utilize tetrahydrobiopterin (BH_4) as cofactor (Figure 1). The activity of both enzymes may be regulated by the availability of cofactor, since the concentration of BH_4 in brain is sub-saturating for either enzyme (3).

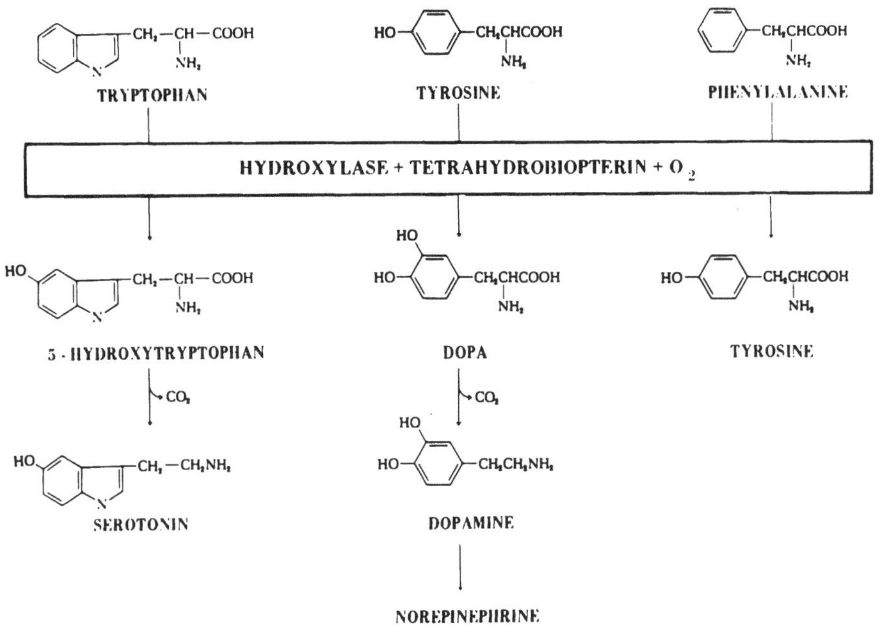

Figure 1. Role of BH_4 in the biosynthesis of serotonin, dopamine and norepinephrine, and in phenylalanine metabolism.

Tetrahydrobiopterin levels are low in the CSF of patients with a variety of neurological disorders, including Parkinson's disease (4), Alzheimer's disease, Shy-Drager syndrome, Steel-Richardson syndrome, Huntington's chorea (5), and dystonia (6). Cerebrospinal fluid BH_4 is also low in a form of depression in which there is decreased activity of central norepinephrine and/or serotonin systems (7). Post-mortem analysis of brains of Parkinson's and Shy-Drager's patients revealed that BH_4 in caudate nucleus is decreased even more than in CSF (3). In Parkinson's patients, brain tyrosine hydroxylase is also low (3). This loss of enzyme and of cofactor which accompanies degeneration of dopaminergic neurons in the nigro-striatal pathway, results in a marked deficiency in dopamine concentration. The current major means of treating Parkinson's patients, by replacement therapy with L-DOPA or L-DOPA/carbidopa, is totally ineffective in 15-25% of patients and, for those that do respond, adverse side-effects are common, often beginning within a few months of therapy. Most Parkinson's patients live for 10 to 20 years after onset of the disease, but full benefit of L-DOPA therapy generally lasts for only 2 to 4 years, and within 5 years the increasing side-effects usually begin to outweigh the remaining benefit of continued treatment (8,9).

The low BH_4 levels associated with Parkinsonism and other neurological diseases, suggest that such disorders might be alleviated by administration of BH_4. Recently, such treatment has been attempted on an experimental basis in patients with Parkinsonism (10,11), dystonia (12), and depression (7,11). Cofactor therapy is also under investigation in patients with variant forms of PKU caused by a deficiency in one of the BH_4 biosynthetic enzymes (13,14,15). In this disorder the activity of all three of the BH_4 dependent hydroxylases is impaired (Figure 1).

Such trials, although resulting in variable success, indicate the potential of cofactor therapy for these disorders. A major set-back has been the very high doses required to produce a clinical response (7,10-15). It has been proposed that BH_4 may be too lipophobic to readily cross the blood brain barrier (16), since after peripheral administration of BH_4, only small increases have been found, either in brain of rats (16,17), or in CSF of humans (15). Thus cofactor analogs are required which reach the brain more efficiently. However, another obstacle to this approach to therapy is the lability of tetrahydropterins, first to oxidation and subsequently to rapid rearrangement to 7,8-dihydropterins.

Tetrahydropterins which function well as cofactors for the hydroxylases are intrinsically unstable, since susceptibility to air oxidation and ability to function as cofactor are closely related. Autooxidation is, therefore, a continous threat to the integrity of tetrahydropterins used for cofactor

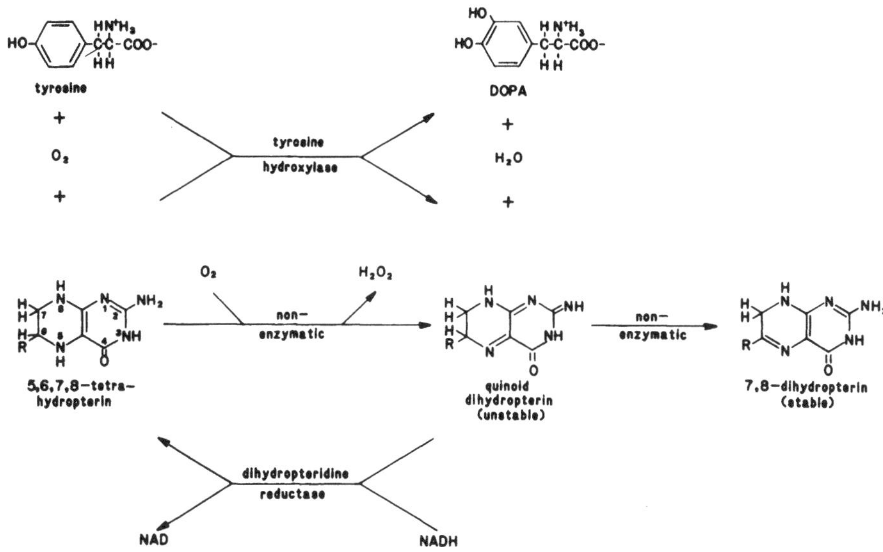

Figure 2. Function of tetrahydropterins and dihydropteridine reductase in the reaction of an aromatic amino acid hydroxylase. BH_4: R = $CH_3CHOHCHOH$-; 6-$MePH_4$: R = CH_3-.

therapy; during storage, during administration, in the gut and peripheral circulation, and finally while in the brain. Dihydropteridine reductase, which is present in large excess in organs such as liver, kidney and brain (18), readily reduces quinoid dihydropterins back to tetrahydropterins. During administration, and in the blood, where dihydropteridine reductase is not present, quinoid dihydropterins rapidly tautomerize to 7,8-dihydropterins which are not substrates for dihydropteridine reductase and, therefore, can no longer be reduced back to the active form of the cofactor (Figure 2). The rate limiting step in the tautomerization is cleavage of the 6C-H bond (19). Replacement of the 6-H by e.g. a methyl group should therefore block the tautomerization, giving the quinoid 6,6-disubstituted-dihydropterin sufficient stability to enter the brain before any significant degradation occurs.

To allow effective treatment at significantly lower doses, we have developed a procedure for synthesizing active cofactor analogs of BH_4 which are disubstituted at the 6-position of the pterin ring (20). While 6,6-disubstituted tetrahydropterins can be labile to auto-oxidation, the stability of their quinoids and their ability to be reduced back to tetrahydropterins, potentiate a greater duration for the effect of each dose and decrease the need to compensate for rapid losses. This stability may be a significant contribution to the successful treatment of neurological disorders related to BH_4 deficiency. The flexibility in the selection of 6-substituents provided by this synthetic method may allow optimization of the properties required of an effective drug. Requirements of BH_4 analogs to be used for replacement therapy include:

1. Simple and inexpensive to synthesize.
2. Overall stability.
3. Good cofactor for aromatic amino acid hydroxylases.
4. Facile regeneration of oxidized cofactor.
5. Ability to enter the brain.

6. Innocuous products.

7. Specificity for target enzyme.

The synthesis of a prototype of this class of compounds, 6,6-dimethyltetrahydropterin (6,6-Me_2PH_4), and its properties, as they relate to the above requirements, are presented here.

1. Synthesis of 6,6-Dimethyltetrahydropterin (6,6-Me_2PH_4)

The synthesis of quinoid 6,6-dimethyldihydropterin is outlined in Figure 3. The commercially available 2-amino-6-chloro-4(3H)-pyrimidinone is purified and then nitrated. Condensation of the nitro pyrimidine (I) with 1,2-diamino-2-methyl-propane occurs regiospecifically, the less hindered amine displacing the halogen. The nitro group of (II) is then reduced, either catalytically or, in higher yield, by dithionite. The resulting 6-substituted derivative of 2,5,6-triamino-4(3H)-pyrimidinone (III) is oxidized and the imine of quinoidal (IV) hydrolyzed via acid catalysis to a 6-substituted derivative of quinoid divicine (V). Upon neutralization, intramolecular Schiff's base condensation forms quinoid 6,6-dimethyl-dihydropterin (q-6,6-Me_2PH_2) (VI). This can be collected, or reduced to the tetrahydropterin, 6,6-Me_2PH_4. This compound has also been synthesized by an alternative method (37).

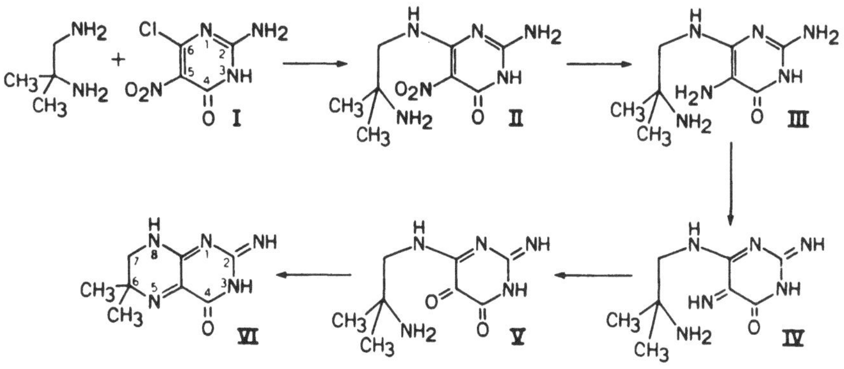

Figure 3. Synthesis of q-6,6-Me_2PH_2 (VI).

2. Stability of Quinoid 6,6-Dimethyldihydropterin (q-6,6-Me_2 PH_2)

Although tetrahydro-6,6-dimethylpterin is as labile to air oxidation as tetrahydro-6-monomethylpterin, the quinoid form of the 6,6-disubstituted compound is significantly more stable than that of the 6-monomethyl. At physiological pH and temperature there is a slow decay of q-6,6-Me_2PH_2 with a half-life of 75 min (Figure 4). Under the same conditions quinoid-6-methyl-dihydropterin (q-6-$MePH_2$) almost completely rearranges in 15 min to the physiologically unusable 6-methyl-7,8-dihydro tautomer. The half life of q-6,6-Me_2PH_2 at 27°C in 0.1M Tris Cl, pH 7.4, is 4 hours (compared to 25 min for q-6-$MePH_2$), and 104 hours at 0°C in 0.1M Tris-Mes, pH 6.8. The rate of decay

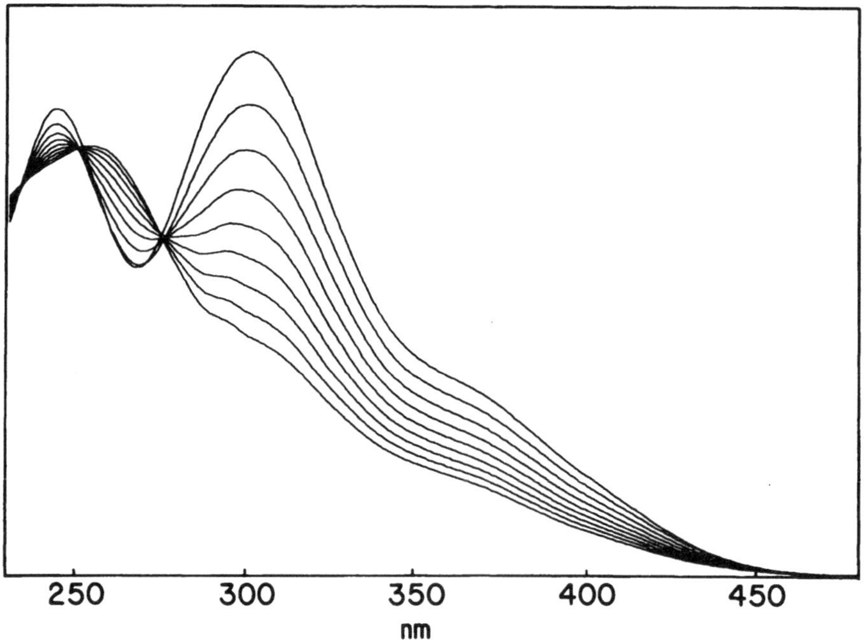

Figure 4. Rate of decay of q-6,6-Me_2PH_2 in 0.1 M Tris Cl, pH 7.4, at 37°. Absorbance spectra were taken at 15 min intervals.

is unaffected by the presence of oxygen, and is slightly accelerated in acid. In non-aqueous solution the quinoid is more stable, having a half life in dimethyl sulfoxide of 48 hours at room temperature and several weeks at -80°C. As a dry powder, q-6,6-Me_2PH_2 has been stored desiccated at -80°C for over a year with no signs of decomposition (Table I).

Table I. Stability of Quinoid-6,6-Me_2PH_2 as a Function of pH, Temperature, Oxygen and Solvent

pH	Temperature		$t_{1/2}$ (hours)
7.4	27°C	air	4
7.4	37°C	air	1.25
7.4	37°C	argon	1.25
0	37°C	air	0.75
0	37°C	argon	0.75
6.8	0°C	air	104
dimethylsulfoxide	25°C	air	48
dimethylsulfoxide	-80°C	air	several weeks
dry powder	-80°C	air	>1 year

The rates of decomposition were determined from the absorption spectra in 1N HCl (pH 0), or Tris-Mes (pH 6.8), or Tris-Cl (pH 7.4) each at 0.1M.

3. Cofactor Properties of 6,6-Me_2PH_4 with Aromatic Amino Acid Hydroxylases

Phenylalanine hydroxylase. 6,6-Me_2PH_4 binds to rat liver phenylalanine hydroxylase with an affinity similar to that of 6-methyltetrahydropterin (6-$MePH_4$), and catalyzes a reaction which is about 4 times slower. An analogous comparison can be made between the cofactor properties of 6R and 6S-BH_4. The unnatural configuration of the 6-substituent in 6S-BH_4 has no effect on Km but slows the rate of reaction four-fold (21)

(Table II). Thus, it appears that the catalytic site of phenylalanine hydroxylase readily accommodates substituents at the 6-position which are either in the natural or unnatural position, or both, but the presence of a substituent in the unnatural position decreases V_{max} about four-fold. In all cases a coupled reaction is catalyzed, one molecule of tyrosine being formed for each cofactor molecule oxidized (Table II).

Table II. Cofactor Properties of 6,6-Me_2PH_4 with Phenylalanine Hydroxylase

Cofactor	Km(mM)	Rel. V_{max}	cofactor oxidized / tyrosine produced
R-BH_4	0.021	1	1
S-BH_4	0.025	0.27	1
6-$MePH_4$	0.1	0.77	1
6,6-Me_2PH_4	0.07	0.17	1

Apparent Km's for cofactor were measured at 1mM phenylalanine and atmospheric oxygen. All reactions were run in 0.1M Tris Cl, pH 7.4 at 27°C and were assayed either spectrophotometrically from the enzyme dependent rate of cofactor oxidation at 340 nm (22), or by analysis for tyrosine production by HPLC with fluorometric detection (23).

Tyrosine hydroxylase. The effect of disubstitution at the 6-position of cofactor has been determined with bovine striatal tyrosine hydroxylase. The second methyl group at the 6-position of 6,6-Me_2PH_4 increases the affinity of cofactor for enzyme when compared to 6-$MePH_4$, but this is counteracted by a correspondingly lower V_{max} (Table III). Constants are expressed as the ratio of V_{max} to K_m, since this represents the relative effectiveness when cofactor concentration is sub-saturating, as will be the case within the brain. Thus at low concentrations, 6-$MePH_4$ and 6,6-Me_2PH_4 are equally good cofactors for tyrosine hydroxylase and three times better than BH_4 (24).

Table III. Reactivity of Cofactor Analogs with Tyrosine Hydroxylase

Cofactor	Rel $\frac{V_{max}}{Km}$
BH_4	1
$6\text{-}MePH_4$	3
$6,6\text{-}Me_2PH_4$	3

Cofactor Km's were measured with saturating tyrosine (0.5mM) and atmospheric oxygen, at pH 7.2 and 37°C. Cofactor was regenerated with either NADH and dihydropteridine reductase, or with ascorbic acid. The same results were obtained with either system. Samples were removed from reactions as a function of time, deproteinized, and analyzed for DOPA by HPLC with fluorometric detection (24).

The values in Table III are from kinetic constants determined with unmodified enzyme. After in vitro phosphorylation there is a significant increase in cofactor affinity and V_{max} of $6,6\text{-}Me_2PH_4$, as well as of $6\text{-}MePH_4$ and BH_4. However, it has been reported that 80% of tyrosine hydroxylase in striatum is in the unphosphorylated form (25).

4. Regeneration of $6,6\text{-}Me_2PH_4$ from Quinoid-$6,6\text{-}Me_2PH_2$

Enzymatic. Quinoid-$6,6\text{-}Me_2PH_2$ is a good substrate for dihydropteridine reductase with a V_{max} double that of quinoid-dihydrobiopterin (q-BH_2). However, the affinity for enzyme is significantly affected by the second methyl group at the 6-position. The Km of q-$6,6\text{-}Me_2PH_2$ is an order of magnitude higher than that for the compound with a single methyl at the 6-position (q-$6\text{-}MePH_2$) and two orders of magnitude higher than that of q-BH_2 (Table IV). Enzyme was from sheep liver or beef liver. The same results were obtained with either enzyme.

Table IV. Quinoid-6,6-Me_2PH_2 as Substrate for Dihydropteridine Reductase

Quinoid Substrate	Km(mM)	Rel. V_{max}
BH_2	0.004	1
6-$MePH_2$	0.03	3
6,6-Me_2PH_2	0.4	2

Kinetic constants were measured in 0.1M Tris Cl, pH 7.4, at 27°C in the presence of saturating NADH (0.1mM). Quinoid substrate was generated in situ either with peroxidase and hydrogen peroxide or with phenylalanine hydroxylase and phenylalanine. Quinoid 6,6-Me_2PH_2 was also produced chemically prior to the reaction.

When q-6,6-Me_2PH_2 is present at low concentrations, its high Km would result in inefficient reduction by dihydropteridine reductase. However, this disadvantage may be compensated for by the large excess of dihydropteridine reductase which is present in hydroxylase containing tissues (18). In addition, the affinity for dihydropteridine reductase might be improved by disubstitution at the 6-position with groups other than methyl.

Non-enzymatic. In the absence of adequate dihydropteridine reductase, such as occurs in patients with a genetic defect in this enzyme, or in the blood of normals, quinoid dihydropterins rapidly tautomerize to 7,8-dihydropterins. The lethality of dihydropteridine reductase deficiency (26) demonstrates that the means of reducing 7,8-dihydropterins to tetrahydropterins, e.g. by dihydrofolate reductase, are inadequate. 6,6-Disubstituted quinoid dihydropterins are sufficiently stable that non-enzymatic reduction to the tetrahydro form by endogenous reducing agents, such as reduced pyridine nucleotides, ascorbic acid, or thiols, may become significant. For example, in rat, ascorbic acid is 1-2 mM in brain and liver (27), NADH is 0.1mM in brain and blood and 0.25mM in liver, NADPH is 0.5mM in

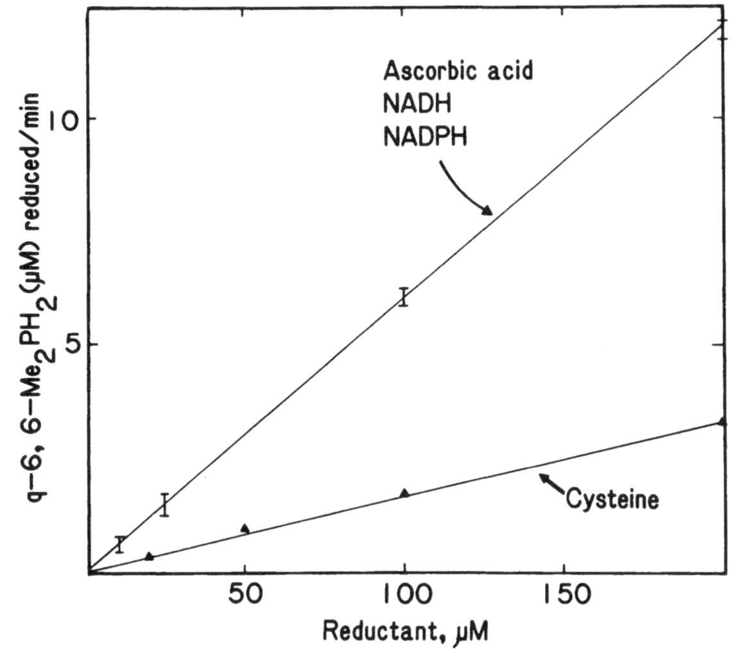

Figure 5. Rate of non-enzymatic reduction of q-6,6-Me_2PH_2 (85 μM) at pH 7.4 and 37°C.

liver (28), and glutathione is in mM concentrations in several tissues. Rates of reduction of q-6,6-Me_2PH_2 by these, or related, reagents were determined at 37° and pH 7.4 (Figure 5). The reactions were first order in each component. With NADH, NADPH and ascorbic acid the rate of reduction was the same, with $k = 660\ M^{-1}\ min^{-1}$. The rate with cysteine was about four times slower, with $k = 160\ M^{-1}\ min^{-1}$. The first order rate constant for the decay of q-6,6-Me_2PH_2 at 37° and pH 7.4 is 0.001 min^{-1}. Thus, brain, for example, has the capacity to reduce q-6,6-Me_2PH_2 non-enzymatically about 1000 times faster than its rate of decay. This difference in rates may be adequate to maintain levels of tetrahydropterin for a significant time, even in the absence of dihydropteridine reductase.

5. Ability of 6,6-Dimethyltetrahydropterin to Enter Brain

Levels of 6,6-Me_2PH_4 reached in brain were compared with those of 6-$MePH_4$. Tetrahydropterin solutions were prepared in 0.1M ascorbic acid, and neutralized with Tris base immediately before injection. Mice were injected i.p. at a dose of 0.1 µmol/g body weight and sacrificed with an overdose of pentobarbital as a function of time after injection. Blood was removed via the heart, and mercaptoethanol added to a final concentration of 0.1M. After centrifugation to separate the cells, the plasma was deproteinized by addition of an equal volume of 1M trichloroacetic acid. Brains were taken out, after perfusion with 0.9% NaCl, and homogenized in 3 volumes of 0.7M trichloroacetic acid, 0.07M mercaptoethanol. Precipitated protein was removed by centrifugation at 20,000g for 5 min. All solutions were argonated, and the brain and blood extracts stored under argon at -80°C until analyzed. Analysis for tetrahydropterins was performed by HPLC with electrochemical detection (29). Three series of animal experiments were run with each pterin. The results of typical experiments with (A) 6,6-Me_2PH_4, and (B) 6-$MePH_4$ are given in Figure 6.

With similar doses in rats the levels of 6-$MePH_4$ reached in brain were 5 nmol/g brain, and 10 times less for BH_4 (16). The levels reached by 6-$MePH_4$ in mouse brain are somewhat lower, and the duration of response much less, than reported for rat. 6,6-Me_2PH_4 reaches a four times higher level and is cleared from the brain more slowly than 6-$MePH_4$ (Figure 6). If the area under the curve is integrated to give a total time vs. concentration response, 6,6-Me_2PH_4 is four to five times more effective than 6-$MePH_4$.

The increased effectiveness of 6,6-Me_2PH_4 compared to 6-$MePH_4$ correlates well with the increased stability of the quinoid 6,6-Me_2PH_2. To determine whether or not the higher levels

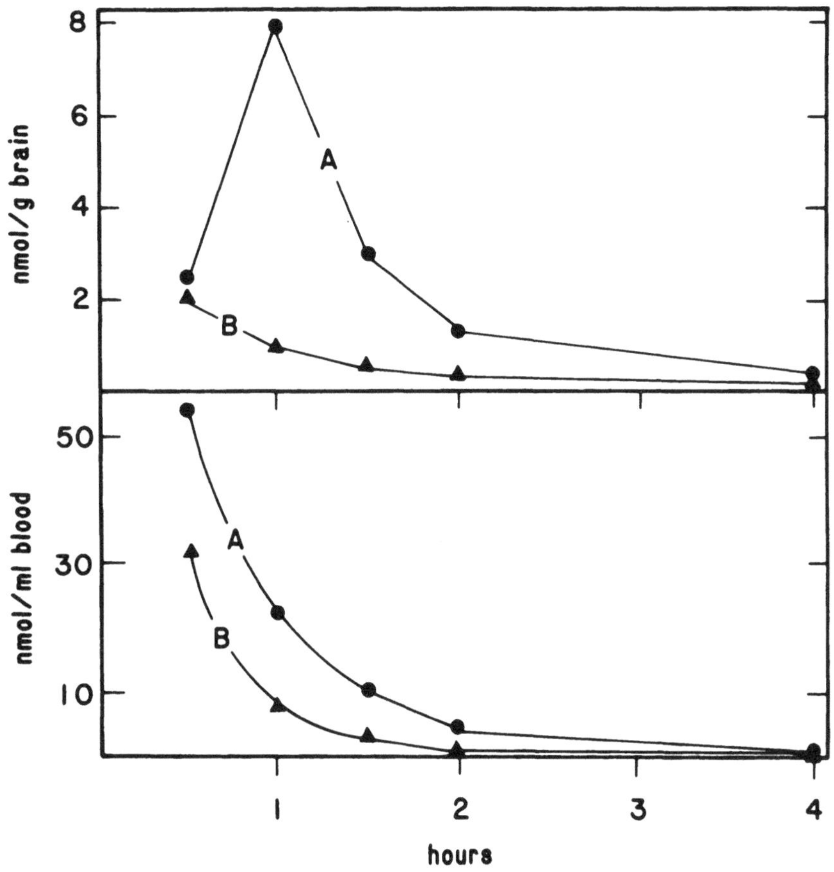

Figure 6. Brain and blood levels of (A): 6,6-Me_2PH_4, and (B): 6-$MePH_4$, as a function of time after administration.

attained with 6,6-Me_2PH_4 actually are due to the contribution of the stable quinoid form of the dihydropterin, mice were injected directly with q-6,6-Me_2PH_2. At a dose of 0.1 µmol q-6,6-Me_2PH_2/g body weight, injected i.p. in dimethylsulfoxide, brain levels reached 2 nmol/g wet weight, measured as 6,6-Me_2PH_4.

6. Innocuous Products

In contrast to quinoid dihydropterins monosubstituted at C6, q-6,6-Me_2PH_2 is unable to tautomerize to a 7,8-dihydropterin. Instead, the disappearance of pteridine-like absorption (Figure 4) suggests that a major ring rearrangement occurs during the decay of q-6,6-Me_2PH_2. Experiments were performed to ascertain whether such degradation products are toxic. Quinoid-6,6-Me_2PH_2 was incubated at pH 7.4 and 37°C for 4 hours, during which time 90% of the material decayed. It was then injected i.p. into mice at a dose of 1 μmole/g body weight. The mice were observed for 5 days and showed no effects of the treatment.

Similar experiments with high doses of q-6,6-Me_2PH_2 were precluded by the limitations in its solubility. The most likely deleterious effect of abnormally high levels of quinoid dihydropterin would be product inhibition of the hydroxylases. Thus, q-6,6-Me_2PH_2 was tested in vitro as an inhibitor of phenylalanine hydroxylase. 6,6-Me_2PH_4 was utilized as cofactor, so that concentrations of q-6,6-Me_2PH_2 would remain constant. No inhibition was observed by concentrations of q-6,6-Me_2PH_2 up to 0.4 mM. In contrast, 6-methyl-7,8-dihydropterin (6-Me-7,8-PH_2) was found to be a competitive inhibitor of phenylalanine hydroxylase, with a K_i of 0.2 mM (Table V).

Table V. Inhibition of Phenylalanine Hydroxylase by Dihydropterins

Cofactor	Cofactor Km (mM)	Inhibitor	K_i (mM)	Type of Inhibition
6,6-Me_2PH_4	0.07	q-6,6-Me_2PH_2	no inhibition up to 0.4 mM	
6-$MePH_4$	0.1	6-Me-7,8-PH_2	0.2	Comp.

Reaction conditions were the same as for Table II, with cofactor concentration varied over a range of 0.05 - 0.3mM.

7. Specificity for Target Enzyme

The biological properties required for a hydroxylase cofactor may be different for each disease. Deficiency in BH_4 biosynthesis or in dihydropteridine reductase both require a structure that is a cofactor for all three hydroxylases. In the latter case, ability to be regenerated by reductase is obviously not necessary, and instead facile non-enzymatic reduction by endogenous reducing compounds is desirable. In contrast, the optimal cofactor for Parkinson's disease may be a molecule that is specific for brain tyrosine hydroxylase, and is inactive with phenylalanine, typtophan, and adrenal tyrosine hydroxylases.

Information on structure-activity relationships of cofactor analogs is accumulating for all three hydroxylases (20,21,29-36). This data, together with the wide range of starting materials that are available for the above synthetic method, may allow production of cofactor analogs with properties optimized for the specific goal.

Acknowledgments

This investigation was supported by Grant CA-31852 awarded by the National Cancer Institute and Grant GM-30368 awarded by the National Institute of General Medical Sciences.

References

1. Levitt, M., Spector, S., Sjoerdsma, A. and Udenfriend, S.: J. Pharmacol. Exp. Ther. 148, 1-8 (1965).
2. Jequier, E., Lovenberg, W. and Sjoerdsma, A.: Mol. Pharmacol. 3, 274-278 (1967).

3. Nagatsu, T., Yamaguchi, T., Kato, T., Sugimoto, T., Matsuura, S., Akino, M., Nagatsu, I., Iizuka, R. and Narabayashi, H.: Clin. Chim. Acta 109, 305-311 (1981).

4. Lovenberg, W., Levine, R.A., Robinson, D.S., Ebert, M., Williams, A.C. and Calne, D.B.: Science 204, 624-626 (1979).

5. Williams, A.C., Levine, R.A., Chase, T.N., Lovenberg, W., and Calne, D.B.: J. Neurol. Neurosurg. Psych. 43, 735-738 (1980).

6. Williams, A., Eldridge, R., Levine, R.A., Lovenberg, W., and Paulson, G.: Lancet, 410-411 (1979).

7. Curtius, H.-Ch., Niederwieser, A., Levine, R.A., Lovenberg, W., Woggon, B. and Angst, J.: Lancet, 657-658 (1983).

8. Schmidt, M.J., in Neuropharmacology of Central Nervous System and Behavioral Disorders (Palmer, G.C., ed.), Academic Press, N.Y., pp. 149-171 (1981).

9. Rose, F.C. and Capildeo, R., eds., Research Progress in Parkinson's Disease. Pitman Medical Ltd., Tunbridge Wells, U.K. (1981).

10. Narabayashi, H., Kondo, T., Nagatsu, T., Sugimoto, T. and Matsuura, S.: Proc. Japan Acad. 58 Ser. B,283-287 (1982).

11. Curtius, H.-Ch., Muldner, H. and Niederwieser, A.: J. Neural. Transmission 55, 301-308 (1982).

12. LeWitt, P.A., Newman, R.P., Miller, L.P., Lovenberg, W. and Eldridge, R.: New Eng. J. Med. 308, 157-158 (1983).

13. Danks, D.M., Schlesinger, P., Firgaira, F., Cotton, R.G.H., Watson, B.M., Rembold, H. and Hennings, G.: Pediat. Res. 13, 1150-1155 (1979).

14. Niederwieser, A., Curtius, H.-Ch., Wang, M. and Leupold, D.: Eur. J. Pediat. 138, 110-112 (1982).

15. Kaufman, S., Kapatos, G., McInnes, R.R., Schulman, J.D. and Rizzo, W.B.: Pediatrics 70, 376-380 (1982).

16. Kapatos, G. and Kaufman, S.: Science 212, 955-956 (1981).

17. Kettler, R., Bartholini, G. and Pletscher, A.: Science 249, 476-478 (1974).

18. Craine, J.E., Hall, S.E. and Kaufman, S.: J. Biol. Chem. 247, 6082-6091 (1972).

19. Archer, M.C. and Scrimgeour, K.G.: Can. J. Bioch. 48, 278-287 (1970).

20. Bailey, S.W. and Ayling, J.E.: Biochemistry 22, 1790-1798 (1983).

21. Bailey, S.W. and Ayling, J.E.: J. Biol. Chem. 253, 1598-1605 (1978).

22. Ayling, J.E., Pirson, R., Pirson, W. and Boehm, G.: Anal. Biochem. 51, 80-90 (1973).

23. Bailey, S.W. and Ayling, J.E.: Anal. Biochem. 107, 156-164 (1980).

24. Ayling, J.E. and Thomas, K.B., manuscript in preparation.

25. Levine, R.A., Miller, L.P. and Lovenberg, W.: Science 214, 919-921 (1981).

26. Danks, D.M., Bartholome, K., Clayton, B.E., Curtius, H.-Ch., Grosse, H., Kaufman, S., Leeming, R., Pfleiderer, W., Rembold, H. and Rey, F.: J. Inher. Metab. Dis. 1, 49-53 (1978).

27. Carr, R.S., Bally, M.B., Thomas, P. and Neff, J.M.: Anal. Chem. 55, 1229-1232 (1983).

28. Lowry, O.H., Passonneau, J.V., Schulz, D.W. and Rock, M.K.: J. Biol. Chem. 236, 2746-2754 (1961).

29. Ayling, J.E., to be published.

30. Kaufman, S.: J. Biol. Chem. 239, 332-338 (1964).

31. Ellenbogen, L., Taylor, R.J. and Brundage, G.B.: Biochem. Biophys. Res. Comm. 19, 708-715 (1965).

32. Ayling, J.E., Boehm, G., Textor, S.C. and Pirson, R.A.: Biochem. 12, 2045-2051 (1973).

33. Nagatsu, T., Mizutani, K., Nagatsu, I., Matsuura, S. and Sugimoto, T.: Biochem. Pharmacol. 21, 1945-1953 (1972).

34. Numata, Y., Kato, T., Nagatsu, T., Sugimoto, T. and Matsuura, S.: Biochim. Biophys. Acta 480, 104-112 (1977).

35. Bailey, S.W. and Ayling, J.E.: Biochem. Biophys. Res. Comm. 85, 1614-1621 (1978).

36. Oka, K., Kato, T., Sugimoto, T., Matsuura, S. and Nagatsu, T.: Biochim. Biophys. Acta 661, 45-53 (1981).

37. Armarego, W.L.F., Waring P.: Aust. J. Chem. 34, 1921 (1981).

3

PTERIDINES IN IMMUNOLOGY

GUANOSINE - TRIPHOSPHATE CYCLOHYDROLASE - DEFICIENCY : ANALYSIS OF THE INFLUENCE ON IMMUNE PARAMETERS IN A GIRL.

Peter W. Joller, Nenad Blau, Manuel Atares, Alois Niederwieser
Department of Pediatrics, University of Zurich
Steinwiesstr. 75, CH-8032 Zurich, Switzerland

J.Cardesa - Garcia
Hospital for Mother and Child, University of Extremadura, Bajadoz, Spain

Introduction

Guanosine-triphosphate (GTP) cyclohydrolase is the first enzyme in the pteridine metabolism and converts GTP to D-erythro-dihydroneopterin-triphosphate (1). The biosynthetic pathway is finally leading to tetrahydrobiopterin which is a cofactor in the synthesis of serotonin and dopamine. Fig.1 (2,3,4,5). The products of this pathway have been shown to play a role not only in signal transmission in the central nervous system but also in immune reactions.

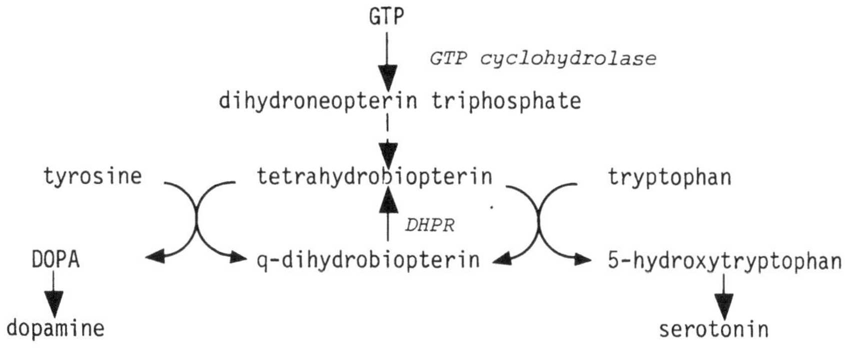

Fig. 1 Biosynthesis of tetrahydrobiopterin, dopamine and serotonin.
DHPR: dihydropteridine reductase

The evidence that pteridines may have a considerable influence on immune mechanisms stems primarily from the findings that increased amounts of pteridines are excreted in the urine of patients suffering from virus infections (6) and also in cancer patients (7). Furthermore one of the earliest signs in kidney graft rejection seems to be an increase in neopterin concentration in urine (8). Most interesting is the report (8) that cyclosporin A, a drug acting on T lymphocytes for the prevention of graft-versus-host disease, is able to suppress neopterin excretion as well. These in vivo data are supported by the in vitro disclosure of increased quantities of pteridines in the supernatant of mixed leukocyte cultures (9).
Taken together a causal relation between the triggering of defence mechanisms and the production of pteridines seems highly possible.

GTP cyclohydrolase deficiency in a patient

A unique possibility to assess the role of pteridines in the immune response was the discovery of a patient with GTP cyclohydrolase deficiency (17,18). Since GTP cyclohydrolase is the key enzyme in the synthesis of all known pteridines our patient is a paradigma for the study of the influence of pteridines on the immune system.

The five year old girl Rocio suffering from GTP cyclohydrolase deficiency shows impressive muscle hypotonia of the trunk and the extremities and is afflicted with severe mental and physical retardation. But there is no history of recurrent infections.

Initially the girl was treated with L-DOPA, 5-hydroxy tryptophane and carbidopa according to the proposal by Bartholomé and Byrd (10), but difficulties in adjusting the dosage and irritability led to discontinuation of this therapy. Since about one year Rocio is given tetrahydrobiopterin at 60 mg per day and improves under this regimen. Two days before drawing blood for our examinations tetrahydrobiopterin therapy was stopped.

For ethical and organisational reasons there was only one possibility to draw blood. Therefore no sequential work up could be envisaged. All tests had to be done simultaneously.

Methods and Results

1.) Parameter relevant for humoral immunity

Protein electrophoresis was carried out on an 2 mm agarose layer, stained with Amidoblack, dried and evaluated by eye and with a densitometer. An elevation of the $\alpha 2$ and β fractions -characteristic for an acute inflammatory infection- was discovered.

For immunoelectrophoresis the micromethod of Scheidegger (11) was used. The most conspicuous finding was an enlarged Immunoglobulin M (IgM) line which in the analysis with monospecific antisera showed no signs of a monoclonal gradient but of an increased polyclonal synthesis as seen in acute infections.

The quantitative analysis of the immunoglobulins, complement components, α 1-antitrypsin and C-reactive protein were done on a Hyland laser nephelometer. IgG and IgA were found to be decreased by 30% and 60% respectively as compared to the mean quantity of normal age matched controls. IgM was elevated by more than 100% over the normal mean. Radioimmunoassayed IgE was 30 IU/ml which suggests that the girl is not suffering from an allergy. C3 value was slightly below the normal range but C4 and C3-proactivator were within normal limits. α 1-antitrypsin proved to be 1.28 g/l which is considerably diminished, despite the fact that isoelectrofocusing revealed a normal (MM) phenotype. In contrast to the expectation in acute infection, C-reactive protein was not elevated.

Serum autoantibodies assayed in immunofluorescence tests on the appropriate substrate proved to be negative for nuclear antigens, DNA, thyreoglobulin, microsomal antigen, mitochondria, smooth muscle and striated muscle.

Hepatitis serology tested with the third generation of radioimmunoassays (ABBOTT) showed no circulating surface or core antigens but antibodies against the surface antigen (HBsAg) were present. This indicates that the girl experienced an infection with hepatitis B virus and made specific antibodies against the viral surface antigen.

α 1-fetoprotein and β human choriogonadotropin were also quantitated in a radioimmunoassay. A slight elevation of 18 IU/ml of α 1-fetoprotein was noticed.

Specific serum antibodies against the vaccination antigens diphtheria and tetanus were looked for in a hemagglutination titration. While anti-diphtheria antibodies could not be shown, anti-tetanus antibodies were present up to a dilution of 1 in 64.

2.) Parameters relevant for cellular immunity

The most accepted in vivo test for cellular immunity is the skin test. We used the intradermal multipuncture multitest (Mérieux) which measures the reactions against the following antigens (in brackets: induration in mm in the patient after 48 hours): tetanus (5), diphtheria (3), streptococcus (0), tuberculin (0), candida (1.5), trichophyton (0), proteus (3) and a glycerine control (0). The test is evaluated in a score which consists of the sum of the indurations and the sum of positive reactions. The alert score is 5/2 and less. Rocio's score is 11/3.

Blood smear analysis of leukocytes and lymphocytes showed 5.9 x 10^9

cells/l and 1.68 x 10^9 cells/l respectively. These figures are within the normal range.

Tests based on lymphocyte surface markers give information about the quantity and balance of the lymphocyte populations and subpopulations. For the enumeration of thymus dependent T lymphocytes we performed the rosette technique with AET treated sheep erythrocytes (12) as well as fluorescence microscopic evaluation of the pan T cells and T lymphocyte subpopulations with monoclonal antibodies of the LEU series (Becton and Dickinson) in the biotin-avidine system (13). The percentage and the absolute number of T cells in both test systems were normal. A slight imbalance in the ratio of helper to suppressor T cells (1.5 instead of 2.0) was noted.

The following B lymphocyte markers were evaluated in fluorescence microscopy: surface membrane immunoglobulins (smIg) surface membrane IgM (smIgM) and histocompatibility antigen HLA-Dr (Ia). Normally we find 5-10% of cells with smIg and smIgM and 1-3% of Ia positive cells. In Rocio the numbers were slightly elevated: smIg 14%, smIgM 15%, Ia 3%.

All the above stated immunocytological marker tests disclose only morphological entities. The function of the different classes and subclasses are revealed exclusively in stimulation experiments with appropriate mitogens like phytohemaglutinin (PHA), pokeweed mitogen (PWM), concanavalin A (ConA) and antigens like candidin, an extract of candida albicans. Reactivity is measured by incubating the Ficoll-separated lymphocytes with the stimulants during 3 to 5 days at 37° C in an atmosphere of 5% CO_2 in air, pulsing the cultures for 16 hours with tritiated thymidine and measuring the incorporated, acid precipitable radioactivity. The results in our patient for all mitogens and the antigen tested showed normal netcounts, a lower than normal incorporation of ^{3}H-thymidin in unstimulated lymphocytes and stimulation ratios (netcounts : unstimulated counts) which were comparable to the ones of the lymphocytes of a healthy control.

Some immune defects are based on the absent activity of certain enzymes of the purin metabolism (14,15). Adenosin deaminase and purin nucleoside phosphorylase activities in the erythrocytes of the patient however were above the normal range.

Terminal deoxynucleotidyl transferase, a lymphocytic enzyme with a proposed action on diversity of the immune reaction (16) was found to be normally active (5.84 microU/mg protein) in Rocio.

Summary of aberrant immunological findings in GTP cyclohydrolase deficiency *
Elevation of: - Acute phase proteins - IgM - alpha-1-fetoprotein - T suppressor lymphocytes (as compared to T helper cells) - B lymphocytes
Diminution of: - IgG - IgA - alpha-1-antitrypsin - Blank value in lymphocyte stimulations

*Every single parameter listed represents only a minor deviation from the normal value. The meaning of the single or the sum of all divergent values is not readily explainable because of the severe impairment of the nervous system.

Discussion

On first sight it is not readily understandable why so many parameters have been evaluated in this patient with GTP cyclohydrolase deficiency despite the fact that gross abnormalities could not be found.
In the clinical immunologists opinion however so many things in the ontogeny, the maturation and the functioning of the immune system are not understood that experiments of nature like the GTP cyclohydrolase deficiency in this girl must be examined with every possible means. We work to learn more about the prerequisites for the faultless operation of one of the most important systems allowing us to survive in a surrounding full of pathogens.
The evidence for a connection between immune reactions and increased synthesis of pteridines stems from in vivo and in vitro experiments. The proven defect in pteridine synthesis in our patient (less than 10% pteridines in the urine compared to healthy controls (18)) and the lack of GTP cyclohydrolase activity in the liver biopsy (less than 4% of the normal (18)) should therefore result not only in the documented severe impairment of the central nervous system but also in immunodeficiency. This is not the case, neither clinically nor based on the laboratory data. So we are left to find an explanation to this fact. A careful analysis of all our data and future additional experiments may probably shed a new light on the whole problem.
We do not think that at the moment it is possible to reconcile the different and differing facts. Therefore we shall only discuss the most provocative interpretations. We hope that final decisions on the basis of further tests and experiments can be made in time to come.

1.) The excretion of large amounts of pteridines during viral infections, in cancer patients and preceding graft rejection may be an epiphenomenon and may not have a direct connection to the proper function of the immune system. This would explain the absence of severe consequences due to the lack of pteridines in our patient.

2.) Pteridines in the supernatant of lymphocyte cultures stimulated allogeneically could be metabolic byproducts concomitant with every cell proliferation. Then it would be obvious that the effects of missing pteridines on the immune reactions would not be impressive.

3.) In contrast to the requirement of large amounts of pteridines for the production of neurotransmitters in the central nervous system, the immune system could be less demanding and could work without impairment with 10% or less of the quantities normally present.

4.) The tetrahydrobiopterin therapy although discontinued for two days prior to examination could have improved not only the neurologic parameters but also minor defiencies in the immune system. In coming years damages due to the slight disfunction of the defence mechanism without this treatment could have accumulated and could have led to severe infections and progressive deterioration of the immune system.

5.) There could be different loosely linked pteridine systems in our body with isoenzymes (analogous to the lactate dehydrogenase system): e.g. liver system, brain system, kidney system, lymphocyte system. In our patient one would propose a mutilated pteridine-synthetizing system in the brain but intact capabilities of pteridine synthesis in the lymphocytes.

Acknowledgements
We wish to thank Dr.Helen Joller for the analysis of autoantibodies, hepatitis serology and α1-globulins and for valuable discussion. We also greatfully acknowledge the expert technical assistance of Mrs. C.Handschin and V.Streit.

References

1.) Yim J.J., Brown G.M. Characteristics of Guanosin Triphosphat Cyclohydrolase I Purified from Escherichia coli. J.Biol.Chem. 251:5087-5094,1976

2.) Curtius H.Ch. Häusermann M., Niederwieser A., Ghisla S. Tetrahydrobiopterin Biosynthesis. In: Biochemical and Clinical Aspects of Pteridines. Vol.I. p 27-50. Wachter H., Curtius H.-Ch., Pfleiderer W. (Eds) Walter de Gruyter, Berlin, New York, 1982

3.) Nichol C.A., Lee C.L., Edelstein M.P., Chao J.Y. Duch D.S. Biosynthesis of Tetrahydrobiopterin by de novo and Salvage pathways in Adrenal Medulla Extracts, Mammalian Cell Cultures and Rat Brain in vivo. Proc. Natl. Acad. Sci. USA 80: 1546-1550 , 1983

4.) Häusermann M. Ghisla S., Niederwieser A. Curtius H.Ch. New Aspects in Biopterin Biosynthesis in Man. FEBS Letters 131: 275, 1981

5.) Kaufman S. Properties of the Pterin-Dependent Aromatic Amino Acid Hydroxylases. In: Aromatic Amino Acids in the Brain. p 81-115, CIBA Found.,Elsevier, New York, 1980

6.) Wachter H., Hausen A., Grassmayr K. Erhöhte Ausscheidung von Neopterin in Harn von Patienten mit malignen Tumoren und mit Viruserkrankungen. Hoppe-Seyler's Z physiol Chem 360: 1957-1960,1979

7.) Dhont J.L., Hayte J.M., Bonneterre J., Adenis L., Demaille A., Ardouin P., Farriaux J.P. Pteridines in the Urine and Serum from Cancer Patients. In: Biochemical and Clinical Aspects of Pteridines. Vol.I. p 133-140. Wachter H., Curtius H.-Ch., Pfleiderer W. (Eds) Walter de Gruyter, Berlin, New York, 1982.

8.) Margreiter R., Spiegelberger M., Fuchs D., Hausen A., Reibnegger G., Wachter H., Huber Ch. Pteridines as a New Marker for the Prediction of Human Allograft Rejection. In: Biochemical and Clinical Aspects of Pteridines. Vol.I. p 233-240. Wachter H., Curtius H.-Ch., Pfleiderer W. (Eds) Walter de Gruyter, Berlin, New York, 1982.

9.) Huber Ch., Margreiter R., Spielberger M., Fuchs D., Hausen A., Reibnegger G., Wachter H. Pteridines in the Assessment of Lymphocyte Activation. In: Biochemical and Clinical Aspects of Pteridines. Vol.I. p 227-232. Wachter H., Curtius H.-Ch., Pfleiderer W. (Eds) Walter de Gruyter, Berlin, New York, 1982.

10.) Bartholomé K., Byrd D.J. L-Dopa and 5-Hydroxytryptophan Therapy in Phenylketonuria with Normal Phenylalanine Hydroxylase Activity. Lancet II: 1980, 1974

11.) Scheidegger J.J. Une micro-méthode de l'immuno-électrophorese. Int Arch Allergy 7: 103, 1955

12.) Pellegrino M.A., Ferrone S., Dierich M.P. Reisfeld R.A. Enhancement of Sheep Red Blood Cell Human Lymphocyte Rosette Formation by the Sulfhydryl Compound 2-Amino Ethylisothiouronium Bromide (AET). Clin Immunol Immunopathol 3: 324-333, 1975

13.) Heggeness M.H., Ash J.F. Use of the Avidin-Biotin Complex for the Localisation of Actin and Myosin with Fluorescence Microscopy. J Cell Biol 73: 783-788,1977

14.) Giblett E.R., Anderson J.E., Cohen F., Pollara B., Meuwissen H.J. Adenosine Deaminase Deficiency in Two Patients with Severely Impaired Cellular Immunity. Lancet II, 1067,1972

15.) Giblett E.R., Amman A.J., Wara D.W. Sandman R., Diamond L.K. Nucleoside Phosphorylase Deficiency in a Child with Severely Defective T-Cell Immunity and Normal B-Cell Immunity. Lancet I: 1010,1975

16.) Baltimore D. Is Terminal Deoxynucleotidyl Transferase a Somatic Mutagen in Lymphocytes ? Nature 248: 409-411,1974

17.) Niederwieser A. Staudenmann W. Wang M.,Curtius H.Ch., Atares M.,Cardesa-Garcia J. Hyperphenylalaninemia with Neopterin Deficency. A New Enzyme Defect Presumably of GTP Cyclohydrolase. Eur.J.Pediatr. 138: 97, 1982

18.) Niederwieser A., Blau N., Wang M., Joller P., Atares M., Cardesa-Garcia J. GTP Cyclohydrolase Deficiency, a New Enzyme Defect Causing Hyperphenylalaninemia with Neopterin, Biopterin,Dopamin and Serotonin Deficiencies, Muscular Hypotonia and Severe Retardation in Development. Eur J. Pediatr. (in press)

PTERIDINES IN THE ASSESSMENT OF LYMPHOCYTE ACTIVATION

Chistoph Huber, Dietger Niederwieser
Department of Internal Medicine, University of Innsbruck

Dieter Schönitzer
Blood Transfusion Center, University Hospital, Innsbruck

Dietmar Fuchs, Arno Hausen, Olga Knosp, Gilbert Reibnegger, Helmut Wachter
Department of Medical Chemistry and Biochemistry, University of Innsbruck

Introduction

In previous publications we have shown that in vitro activation of human peripheral blood mononuclear cells is associated with release of two pteridines: one was identified as neopterin, whereas the structure of the other has not yet been unraveled (1-3). These results demonstrated the theoretical basis of elevated urinary neopterin levels in patients suffering from viral infections and various malignancies (4,5). In this report we present further evidence suggesting the existence of a complex regulatory network controlling in vitro pteridine release from activated mononuclear blood cells: we show that T-cells under the genetic control of the major histocompatibility complex (MHC) produce a factor which enables macrophages to release the above mentioned two pteridines specificities.

Materials

Pteridines in culture supernatants were assessed by means of HPLC as previously described (6). Results are expressed in nmol/l for neopterin and for the "4.2" specificity in arbitrary fluorescence units (1 a.f.u. 353 nm/438 nm $\approx$1 nmol/l neopterin). Mixed lymphocyte cultures (MLC) were performed using standard techniques (7). Purification of T-cells was achieved by the combined use of density gradient separation and passage

Biochemical and Clinical Aspects of Pteridines, Vol. 2

over Sephadex G 10 columns (8); monocytes were isolated by flotation on Percoll and subsequent surface adherence (8).

Results

Activation of T-cells induced by a variety of stimuli such as allogeneic peripheral blood mononuclear cells, autologous virus-modified cells and polyclonal activators is associated with increased release of two pteridines into the culture supernatants. One of these pteridines was identified as neopterin; the structure of the other, which operationally according to its retention time on HPLC was designated as "4.2"-pteridine, is not yet unraveled. A composition of these data is given in Table 1.

Table 1: Activation signals inducing release of marker pteridines into the culture supernatants of human peripheral blood mononuclear cells.

signal for T cell activation	experiments	pteridine "4.2"	specificity neopterin+)	proliferation o)
histoincompatible HLA-DR	8	539 (41)	37 (8)	37 (4)
EBV modified self	4	116 (36)	14 (3)	16 (4)
PHA (1 μg/ml)	13	212 (36)	13 (2)	14 (2)
Con A (10 μg/ml)	13	78 (14)	6 (1)	32 (4)
none	29	3 (2)	< 1	<1

+) nmol/l, mean and mes (in brackets)
o) ^{3}H-TdR incorporation, cpm x 10^{-3}

Release of the two marker pteridines into supernatants of MLC was clearly controlled by genes coded for by the major histocompatibility complex. These genes were found to be closely linked to or identical with the HLA-D region of this complex. In this respect pteridine release clearly resembled induction

Table 2: Release of marker pteridines in family studies of human MLC is genetically controlled by genes coded for by the major histocompatibility complex (HLA)

HLA-identity	no. of tests	proliferation+)	pteridine specificity "4.2"	neopterin°)
none	8	37 (4)	539 (41)	37 (8)
complete genotypical identity at all loci	8	< 1 (< 1)	67 (30)	< 1 (< 1)
phenotypical identity for HLA-DR	8	5 (2)	141 (38)	7 (1)

+) ^{3}H-thymidine incorporation x 10^{-3} mean (SE)   o) nmol/l, mean (SE)

of cellular proliferation. Results are composed in Table 2. The control of cellular functionsby HLA genes identical with or closely linked to the D-region of the HLA complex was so far exclusively a feature of T lymphocytes of the helper subtype. Further support for the view that activation of T-cells is essential for induction of in vitro pteridine release was obtained from blocking studies with cyclosporin A (results not shown). This substance so far is not known to interfere with any other cellular function except with the activation of T-cells (9). When, however, highly purified human T cells derived from T-cell lines or obtained by extensive purification procedures from the peripheral blood were investigated, a complete lack of pteridine release was observed. Results are summarized in Table 3.

This strongly suggested that in addition to T-cell activation accessory cells are required to release neopterin and the 4.2-pteridine. Experiments in which highly purified human T-cells after activation were tested in the presence or absence of monocytes supported this view. A typical experiment is given in Table 4.

Table 3: Failure to detect marker pteridines in supernatants of highly purified human T-cells.

cell type	no. of different suspensions tested	activation+)	pteridine specificity "4.2"	neopterin°)
thymocytes	1	-	o	o
		+	o	o
peripheral blood T cells	8	-	o	o
		+	o	o
IL-2 dependent T-cell lines x)	10	-	o	o
		+	o	o
malignant T-cell lines ++)	4	-	o	o
		+	o	o

+)induced by allogeneic lymphoblastoid cell lines; o)nmol/l; x)cytotoxic T cell lines derived from MLC;++)MOLT-4, CCRF-HSB 2, CCRF-CEM, JM.

Table 4: Release of marker pteridines in human MLC requires both activation of T-cells and presence of accessory cells

responder	stimulator	proliferation (cpm)	pteridine release "4.2"	neopterin+)
PBMNC	O	402	O	O
T	O	912	O	O
MØ	O	312	O	O
PBMNC	LCL++)	190 031	145	5
T	LCL	195 345	21	O
MØ	LCL	835	128	O
T + MØ	LCL	105 502	288	24

+) nmol/l; ++)allogeneic lymphoblastoid cell lines.
PBMNC – peripheral blood mononuclear cells
T – T-cells, MØ - macrophages

Preliminary experiments in which supernatants derived from alloantigen stimulated human T-cells were tested to induce pteridine release from highly purified cultures of macrophages indicated that such a cooperation is in fact operative (results not shown). We thus would state that release of pteridines from antigenically activated human mononuclear cell suspensions is mediated via cooperation of T-cells and macrophages where the first cell upon antigenic stimulation produces a factor which renders macrophages susceptible to release pteridines.

Discussion

Since the last Meeting further progress has been achieved in the understanding of the mechanism underlying the release of distinct pteridine specificities into the supernatants of cultures from activated mononuclear blood cells. We now demonstrate that two prerequisites must be fulfilled to induce this phenomenon: (1) the activation of T-cells: this is indicated by the lack of resting T cells to induce this release, as well as by the genetic control via the HLA-D region of the human major histocompatibility complex and by the capacity of cyclosporin A to inhibit this release. (2) the presence of accessory cells of the monocyte/macrophage lineage. Their requirement is indicated by the failure of highly purified T-cells to release the two marker pteridines as well as by the results of reconstitution experiments with this cell type. Preliminary results of our laboratories suggest that in this T-lymphocyte/ macrophage interaction the T-cell produces a factor which renders macrophages capable of releasing neopterin and the 4.2-pteridine specificity.

From the immunologist's point of view this intimate cooperation between T-cells and macrophages is not surprising: similar observations have been made in context with almost every antigen inducing cellular immune responses (10): macrophages have been shown to present antigens to inducer-T cells and to produce factors such as interleukin-1; T-cells on the other hand, subsequent to recognition of the antigen, release other mediators such as the migration-inhibitory-factor or the macrophage-activating-factor to immobilize these cells at the site of inflammation and to enhance their functional capacities for phagozytosis, degradation and killing (10). Our in vitro data of release of marker pterdines controlloed by a cooperation between activated T-cells and macrophages further support the view that such molecules can be used for biochemical markers to assess the activation of the T-cell/macrophage axis (1).

References

1. Huber, Ch., Margreiter, R., Spielberger, M., Fuchs, D., Hausen, A., Reibnegger, G., Wachter, H.: Biochemical and Clinical Aspects of Pteridines. Vol. 1, pp 227, Walter de Gruyter, Berlin-New York, 1982.

2. Huber, Ch., Fuchs, D., Hausen, A., Margreiter, R., Reibnegger, G., Spielberger, M., Wachter, H.: J. Immunol. 130, 1047 (1983)

3. Fuchs, D., Hausen, A., Huber, Ch., Margreiter, R., Reibnegger, G., Spielberger, M., Wachter, H.: Hoppe-Seyler's Z. Physiol. Chem. 363, 661 (1982)

4. Wachter, H. Hausen, A., Graßmayr, K.: Hoppe-Seyler's Z. Physiol. Chem. 360, 1957 (1979)

5. Hausen, A., Wachter, H.: J. Clin. Chem. Clin. Biochem. 20, 593 (1982)

6. Hausen, A., Fuchs, D., König, K., Wachter, H.: J. Chromatography 227, 61 (1982)

7. Huber, Ch., Fink, U., Leibold, W., Schmalzl, F., Peterson, P.A., Klareskog, L., Braunsteiner, H.: J. Immunol. 127, 726 (1981)

8. Huber, Ch., Merkenschalger, M., Gattringer, C., Royston, I., Fink, U., Braunsteiner, H.: J. Exp. Med. 155, 1222 (1982)

9. Borel, J.F.: Transplant. Proc. XIII, 344 (1981)

10. Rosenthal, A.S.: New Engl. J. Med. 303, 1153 (1980)

Acknowledgements

This work was financially supported by the Austrian Research Funds "Zur Förderung der wissenschaftlichen Forschung", P. 4905. We are further indebted to Mrs. A.M. Födinger, J. Pichl and I. Katholnig for skilful technical assistance and to Mrs. J. Gschnitzer for secretarial help.

PTERINS AS FUNCTIONING CONSTITUENTS OF THE LYMPHOKINE SYSTEM

Irmgard Ziegler, Ulrike Hamm
Institut für Toxikologie und Biochemie der Gesellschaft für Strahlen- und Umweltforschung, Abteilung Zellchemie; D - 8000 München 2, Germany

Jürgen Berndt
Institut für Toxikologie und Biochemie der Gesellschaft für Strahlen- und Umweltforschung, Abteilung Zellchemie; D - 8000 München 2, Germany

Introduction

Earlier experiments had shown that biopterin is accumulating in the blood cells during normal (1) and especially, during leukemic hemopoiesis (2) and that neopterin is excreted by alloantigen-induced lymphocytes (3). The follow-up of blood pteridines during clinical cases (4) gave us the notion that intermediates of pterin metabolism may not only represent lymphocytic products but may be regulators of the immune response.

A screening of 15 of the major intermediates of pterin metabolism for their action during concanavalin (Con A) induced lymphocyte activation showed that most of them had no effect. Sepiapterin (SP), dihydrobiopterin (BH_2) and tetrahydrobiopterin (BH_4) however, proved to be co-stimulators, whereas xanthopterin and isoxanthopterin were proliferation inhibitors. Both the co-stimulating and the inhibiting effect were highly dependent on the structural and optical isomers (5).

The present investigation should further clarify the view that those pterins are functioning constituents of the lymphokine system.

Biochemical and Clinical Aspects of Pteridines, Vol. 2

Materials and Methods

The mouse spleen lymphocytes were obtained as described (6). The cells were grown in RPMI 1640 tissue culture medium with 5% inactivated fetal calf serum (FCS) and 10^{-4}M 2-mercapto - ethanol at 5% CO_2 and 37°. The synthesis of DNA and RNA was determined by incorporation of ^{3}H-thymidine (^{3}H-TdR) and ^{3}H-uridine (^{3}H-U), respectively. The Microtiter assay system in combination with a semiautomatic cell harvester (Flow, Meckenheim) has been used.
Line L 1210, derived from a lymphocytic mouse leukemia (see 7) and line LS-2, established 1974 by J.Lange (Stuttgart) from a patient with acute lymphocytic leukemia were maintained in RPMI 1640 + 5% FCS + 10^{-4}M 2-mercaptoethanol; whereas line L 5222, derived from a myelomonocytic rat leukemia (8) re - quires 20% FCS for long term maintenance. L 1210 and L 5222 were supplied by the Deutsches Krebsforschungszentrum, Heidelberg; LS-2 was supplied by the Institut für Klinische Hämatologie der Gesellschaft für Strahlen- und Umweltforschung, München.

Results

1) The co-stimulating effect of SP, BH_2 and BH_4 during T-cell activation. Con A is known to activate T-lymphocytes (see 6). SP, BH_2 and BH_4 promote this activation. The ex - tent of the promotion is dependent on the concentration of both partners. On the one hand it is dependent on the activation brought about by the lectin; it is small at optimum Con A concentrations but may double the rate at suboptimal and supraoptimal concentrations of Con A (fig.1).
On the other hand, by example of SP it is shown that the co-stimulating effect at a constant lectin concentration is dependent on the concentration of the pterin (fig.2). It peaks at 30 µM SP in the medium; even though the immediate effective concentration may be lower. The need for

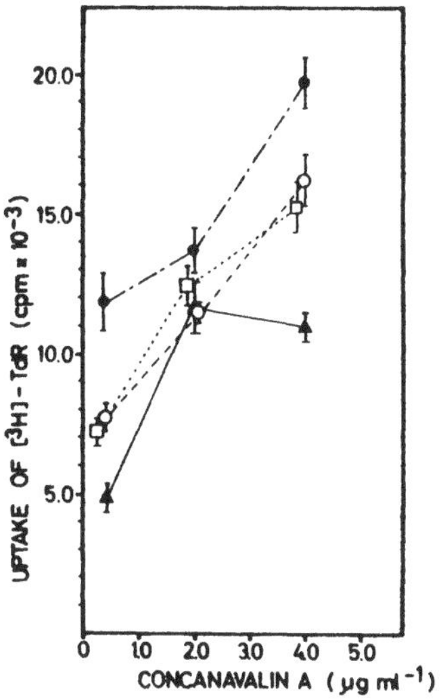

Fig.1 Effect of SP, BH_2 and BH_4 (1.4×10^{-5}M) on the incorporation of ^{3}H-TdR into DNA of mouse spleen lymphocytes at various Con A concentrations.

▲——▲ control

○- - -○ sepiapterin

□· · · · □ dihydrobiopterin

●-·-·● tetrahydrobiopterin

= S.D.

permeation into the cell and a further metabolic conversion during the 72 h incubation period may shift the optimum to higher pterin concentrations than are needed at its molecular target.

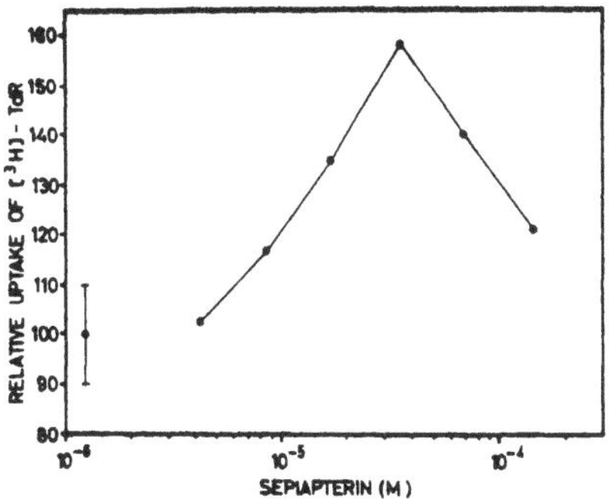

Fig.2 Effect of SP on the incorporation of ^{3}H-TdR into DNA of Con A (0.4 µg/ml) stimulated mouse spleen lympho - cytes.

= S.D. of control

2) The co-stimulating effect of BH_4 during B-cell activation.

Lipopolysaccharide E endotoxin (LPS) is known to stimulate B-cells (see 6). SP and BH_2 fail to promote this B-cell activation. SP is even inhibiting at concentrations $> 5 \times 10^{-5}$M. Only BH_4 proves to be a co-stimulator (fig.3). Its co-stimulating activity peaks at a concentration of 15 µM in the medium and is therefore similar to the pterin concentrations needed for the enhancement of T-cell activation. In no experiment a promotion as high as obtained during T-cell activation has been found yet.

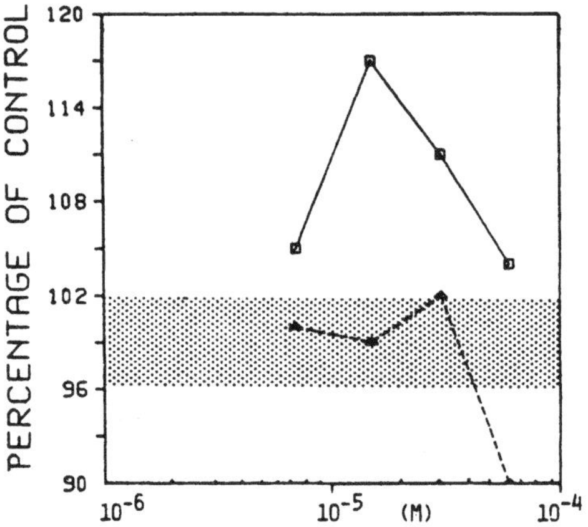

Fig.3 Effect of SP and BH_4 on the incorporation of ^{3}H-U into RNA of lipopolysaccharide E (1.5 µg/ml) stimulated mouse spleen lymphocytes.

△- - -△ sepiapterin

□——□ tetrahydrobiopterin

Dotted area: 95% probability level of control

4) The inhibiting action of xanthopterin and its molecular target. As previously shown (5), both catabolites xanthopterin and isoxanthopterin inhibit the proliferation of activated lymphocytes as well as of lymphoid cell lines (LCL). Half maximum inhibition of lympho -

cyte activation was found at 18 µM xanthopterin (5). Isoxanthopterin seems to be trapped by the alpha$_1$-acid glycoprotein present in the FCS of the medium (5,9). Therefore, its action is harder to rationalize and further investigations concentrated on xanthopterin.
Fig.4 demonstrates that a LCL, exposed to various concentrations of xanthopterin maintains its initial DNA synthesis for about 20 h; a marked inhibition coincides with the onset of the main period of DNA synthesis which in turn, is based on a preceding RNA formation.

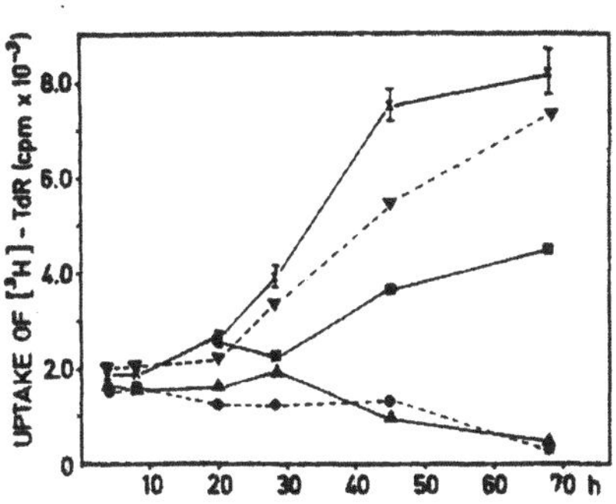

Fig.4 Effect of xanthopterin on ^{3}H-TdR incorporation into the DNA of a LS-2 cell culture.

x——x control; $\bar{x}$ = S.D.
●--- ● 1.4 x 10^{-4}M xanthopterin
▲——▲ 7.0 x 10^{-5}M xanthopterin
■——■ 3.5 x 10^{-5}M xanthopterin
▼---- ▼ 1.75 x 10^{-5} xanthopterin

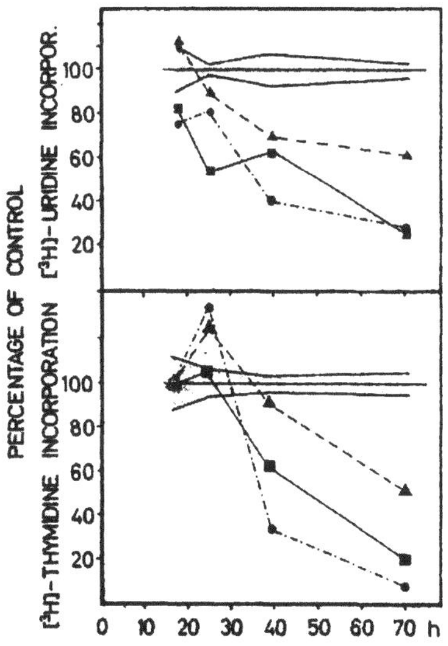

Fig.5 Effect of xanthopterin on the RNA and DNA synthesis in Con A (1.25 µg/ml) stimulated mouse spleen lymphocytes.

○-·-·-○ 7.0 x 10^{-5}M xanthopterin
■——■ 3.5 x 10^{-5}M xanthopterin
▲- - - ▲ 1.75 x 10^{-5}M xanthopterin

Dark area: 95% probability level of the control

In the case of Con A induced lymphocyte stimulation the RNA synthesis was found to be reduced prior to the inhibition of ^{3}H-TdR incorporation into DNA; moreover, it proved to be more sensitive at low xanthopterin concentrations (fig.5). In contrast to DNA synthesis however, about one third of the RNA synthesis remains unaffected even at high inhibitor concentrations and after prolonged incubation time.

These data indicate that the molecular target of the inhibiting pterin is the synthesis of RNA rather than of DNA and among this, of only one of the RNA species.

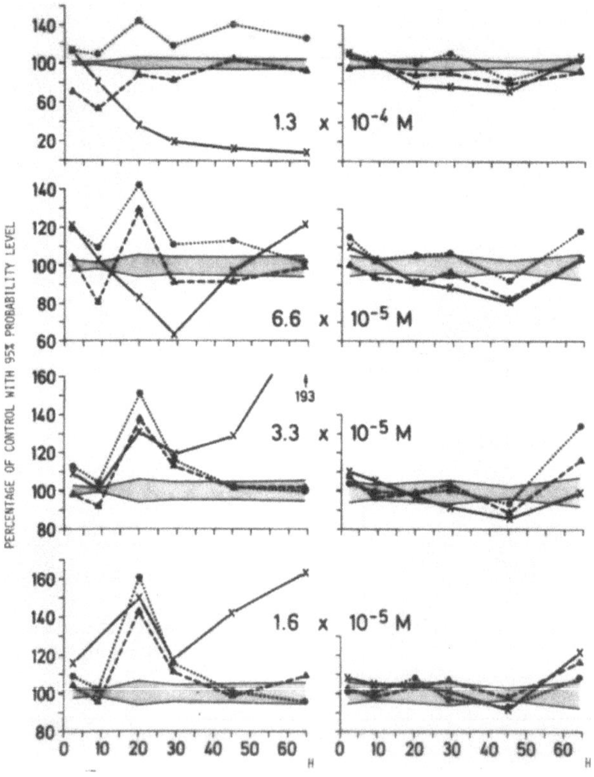

Fig.6 Effect of SP, BH_2 and BH_4 on ^{3}H-TdR incorporation into the DNA of L 5222 (myelomonocytic rat leukemia) and L 1210 (lymphocytic mouse leukemia). 5% FCS medium.

x——x sepiapterin; ●····● dihydrobiopterin; ▲---- ▲ tetrahydrobiopterin

5) The effect of SP, BH_2 and BH_4 on leukemic cell lines.

The cell lines L 1210, LS-2 and L 5222 were exposed to 5% FCS medium. This FCS content is sufficient for both lymphoblastoid cell lines (L 1210 and LS-2) but not sustaining continous growth of the myelomonocytic line (L 5222). The addition of SP and to a lesser extent of BH_4, are inhibiting at concentrations > 3.3×10^{-5}M (fig.6; LS-2 showed identical results with L 1210 and therefore, the data are omitted). The gradual recovery with time indicates that the inhibitor is metabolically removed as long as its concentration is not too high. The increased sensitivity of L 5222 as compared to both lymphoblastoid lines may be interpreted by a lower SP metabolizing activity.

If the pterins are present at concentrations < 3.3×10^{-5}M, a difference between both cell types is most obvious (fig.6). SP, BH_2 and BH_4 stimulate the DNA synthesis in the myelomonocytic line, whereas the incorporation rates of ^{3}H-TdR in both lymphoblastoid cell lines are unaffected (data of LS-2 identical with L 1210). The enhancement of DNA synthesis in L 5222 proved to be ephemeral with BH_2 and BH_4, whereas SP sustained an increased rate for at least 60 h.

Discussion

In 1969 the criteria for lymphokines were defined by Dumonde et al. (10). Subsequently those factors were described manyfold but largely remained unidentified yet.They are (i) soluble factors; (ii) they are produced by antigen-activated lymphocytes; (iii) they are different from classical antibodies; (iv) they are amplifiers, regulators and mediators of the immune response.

The criteria (i) through (iii) are self-evident or were shown

to be true for pterins in earlier investigations (2,3).
The amplification rates, as required by criterion (iv) and as obtained by SP, BH_2 and BH_4 correspond to those usually reported for natural lymphokines (11).
The regulatory function of lymphokines implies that the lymphocyte stimulation is finally terminated and cellular maturation takes place (cf. 6,11). This emphasizes the regulatory aspect of pterin metabolites, in that some anabolites are co-stimulators, but some of the catabolites are inhibitors of cell proliferation. Thus the control system is governed by the equilibria present between the various anabolic and catabolic steps. Since xanthopterin and isoxanthopterin are inhibitors, but dihydroxanthopterin, leucopterin and the lumazines are ineffective (5), the catabolic route underwent by the respective cell lines may exert a decisive control function. Its molecular target seems to be part of the RNA synthesis.

A first approach to the mediating role of pterin metabolites is indicated by the experiments described above. T-lymphocytes are known to assist B-cell response (**cf.** 6). Whereas T-cells are co-stimulated by SP, BH_2 and BH_4 either, B-cell co-stimulation is dependent on the delivery of the end product BH_4. This supply readily may be done by the mediating T-cells.

In the last decades the supply of lymphokines by the "conditioned medium" became well known. It results from a previous growth of activated T-lymphocytes (cf.6). Moreover, lymphokines are known to represent part of the growth factors supplied by the FCS and are needed to various extents by the different cell lines. The experiments demonstrate that SP, BH_2 and BH_4 are among those factors which may substitute for FCS.

In conclusion the classical and subsequent criteria for lymphokines are fulfilled by some intermediates of pterin metabolism. Its detailed study during the immune response is needed for further development of these initial results.

References

1. Ziegler,I., Kolb, H.J., Bodenberger,U., Wilmanns,W.: Blut 44, 261 - 270 (1982).

2. Ziegler,I., Fink,M., Wilmanns,W.: Blut 44, 231-240 (1982)

3. Fuchs, D., Hausen,A., Huber,C., Margreiter,R., Reibn - egger,G., Spielberger,M., Wachter,H.: H.S.Zeitschr.f. Physiol.Chem. 363, 661 - 664 (1982).

4. Fink, M., Ziegler,I., Maier,K., Wilmanns,W.: Cancer Res. 42, 1574 - 1578 (1982).

5. Ziegler,I., Hamm,U., Berndt, J.: In Chemistry and Biology of Pteridines (Ed. J.Blair); W.de Gruyter, Berlin (in press).

6. Ling,N.R., Kay,J.E.: Lymphocyte stimulation, second ed., North Holland Publ.Comp., New York, 1975.

7. Habs,M.: in Maligne Tumoren (Ed.D.Schmähl). 3.Aufl., pp.455 - 499. Edition Cantor Aulendorf (1981).

8. Ivankovic,St., Zeller,J.: Blut 28, 288 - 292 (1974).

9. Ziegler,I., Armarego,W.L.F., Fink,M.: In Chemistry and Biology of Pteridines. (Ed.J.Blair). W. de Gruyter, Berlin (in press).

10. Dumonde, D.C., Wolstencroft, R.A., Panayi,G.S., Matthew, M., Morley,J., Howson, W.T.: Nature 224, 38 - 44 (1969).

11. Sanderson,R., Vesole,D., Jakway,J., Talmage,D.: Immunol. Rev. 51, 177 - 191 (1982).

Acknowledgement

Dr.M.Berger, Deutsches Krebsforschungszentrum Heidelberg, supplied us with the cell lines L1210 and L5222 and relevant information for which we are very grateful.

The work was supported by the Deutsche Forschungsgemeinschaft.

NEOPTERIN EXCRETION DURING ALLOGENEIC, SYNGENEIC AND AUTOLOGOUS BONE MARROW TRANSPLANTATION

D. Niederwieser, Ch. Huber,
Dept. of Internal Medicine, University of Innsbruck, Austria

A. Gratwohl, P. Bannerth,D. Speck,
Division of Hematology, Dept. of Internal Medicine,
Kantonspital Basel, University of Basel, Switzerland

D. Fuchs, A. Hausen, G. Reibnegger, H. Wachter,
Dept. of Medical Chemistry and Biochemistry, University of Innsbruck, Austria.

Introduction

As already discussed in our previous publications lymphocyte activation is invariably associated with increased excretion of neopterin (1,2). As a logical extension of this observation we have searched for the use of this biochemical marker to predict immunological complications such as acute rejection of renal, hepatic or pancreatic allografts (3,4). Our presentation deals with the utility of the determination of urinary neopterin levels for the clinical monitoring of bone marrow transplant recipients. Results are presented suggesting that the neopterin marker for this population of patients represents a most useful tool to predict both immunological complications as well as haemolymphopoetic reconstitution.

Material and Methods

<u>Patients</u>: So far 7 recipients of allogeneic bone marrow grafts

Biochemical and Clinical Aspects of Pteridines, Vol. 2

from HLA identical and MLC negative sibling donors as well as two recipients transplanted with syngeneic or autologous marrows were evaluated. Their clinical characteristics are summarized in Table 1. Details of the management of these patients were previously published (5).

Prior to transplantation all patients received high dose cyclophosphamide followed by total body irradiation. Immunosuppression for prophylaxis of rejection and graft versus host disease consisted of Cyclosporin A except for patients transplanted with syngeneic bone marrow grafts.

TABLE I

Clinical characteristics of bone marrow transplant recipients

patient	diagnosis[1]	bone marrow donor	TREATMENT pretransplant		TREATMENT posttransplant	complication	out-come
M.O.	CML	allogeneic	CTX[2]	TBI[3]	Cy A[4]	GvHD[5]	dead
A.F.	ALL 3. rem.	allogeneic	CTX[2]	TBI[3]	Cy A[4]	none	alive
O.J.	AML 1. rem.	allogeneic	CTX[2]	TBI[3]	Cy A[4]	GvHD, herpes simp.	alive
K.B.	ALL 2. rem.	allogeneic	CTX[2]	TBI[3]	Cy A[4]	relaps	dead
K.M.	ALL 1. rem.	allogeneic	CTX[2]	TBI[3]	Cy A[4]	herpes simplex	alive
H.R.	CML	allogeneic	CTX[2]	TBI[3]	Cy A[4]	GvHD, herpes simp.	alive
W.F.	ALL no rem.	allogeneic	CTX[2]	TBI[3]	Cy A[4]	none	alive
P.P.	ALL 1. rem.	syngeneic	CTX[2]	TBI[3]	----	septic., herpes s.	alive
G.E.	metastatic breast cancer	autologous	CTX[2]	TBI[3]	Cy A[4,6]	herpes simplex	alive

1. AML=acute myeloid leukemia
   ALL=acute lymphoblastic leukemia
   CML=chronic myeloid leukemia
2. high dose cyclophosphamide (6o mg/kg x 2 days)
3. total body irradiation with looo rad
4. cyclosporin A
5. graft versus host disease
6. this patient has previously received a liver graft

Laboratory investigations: Neopterin levels were measured in urine samples by high pressure liquid chromatography (HPLC). Details of this method were described previously (6). Neopterin levels are expressed in µmol/mol of urinary creatinine.

Results

1. Results of urinary neopterin excretion in patients transplanted with allogeneic bone marrow grafts

Neopterin excretion values of a representative case in relation to the various clinical parameters are demonstrated in figure 1. As shown, induction of haemopoetic aplasia following treatment with high dose cyclophosphamide and TBI is accompanied by a reduction of neopterin excretion, which reached its maximal suppression 5 days after transplantation. This time also corresponded with the time of maximal depletion of circulating lymphocytes. Subsequently neopterin levels increased rapidly reaching peak values 3 days after haemolymphatic reconstitution, defined as the time at which more than 1000 circulating leukocytes were detectable. Thereafter neopterin excretion declined reaching levels in the upper range of normal controls. On day

Figure 1

Neopterin excretion in relation to peripheral blood lymphocyte number in a bone marrow transplant recipient with uncomplicated clinical course

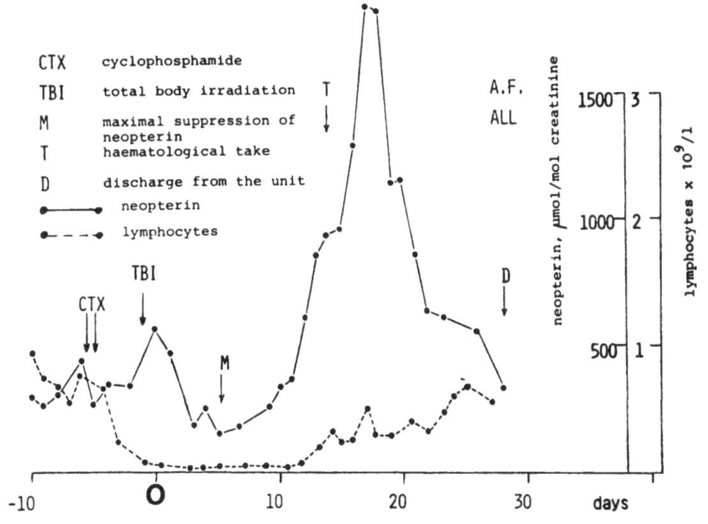

FIGURE 2
NEOPTERIN EXCRETION PATTERNS IN ALLOGENEIC BONE MARROW TRANSPLANT RECIPIENTS LACKING MAJOR IMMUNOLOGICAL COMPLICATIONS

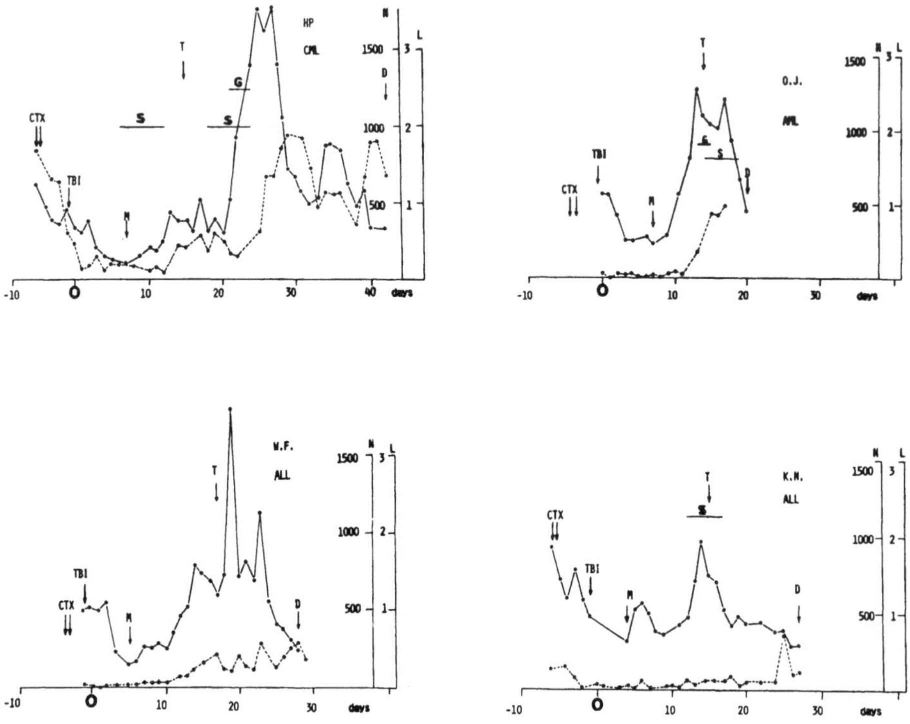

N neopterin, µmol/mol creatinine •——•
L lymphocytes x 10^9/l •------•
CTX high dose cyclophosphamide
TBI total body irradiation

M maximal suppression of neopterin
T haematological take
D discharge from the unit
G graft versus host disease
S herpes simplex infection

29 after transplantation the patient was discharged from the unit without evidence of either complicating infections or graft versus host disease.

Essentially the same neopterin patterns were observed in four additional allograft recipients lacking major immunological complications (Fig. 2). Two further cases developed most severe and, in one of these patients, lethal graft versus host disease after haemopoetic recovery. The course of one of these patients is depicted in figure 3. As shown neopterin patterns during the time before development of graft versus host disease resembled those of the uncomplicated cases. Subsequently despite intensification of immunosuppression neopterin levels increased again and the patient died due to progressive graft versus host disease.
(P ≙ 1000 mg Prednisolon).

Figure 3

Neopterin excretion and lymphocyte patterns in allogeneic bone marrow transplant recipients with most severe and lethal graft versus host disease

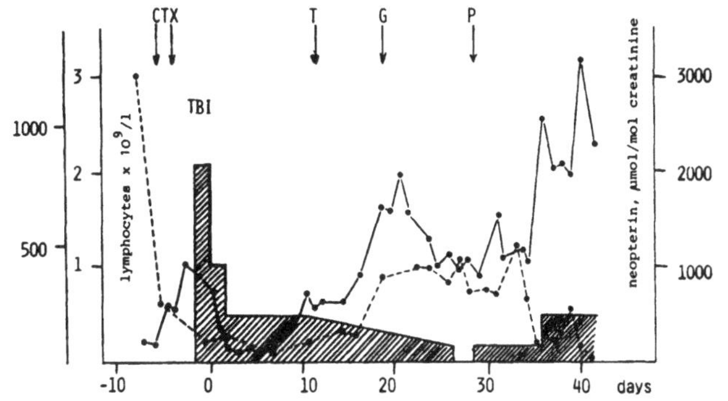

CTX cyclophosphamide

TBI total body irradiation

T haematological take

G graft versus host disease

cyclosporin A

neopterin

lymphocytes

2. Results of urinary neopterin excretion in patients transplanted with autologous or syngeneic bone marrow

Also investigated were neopterin values of two patients transplanted with autologous or syngeneic bone marrow. These combinations enabled to study neopterin levels in the absence of immunostimulation by histoincompatible tissues. Results of the autologous transplant are depicted in figure 4. As demonstrated a pattern virtually identical with those obtained in allogeneic combinations was observed.

Figure 4

Neopterin excretion pattern in a autologous transplant recipient

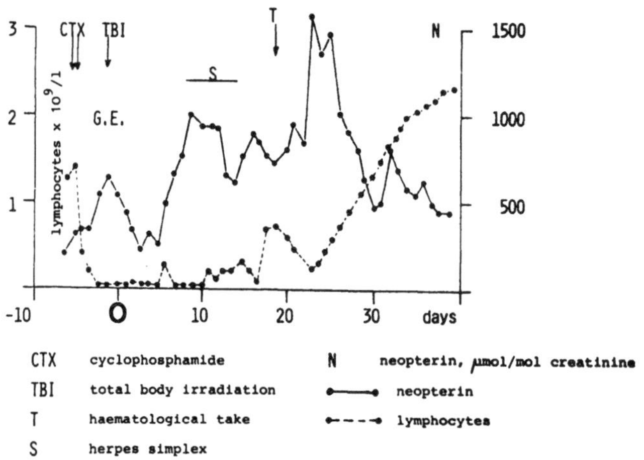

Conclusions

Our preliminary results of evaluation of urinary neopterin values in bone marrow transplant recipients may be summarized as follows:

1. Haemolymphatic aplasia induced by high dose cyclophosphamide and TBI was associated with reduced neopterin excretion. This, together with previous in vitro findings (1, 4), suggested that neopterin synthesis is mediated by a CTX and/or irradiation sensitive haemolymphatic producer cell.

2. Haemolymphatic reconstitution after autologous syngeneic or allogeneic BMT was invariably associated with and preceded by rapidly increasing neopterin values. This observation further points to the crucial role of the haemolymphatic tissues for production of neopterin.

3. In the absence of severe immunological complications such as infection or graft versus host disease urinary neopterin release returned to normal levels. In contrast, immune stimulation by viral disease or severe graft versus host disease were accompanied by extremely high neopterin values.

4. We would thus state that the monitoring of neopterin patterns in bone marrow recipients enables both prediction of bone marrow engraftment and of subsequent immune complications as a consequence of infection or graft versus host disease.

References

1. Huber, Ch., Fuchs, D., Hausen, A., Margreiter, R., Reibnegger, G., Spielberger, M., Wachter, H.: J. Immunol. 130, 1047-1050 (1983).
2. Fuchs, D., Hausen, A., Huber, Ch., Margreiter, R., Reibnegger, G., Spielberger, M., Wachter, H.: Hoppe-Seyler's Z. Physiol. Chem. 363, 661 (1982).
3. Margreiter, R., Fuchs, D., Hausen, A., Huber, Ch., Reibnegger, G., Spielberger, M., Wachter, H.: Transplantation, in press.

4. Niederwieser, D., Huber, Ch., Wachter, H.: Wien. klin. Wschr. 5, 161-164 (1983).

5. Speck, B., for the Basel bone marrow transplant team: Anticancer Research 1, 249-254 (1981).

6. Fuchs, D., Hausen, A, Reibnegger, G., Wachter, H.: In: Biochemical and Clinical Aspects of Pteridines, Vol. 1; Wachter, H., Curtius, H.Ch., Pfleiderer, W. (eds), Walter de Gruyter: Berlin, New York 1982 pp 67-80.

Acknowledgement

This study was financially supported by the Austrian Research Funds "Zur Förderung der wissenschaftlichen Forschung", project P 4905.

THE DYNAMICAL BEHAVIOUR OF URINARY NEOPTERIN EXCRETION IN ALLOGRAFT RECIPIENTS - A HIGHLY INFORMATIVE PARAMETER FOR ASSESSMENT OF IMMUNOLOGICAL COMPLICATIONS

G. Reibnegger, D. Fuchs, A. Hausen, O. Knosp, H. Wachter
Institute of Medical Chemistry and Biochemistry, University of Innsbruck

R. Margreiter, M. Spielberger
Dept. of Surgery I, University of Innsbruck

Ch. Huber
Dept. of Internal Medicine, University of Innsbruck

Introduction

A simple and reliable biochemical marker of increased lymphocytic activity would undoubtedly be a tremendous asset in immunological monitoring where, due to the complexity of the assays used, only few studies have been able to employ daily testing. The development in human allotransplantation surgery represents a particularly urgent challenge for the search of a marker of that kind: although a remarkable progress has been achieved in matching for major histocompatibility complex (MHC) antigens as well as in establishing various immunosuppressive modalities, acute graft rejection remains the predominant cause for graft loss.

A timely and reliable diagnosis of acute graft rejection is generally believed to facilitate an appropriate medication and thus to improve the probability of graft survival. Acute rejection of allografts is preceded by rapid proliferation of alloantigen reactive lymphocytes (1), representing the only

Biochemical and Clinical Aspects of Pteridines, Vol. 2

specific effector cell population mediating allograft rejection (2-4).

We have recently reported on the obvious association between the excretion of neopterin (and a further, hitherto not fully identified pterin derivative) and activation of cellular immunity, both in vitro and in vivo, and small pilot studies on the applicability of this observation for monitoring lymphocytic activity in allograft recipients have been most encouraging (5-8).

The aim of this report is twofold:

(a) We summarize our clinical experience with neopterin as a biochemical marker for allograft rejection in 1oo consecutive recipients.

(b) We present in some detail a simple method for the statistical evaluation of the temporal courses of neopterin excretion in this patient group which takes into account not only one single determination per day, but rather includes the dynamical characteristics of neopterin excretion over a longer period of time.

Patients and Methods

Since November 1981 until the end of November 1982, urinary neopterin levels were determined daily in 96 consecutive cadaveric kidney, 3 liver and 1 pancreas recipient(s). Urine sampling was initiated from day 3 post-transplantation after removal of the urine catheter and was continued until discharge from the transplantation unit. On average a total of 21.6 (range from 7 to 51) samples per patient were evaluated.

Post-transplant immunosuppression consisted of either Prednisone and Azathioprine or Cyclosporin A (CyA). Diagnosis of rejection episodes was based on clinical symptoms, blood chemistry, renal scanning and fine needle aspiration cytology.

Viral infections were diagnosed on the basis of clinical symptoms and repeated serological testing.

Rejections were treated with gramdoses of methylprednisolone up to a maximum of 6 grams on consecutive days.

Evaluation of urinary neopterin was performed employing an optimized high performance liquid chromatography (HPLC) procedure which has already been described (9). Urinary neopterin levels have been related to simultaneously detected urinary creatinine (10) in order to avoid errors due to physiological variations of urine concentration. No chemical manipulations (e.g. oxidation of the urine specimens) preceding analysis have been performed (9). Samples of the 24 hour urines have been analysed.

Results

According to their clinical symptoms, the patients distributed themselves into 3 main groups: Group A consisted of 29 patients showing an uncomplicated post-transplant course. Group B comprised 63 cases presenting with immunological complications: acute graft rejections (n=38), viral infections (n=17) and acute rejection episodes associated with viral diseases (n=8). Finally, in group C patients (n=4), prophylactic immunosuppression with Cy A had to be withdrawn due to serious drug toxicity, although the patients did not exhibit immunological problems.

a) Patients with an uncomplicated post-transplant course

In a total of 474 urine samples from the 29 patients comprising group A neopterin levels were determined. The mean level

Table 1

Urinary neopterin levels in healthy subjects and in patients with renal allografts

	No.of subjects	No. of determinations	mean level[a]	SD[a]
healthy subjects	417	417	128	38
group A[b]	29	474	247	195
group B[b]	63	1495	662	874
group C[b]	4	107	422	354
acute rejections on the day of diagnosis	19	19	607	288

a) μmol neopterin/mol creatinine

b) see text for explanation

and the standard deviation, compared with the results of a previous study on 417 healthy individuals (11) and with the respective results of the other patient groups of the present study are reported in table 1. The distribution characteristics of the results, expressed in relative frequencies versus neopterin levels, are given in figure 1. Also indicated in the figure are the upper normal limit, observed in the group of healthy persons (11), and the range of neopterin levels containing about 95% of the determinations in group A patients. Obviously, group A patients exhibited higher neopterin levels in urine compared to the healthy controls, though they were lacking signs of immunological complications.

b) Patients with acute graft rejections, without viral disease

From the 38 post-transplant courses of the group B patients, 19 were not evaluable statistically due to delayed initiation of neopterin sampling: rejection episodes started either during the first 3 post-operative days or after discharge from the hospital. On the day of the clinical manifestation of acute rejection, the remaining patients exhibited on average higher neopterin levels than group A patients (table 1). However, when judging the results in detail, a problem appeared: though 19/19 (100%) of the urinary neopterin values on the day of rejection diagnosis (=d0) were above the upper limit of tolerance of the healthy control group (11), 8/19 (42%) were within the 95% range of group A patients (500 µmol neopterin/mol creatinine; figure 1):

n = 8 : range 257 to 450 µmol neopterin/mol creatinine
n = 11 : range 532 to 1360 µmol neopterin/mol creatinine

Obviously the discriminating power of one single urinary neopterin determination per day is too small to permit a clear

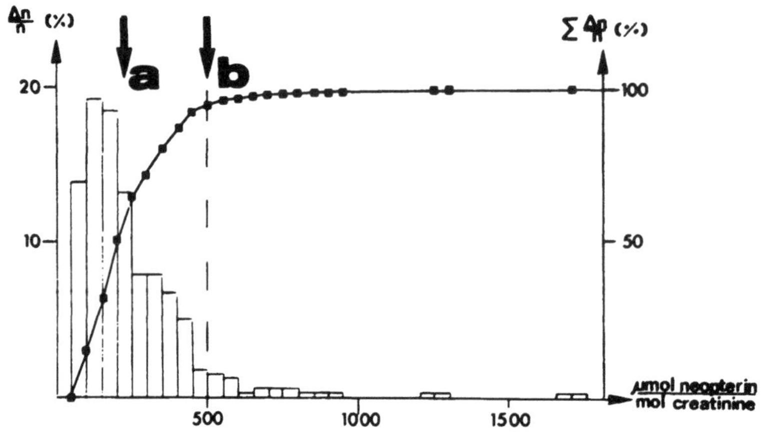

Figure 1 Distribution of relative frequencies of urinary neopterin levels (umol/mol creatinine) in 29 renal allograft recipients with an immunologically uncomplicated post-trans - plantation course. The figure refers to 474 single determinations.

$\frac{\Delta n}{n}$ relative frequency

$\sum \frac{\Delta n}{n}$ cumulative relative frequency

a) upper limit of tolerance of urinary neopterin levels observed in 417 healthy individuals (ref.11)

b) about 95% of the determinations are contained in the range below 500 μmol neopterin/mol creatinine cut between group A patients and cases with acute rejection episodes. An improvement of this unsatisfactory situation could be achieved by taking into account the dynamical properties of the neopterin excretion to some extent.

Figure 2 presents 4 typical temporal courses of the neopterin excretion observed in representative individuals. Normal controls revealed stable and low neopterin excretion patterns (fig. 2a). Group A cadaveric kidney recipients exhibited decreasing and finally normalizing values (fig. 2b). Contrarily, immunological complications were associated with rapidly increasing neopterin levels, usually preceding clinical manifestation of the underlying event (fig. 2c), and withdrawal of CyA caused a rapid and transitory increase in urinary neopterin excretion (fig. 2d).

We have developed a simple method of introducing information concerning dynamical properties: let N_i be the urinary neopterin level on day i. One can define moving averages M_i of relative change on N_i on day i, generally of order m:

$$M_i(m) = \left(\frac{N_i - N_{i-1}}{N_{i-1}} + \frac{N_{i-1} - N_{i-2}}{N_{i-2}} + \dots + \frac{N_{i-m+1} - N_{i-m}}{N_{i-m}}\right)\frac{1}{m}$$

This procedure provides a smoothing effect on the time course of the relative variations in neopterin levels from day to day. However, the larger the order m of averaging, the more detailed information is lost. Hence, we have decided to restrict the process of averaging to order m = 2:

$$M_i(2) \overset{!}{=} M_i = \left(\frac{N_i - N_{i-1}}{N_{i-1}} + \frac{N_{i-1} - N_{i-2}}{N_{i-2}}\right)\frac{100}{2} \quad (\%)$$

The moving averages of relative change in the neopterin excretion of our 29 group A patients, plotted against the respective neopterin levels, are shown in figure 3. (Due to the ave-

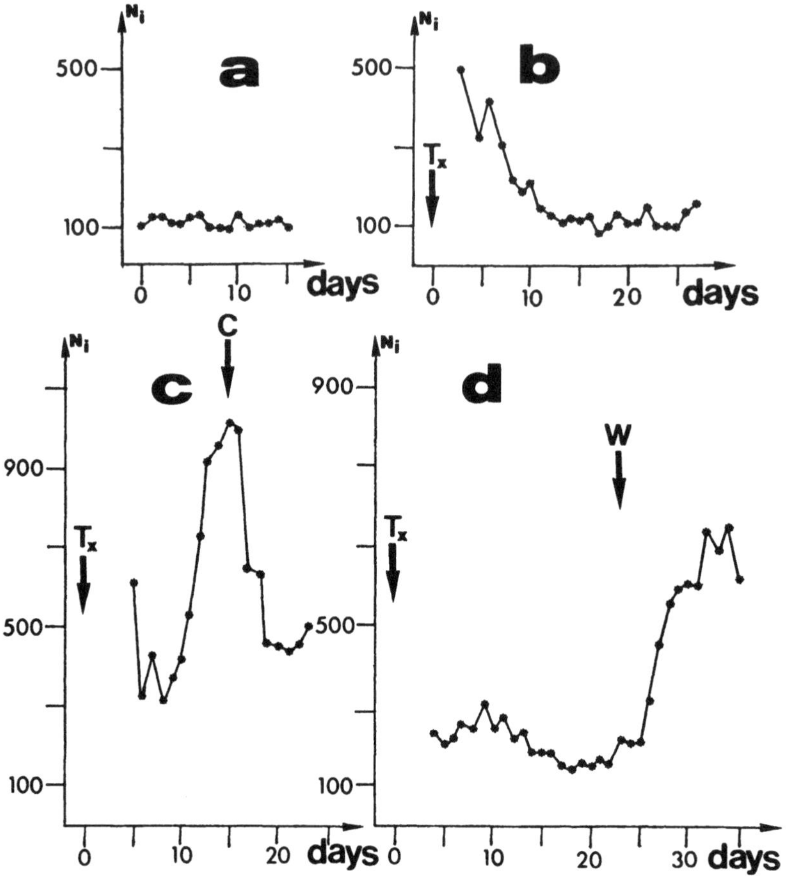

Figure 2 Urinary neopterin excretion of healthy subjects and of renal allograft recipients

a) healthy individual
b) immunologically uncomplicated post-transplantation course
c) post-transplantation course with acute rejection episode
d) post-transplantation course with withdrawal of CyA

N_i neopterin, µmol/mol creatinine
T_x day of transplantation
C clinical diagnosis of rejection
W withdrawal of CyA

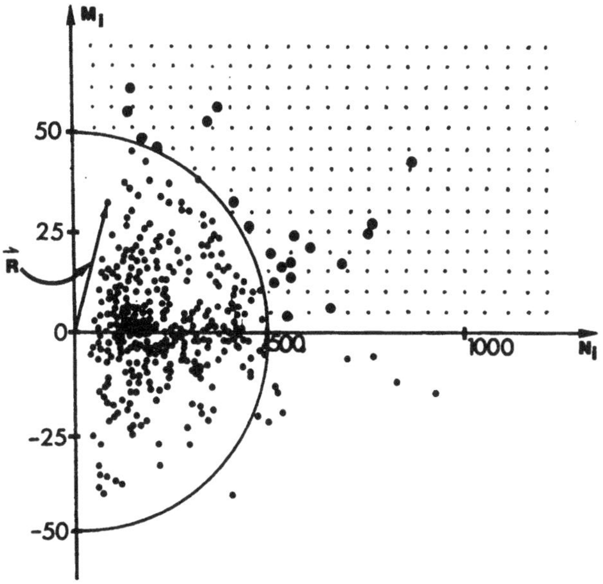

Figure 3 Moving averages M_i of the relative changes of neopterin excretion versus urinary neopterin N_i, observed in 29 renal allograft recipients with an immunologically un - complicated post-transplantation course. The figure refers to 416 single determinations.

M_i moving average on day i (%)

N_i neopterin, µmol/mol creatinine

$\vec{R}$ positional vector of a (N_i,M_i) determination

The dotted area contains about 5% (21/416) of the determinations with an elevated R-value (R +5o).

raging process, the number of determinations is reduced to 474 - mn = 474 - 2x29 = 416). The figure demonstrates that the 416 determinations (N_i, M_i) are forming a cluster of points, distributed approximately semi-circularly in the scale chosen. This observation suggests the definition of a positional vector $\vec{R}_i$ for each point (N_i, M_i), the length of which is given by

$$\left| \vec{R}_i \right| = R'_i = \sqrt{\left(\frac{N_i}{10} \right)^2 + M_i^2}$$

The division of N_i by 10 is made solely to take into account the different numerical scale factors of the two axes in figure 3.

Since the closer inspection of the time series of neopterin excretion in patients with acute rejection reveals increasing N_i values ($M_i \geq o$) before and during rejection and decreasing N_i levels ($M_i < o$) coupled with improvement of the immunological complications (fig. 2), it seemed reasonable to separate the increasing and decreasing branches of the time series. Mathematically this can be achieved most easily by multiplying R'_i by the sign of M_i:

$$R_i = R'_i \text{ sign } (M_i); \qquad \text{sign } (M_i) = \begin{cases} +1 \text{ for } M_i \geq o \\ -1 \text{ for } M_i < o \end{cases}$$

The distribution of the relative frequencies of R_i, plotted against R_i, for group A patients (416 determinations) is given in figure 4. The application of a 95% criterion yields the following result: about 95% of the 416 determinations on group A patients reveal $R_i < +50$; only 5% are "false positive" and are in the "critical region" in figure 3, which is indicated by dotting.

In the fashion of a "movie", figure 5 shows the situation in

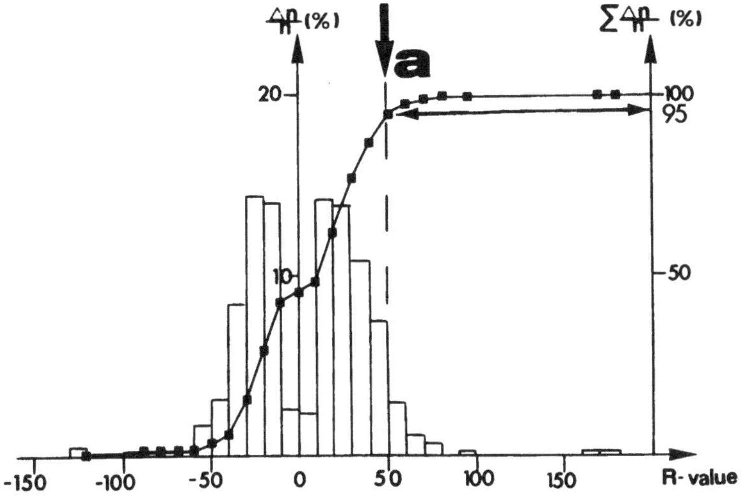

Figure 4 Distribution of the relative frequencies of the R-values observed in 29 renal allograft recipients with an immu - nologically uncomplicated post-transplantation course. The figure refers to 416 determinations.

$\frac{\Delta n}{n}$ relative frequency

$\sum \frac{\Delta n}{n}$ cumulative relative frequency

a) about 5% of the determinations are above R = +5o.

patients suffering from acute rejection 1 and 2 days before d0,on d0,1 and 2 days after d0 and 3 and 4 days after d0 applying the reported scheme of evaluation. (The apparent discrepancy in the numbers of determinations in figure 5a,b stems from the fact, that computation of M_i is possible only after determination of at least 3 N_i values).

The figure reveals the substantial gain in discriminating power of the evaluation scheme explained above: on d0 all R_i values are positive and only 2/17 (12%) are in the "normal range", represented by the semicircle. 1 of these 2 cases showed an elevated R-count ($R > 70$) one day after d0 - indicating that the clinical diagnosis had been made very early in this patient. The second of the 2 "false negative" cases remained within the "normal" semicircle throughout the following fortnight, finally the graft had to be ectomized due to a peracute rejection which had taken place during the very first days after transplantation. We have no explanation at hand for the normal neopterin levels in this patient, but we tentatively conclude, that the peracute rejection might have destroyed the organ to such an extent that even the immune response had ceased, at least that phase which appears to be associated with enhanced neopterin excretion.

The portion of the determinations on d0 with $N_i < 500$ µmol neopterin/mol creatinine, but $R_i > 50$, is particularly marked in figure 5b: on the basis of single determinations of N_i only, these specimens would be "false negative", according to the extended scheme of evaluation, however, they are classified correctly.

According to figure 5a, even 1 and 2 days before d0 an appreciable part of the determinations (14/24; 58%) is already in the "critical region".

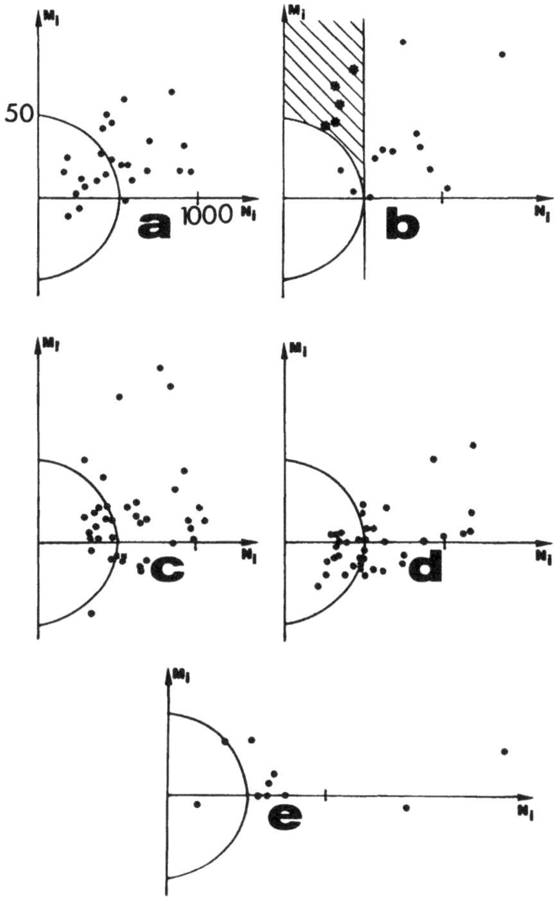

Figure 5 Temporal behaviour of the urinary neopterin excretion during rejection episodes and on the day of nephrectomy

a) 1 and 2 days before d0 (= day of clinical rejection diagnosis) b) on d0; c) 1 and 2 days after d0; d) 3 and 4 days after d0; e) on the day of nephrectomy (irreversible rejections)

The hatched area contains the values with N<5oo μmol neopterin/mol creatinine, but R> +5o: They are classified correctly only on the basis of both N_i and M_i.

N_i-neopterin, μmol/mol creatinine; M_i=moving averages (%)

After the day of clinical manifestation of the rejection, the time courses exhibited different features: partly neopterin excretion returned to the "normal range", partly it increased further and immunosuppression was achievable only by high doses of 6-methylprednisolone and in 10/38 cases the rejections proved to be irreversible, resulting in nephrectomy. The respective determinations on the day of nephrectomy in these patients is given in Fig. 5e.

c) Patients with viral disease with/without concomitant acute rejection

17 patients suffered from different viral diseases (cytomegalovirus, herpes simplex virus, Epstein-Barr virus, mycoplasma) without signs of graft rejection, whereas in 8 patients both immunological events occurred simultaneously.

Figure 6 shows, analogue to figure 5, the N_i,M_i values on the day of maximal neopterin levels in the urine of these 25 patients.

Obviously the values during viral infections are elevated, as is already known (12), and even higher than during rejections. However, similarly to reversible rejection episodes, also in cases of virus disease urinary neopterin levels normalize during clinical consolidation.

Summary

Analysis of the temporal course of urinary neopterin excretion subsequent to allograft transplantations enables a reliable and accurate prediction and/or verification of immunological complications such as rejection episodes or viral infections.

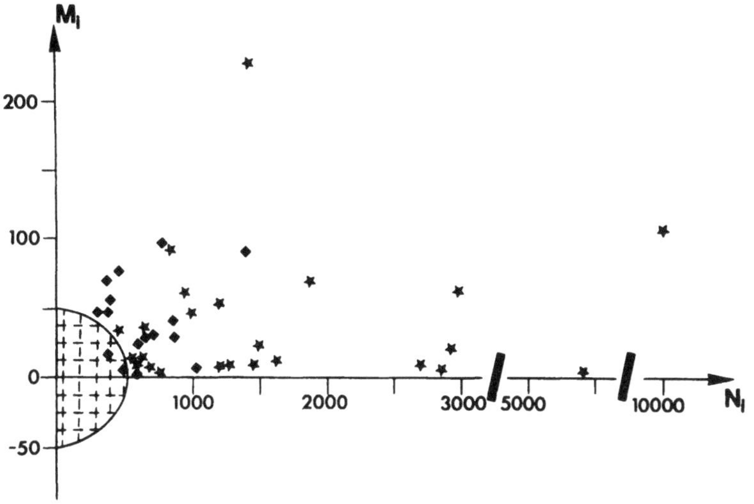

Figure 6 Moving averages M_i (%) of the relative changes in neopterin excretion versus urinary neopterin levels (μmol/mol creatinine) N_i in renal allograft recipients with immunologically complicated post-transplantation course

◆ acute graft rejection, determined on the day of clinical diagnosis of rejection

★ viral infection, determined on the day of maximal neopterin excretion

The semicircle is drawn accordingly to fig.3

Normal ranges of the neopterin levels established on healthy subjects appeared to be of little value in the assessment of allograft recipients; rather, comparison with the results observed in allograft recipients without immunological complications proved to be a far better suited and reliable procedure: introduction of moving averages of the relative change of day-to-day determinations substantially increased the discrimination potency of the method.

The method is by no means restricted to renal allotransplantation: liver and pancreas recipients exhibited quite similar neopterin excretion patterns. Due to the small number no statistical evaluation has been hitherto performed.

We believe the promising results with neopterin as a biochemical marker to be of considerable importance in monitoring allograft recipients. The applied HPLC-procedure allows for daily determination of neopterin levels in urine also for large scale clinical studies.

References

1. Bach, F.H., Bach, L.M., Sondel, P.M.: Nature 259, 273 (1977).
2. Mitchison, N.A.: Proc. Roy. Soc. Lond. (Biol.) 142, 72 (1954).
3. Brunner, K.T., Mauel, J., Cerottini, J.C., Chapius, B.: Immunol. 14, 181 (1968).
4. Häyry, P., Defendi, V.: Science 168, 133 (1970).
5. Fuchs, D., Hausen, A., Huber, C., Margreiter, R., Reibnegger, G., Spielberger, M., Wachter, H.: Hoppe-Seyler's Z. Physiol. Chem. 363, 661 (1982).
6. Huber, C., Margreiter, R., Spielberger, M., Fuchs, D., Hausen, A., Reibnegger, G., Wachter, H.: In: Biochemical and Clinical Aspects of Pteridines, Vol. 1, H. Wachter, H.Ch. Curtius, W. Pfleiderer (eds), Walter de Gruyter, Berlin-New York (1982).

7. Margreiter, R., Spielberger, M., Fuchs, D., Hausen, A., Reibnegger, G., Wachter, H., Huber, C.: In: Biochemical and Clinical Aspects of Pteridines, Vol. 1, H. Wachter, H.Ch. Curtius, W. Pfleiderer (eds), Walter de Gruyter, Berlin-New York (1982).

8. Huber, C., Fuchs, D., Hausen, A., Margreiter, R., Reibnegger, G., Spielberger, M., Wachter, H.: J. Immunol. 130, 1047 (1983).

9. Hausen, A., Fuchs, D., König, K., Wachter, J.: J. Chromatogr. 227, 61 (1982).

10. Hausen, A., Fuchs, D., Wachter, H.: J. Clin. Chem. Clin. Biochem. 16, 357 (1981).

11. Hausen, A., Fuchs, D., Grünewald, K., Huber, H., König, K., Wachter, H.: Clin. Biochem. 15, 34 (1982).

12. Wachter, H., Hausen, A., Grassmayr, K.: Hoppe-Seyler's Z. Physiol. Chem. 360, 1957 (1979).

Acknowledgement

This study was financially supported by the Austrian Research Funds "Zur Förderung der wissenschaftlichen Forschung", project p 4905

Appendix

The following two examples are aimed to numerically demonstrate the evaluation procedure described in the present paper.

CASE 1 : Suppose the initial 3 determinations of urinary neopterin (µmol/mol creatinine) to yield:

$N_1 = 178$

$N_2 = 25o$ (The index 1,2,3,... denotes the

$N_3 = 463$. day.)

M_i is, per definition, evaluable only after at least 3 determinations. Hence,

$$M_3 = \left(\frac{25o - 178}{178} + \frac{463 - 25o}{25o} \right) \cdot \frac{1oo}{2} = 62.8 \ (\%)$$

$$R_3 = \sqrt{62.8^2 + (463/1o)^2} = 78.o$$

Although the data on day 1 and 2 are only moderately high, the tendency is strongly increasing, on day 3 we would conclude from the elevated R-value that an immunological complication is highly probable.

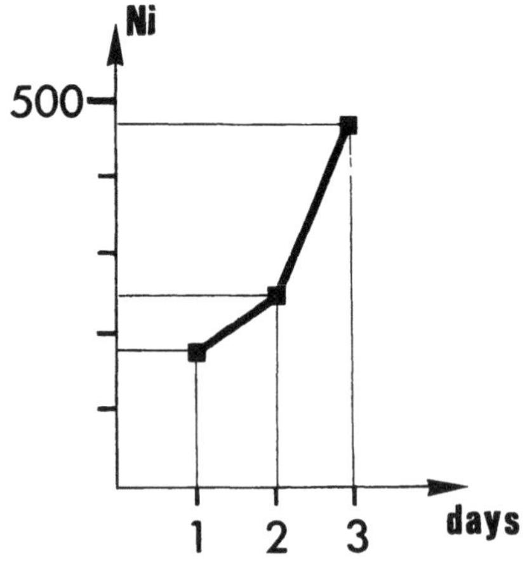

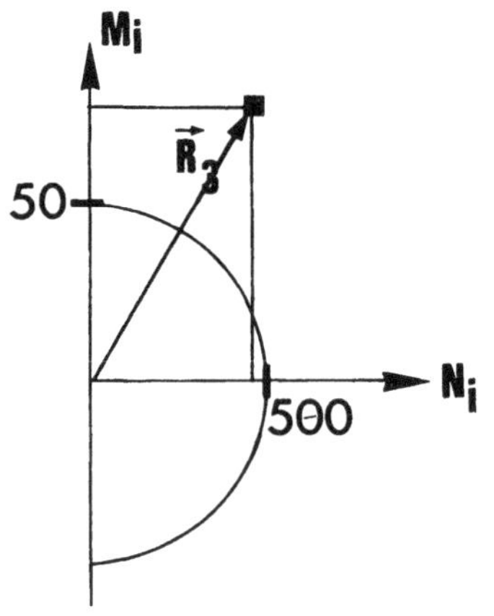

N_i = neopterin, µmol/mol creatinine, M_i = %

CASE 2 : Let the initial 3 values be: (neopterin, μmol/mol creatinine)

N_1 = 852

N_2 = 6o2

N_3 = 478 . (Again, the index 1,2,3 denotes the day)

$$M_3 = \left(\frac{6o2 - 852}{852} + \frac{478 - 6o2}{6o2}\right) \cdot \frac{1oo}{2} = -25.o\ \%$$

$$R_3 = -\sqrt{(-25.o)^2 + (478/1o)^2} = -53.9$$

The neopterin levels are appreciable higher than in the previous case. However, the tendency is decreasing, resulting in a negative R-value.
We would conclude this course to be immunologically uncomplicated.

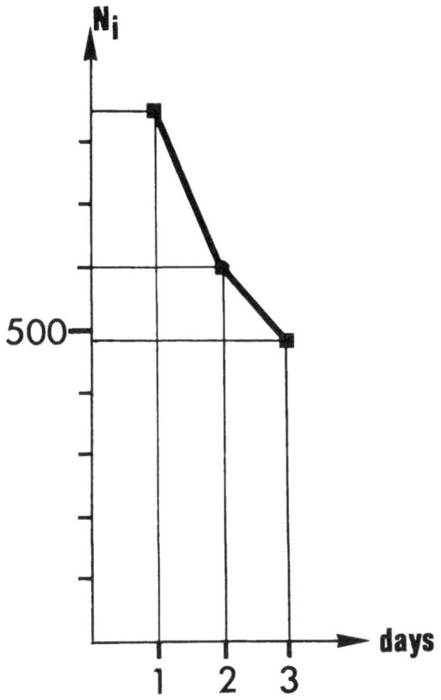

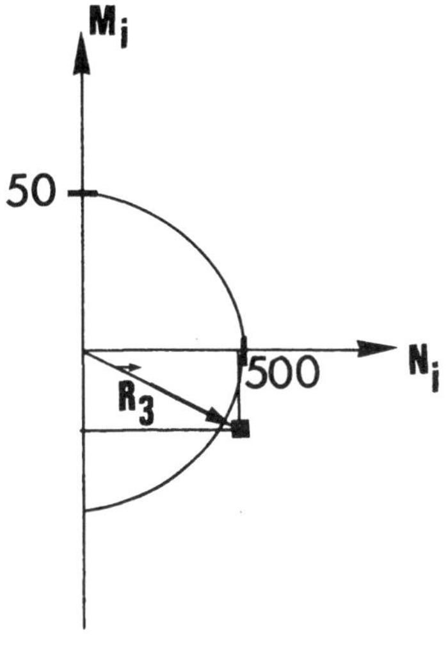

N_i = neopterin, μmol/mol creatinine

M_i = %

NEOPTERIN IN PLASMA AND URINE AND BIOPTERIN IN BLOOD FRACTIONS AFTER BONE MARROW TRANSPLANTATION

Michael Fink, Ulrich Jehn, Wolfgang Wilmanns,
III. Medizinische Klinik der Ludwig Maximilians Universität,
Klinikum Großhadern,
D-8000 München 7o, Germany

Hartmut Rokos,
Forschungslabor Henning Berlin GmbH,
D-1000 Berlin 42, Germany

Irmgard Ziegler,
Institut für Toxikologie und Biochemie der Gesellschaft für
Strahlen- und Umweltforschung, Abteilung Zellchemie;
D-8000 München 2, Germany

Introduction

Allogeneic bone marrow transplantation (BMT) is an established procedure in the treatment of acute leukemia and aplastic anemia. Before transplantation, the malignant cells of the patient must be eradicated, and the bone marrow conditioned for acceptance of the graft, using cyclophosphamide and total body irradiation. The donor's hemopoietic bone marrow cells are aspirated from the anterior and posterior iliac crests and then transfused intravenously into the recipient. A take can be diagnosed by a rise of reticulocytes, leucocytes, and thrombocytes in the peripheral blood, usually within 2-4 weeks. Failure of engraftment has been observed in about 1% of leukemic

Abbreviations:

BMT	= Bone Marrow Transplantation
GVHD	= Graft-Versus-Host-Disease
HLA	= Human Leucocyte Antigen
MLC	= Mixed Lymphocyte Culture

and 5% of aplastic recipients (review see 1). Since also immunocompetent cells are transplanted into an immunosuppressed recipient, not only the graft, but mainly the patient can be attacked by immunological interactions. This is called graft-versus-host-disease (GVHD) and its target organs are skin, liver and intestine. Considerable GVHD occurs in about 2o% of HLA-identical recipients, with a fatal outcome in approximately 8% (1). Patients older than 3o years have a higher incidence of GVHD.
In earlier studies, increased stem- cell proliferation occuring during leukemias (2) and experimental bone marrow transplantation (3) was found to result in increased biopterin levels of the respective peripheral blood cells. Furthermore, highly elevated biopterin levels were demonstrated in the whole blood during the early phase after BMT in man (4). In renal transplantations the urinary neopterin levels have shown to rise 1-2 days before the clinical manifestation of graft rejection (5,6). These data prompted an investigation of biopterin in fractionated blood as well as neopterin in plasma and urine and their correlation with engraftment and GVHD after BMT in man.

Material and Methods

Venous blood was drawn into sodium citrate (1:10). The separation of the blood samples into plasma and cellular fractions has been described elsewhere (3). Lymphocytes were isolated by Ficoll-Hypaque density gradient centrifugation. Biopterin was determined using the Crithidia assay (7). The procedure was adapted to a micromethod and performed in the usual Microtiter assay system. Growth was measured by incorporation of ^{3}[H]-thymidine into the DNA of Crithidia fasciculata. This assay allows the biopterin determination in the fmol range. Urinary neopterin was measured in the morning urine. The radioimmunassay for the determination of neopterin in plasma and oxidated neopterin in urine is described elsewhere (8).

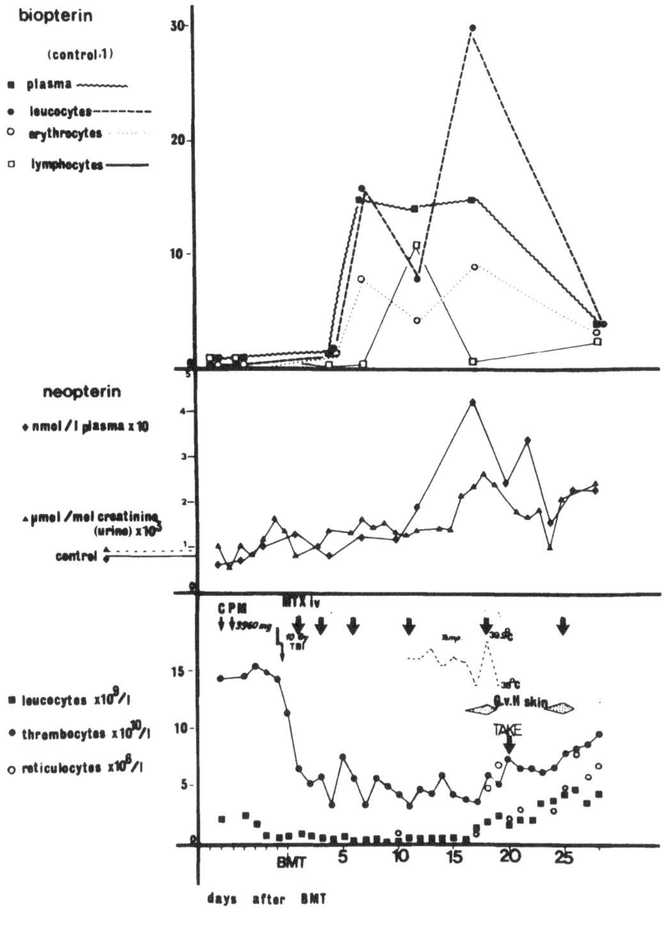

Fig.1 Biopterin and neopterin levels at the time of take and mild GVHD in Pat. H.H. The rise of neopterin is preceding GVHD.
Take = engraftment proven by aspiration of bone marrow
Biopterin values are related to normal control [control =1].

Results

BMT of patient H.H., 32 yrs, had been performed in a third complete remission of a T-ALL from his HLA- and MLC-identical brother. For prophylaxis against interstitial pneumonia,trimethoprim and sulfamethoxazol were given; for prophylaxis of GVHD the patient received methotrexate in weekly intervals. The patient developed a temperature of 39,9^{o} C 11 days after BMT, which was treated with different combinations of antibiotics including an aminoglycoside (Tobramycin) between day 11 and 25. Renal function was normal during this episode. The temperature returned to normal on day 2o after BMT, when a complete take was diagnosed. Mild GVHD of the skin, diagnosed by histology, was evident day 17 and 24, which disappeared under the weekly dose of methotrexate. Biopterin values were low in the first days and displayed a sharp rise about day 5 after BMT in all blood cell fractions; a value as high as 3o-fold of normal has been measured on day 17 in the leucocytes. A change in the distribution of biopterin in blood cells was observed at day 12, but a clear-cut interpretation would be premature; technical errors in dividing blood fractions cannot be excluded. When a complete take was indicated by the clinical data, a sharp decline of biopterin in each cell fraction had been observed. Neopterin-concentrations in plasma and urine rose during the take and GVHD.
Beginning at day 42, the patient developed a second GVHD of the skin which began as a mild facial edema and rush. It disappeared under corticosteroids around day 6o. In spite of this treatment, the patient suffered from a severe, localised GVHD of the skrotal skin, which started at day 5o, reaching a peak between day 6o-7o, and then healed. The neopterin levels in plasma and urine preceded this GVHD with maximum values between day 35 and 55. Even more impressive was the rise and fall of the lymphocytic biopterin, which was also preceding the clinical course.(Fig.2) After the skrotal GVHD had disappeared, the patient developed another episode of elevated temperature of unknown origin,which

was treated with Tobramycin (Fig.3). This time the renal function was moderately impaired with a creatinine of 2 times the normal value. Day 99 and 100, corticosteroids were increased because of a pruritus, which can also be a symptom of dermal GVHD. The decrease of the cell counts and the histologically proven hypoplasia of the bone marrow during this episode remains unknown. Fig.3 demonstrates the correlation between the elevated neopterin in the plasma and the treatment with Tobramycin, which seems to be closer correlated than with the temperature. Day 138, the patient had a septicemia and again was treated with Tobramycin. The temperature normalized promptly, but creatinine rose from 115 to 248 µmol/l (1,3-2,8 mg%) in the serum within one day; the urinary neopterin levels between day 137 and 139 were 822, 1261 and 1889 µmol/mol creatinine respectively. This observation is in good correlation with a report on elevated neopterin-levels in plasma and urine of patients with impaired renal function (17,18), but additionally an influence of neopterin levels by pharmacological interaction seems to be possible in this case.

Pat. V.E., 23 yrs., suffered from acute myeloblastic leukemia, which was not sufficiently controlled by conventional chemo - therapy. Therefore bone marrow was transplanted from his HLA- and MLC- identical father. Trimethoprim-sulfamethoxazol and methotrexat were given as in the other patient. After day 23, the patient had a mild diarrhoea with 1-2 stools per day around day 35. Day 38, GVHD of the small intestine developed, which was diagnosed microscopally and was accompanied by a stool frequency of 8 and more per day, partially with a considerable amount of blood admixed. The neopterin values in plasma and urine rose from day 23 to values as high as five-fold of normal day 4o and exhibited a rapid fall after the beginning of a high-dose corticosteroid-treatment. The withdrawal of steroids was followed by another sharp rise of neopterin.

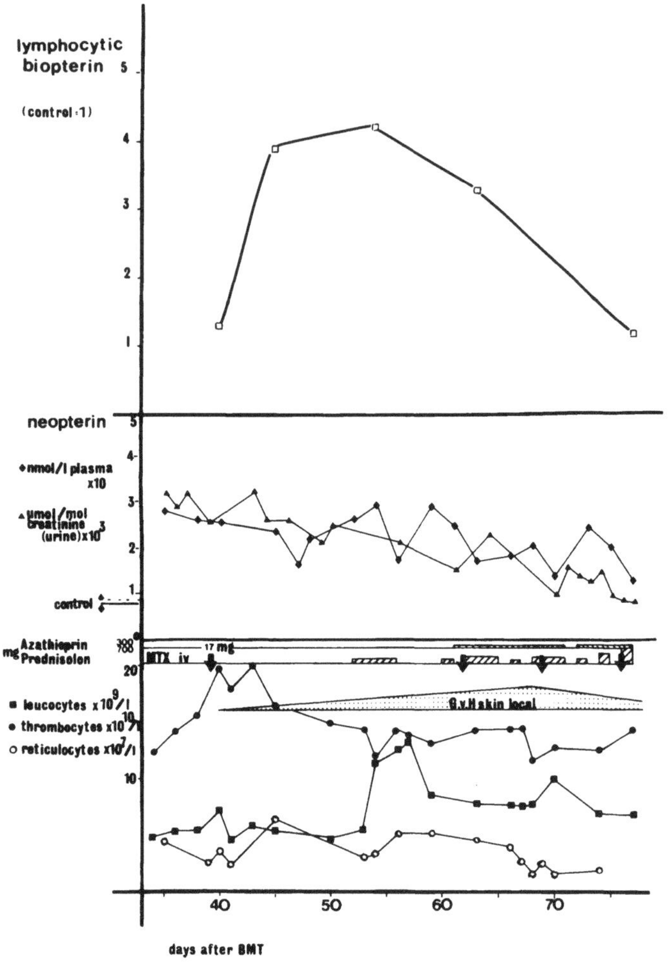

Fig.2 Biopterin in blood lymphocytes and neopterin in plasma and urine in a localized dermal GVHD.
Details see text.
Biopterin values are related to normal control [control =1]

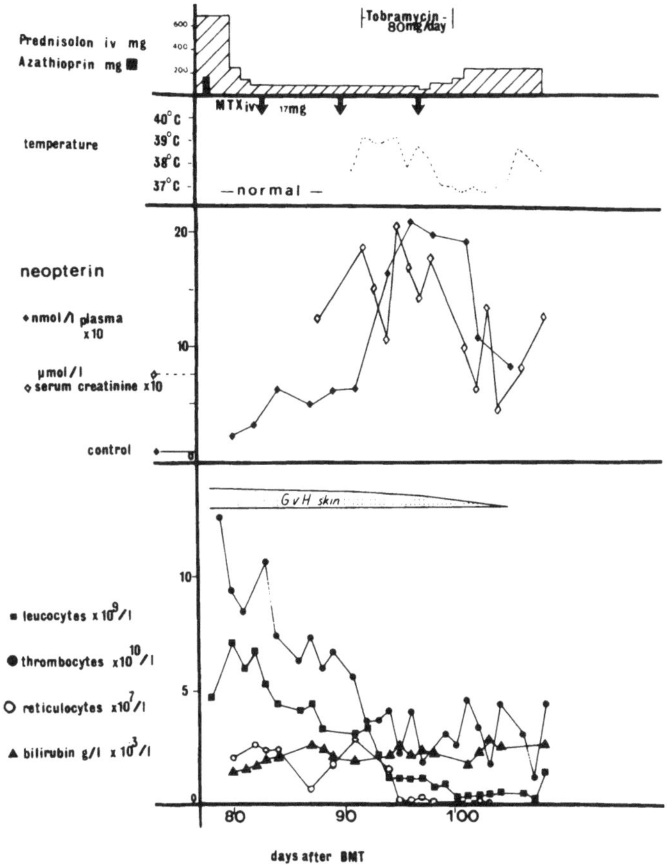

Fig.3 Elevation of plasma-neopterin during the time of f
impaired renal function and treatment with Tobramy
Details see text

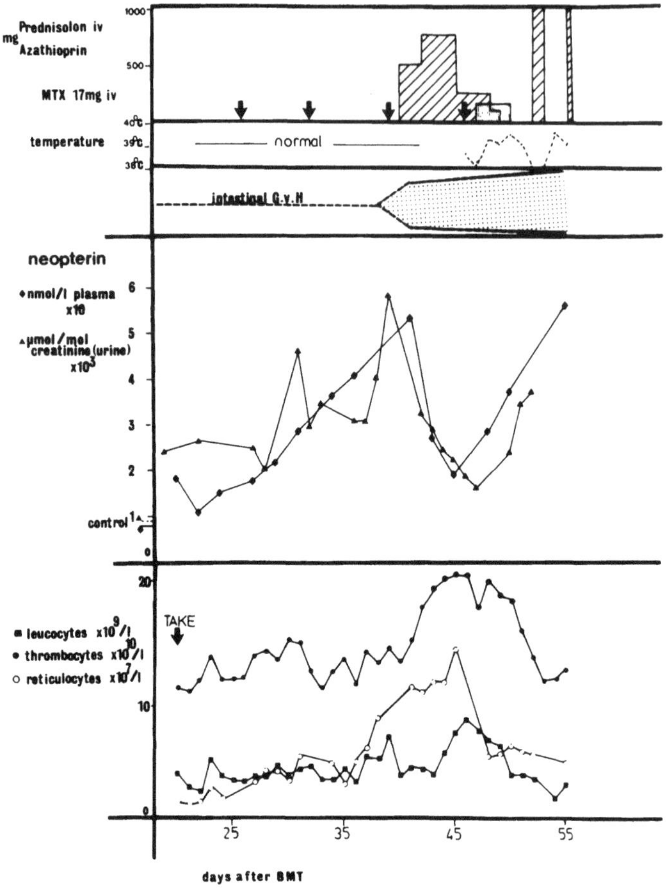

Fig.4 Rise of neopterin-levels in plasma and urine before clinical manifestation of severe intestinal GVHD. Details see text.

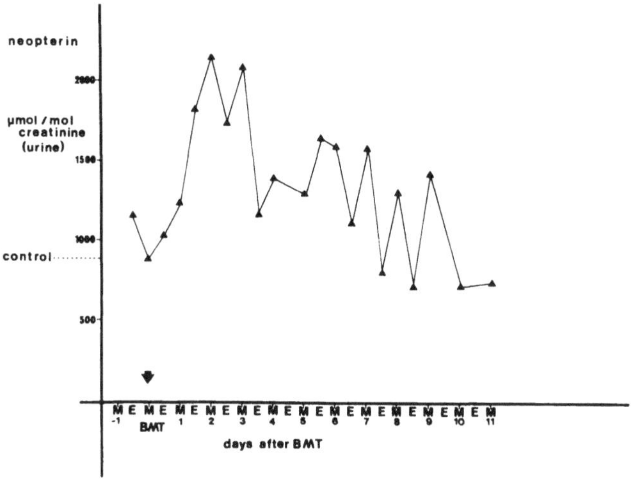

Fig. 5 Neopterin levels in the urine after BMT in a patient with paroxysmal noctural hemoglobinuria.
M = morning urine; E = evening urine
Details see text.

In paroxysmal noctural hemoglobinuria, bone marrow can be transplanted without prior eradication of the patient's marrow (10). In the third patient, a BMT was performed in this way, however, without success. The neopterin values in the urine exhibited a short rise during the first days after transplantation (Fig.5). No complication or improvement of the patient's clinical status was observed.

Discussion

Failure of engraftment is a typical problem of BMT, especially in patients with aplastic anemia. In Beagle dogs, the biopterin concentration in peripheral blood cells after autologus BMT is in good correlation with the engraftment but does not precede the number of the respective cell population (3). During allogeneic BMT in man a considerable rise of biopterin blood levels has been found during the early phase, while other blood parameters showed no sign of proliferation (4). The circulation of immature transfused bone marrow cells with a high biopterin content may be considered as a cause of the observed elevated biopterin levels, although in our patients, only few blood cells were immature.

In contrast, highly elevated biopterin levels during the first days after BMT have been found in whole blood of a patient,who died 2o days after BMT without sufficient signs of a take (4). In this case, the assumption of engraftment was based on a maximum of $5^{0}/00$ reticulocytes 2 weeks after BMT. Altogether, a full explanation for the early and marked rise of biopterin level after BMT cannot yet be given. Experimental data indicate that biopterin may participate in the control and stimulation of the engraftment (13).

Another problem of BMT is GVHD, since it can be a life-threatening complication. However, this complication also displays an antileukemic effect and thus, in its milder variant, may prolong the life of the patient (11). A predictive test for the severity of GVHD would improve timing and intensity of the immunosuppressive therapy. Pteridines could be useful in this regard, since they seem to be indicators (6,12) and active components (13) of the immune response.

Neopterin has been shown to be of prognostic value in renal graft rejections (5,14). In our patient V.E., it exhibited highly elevated plasma- and urine levels more than one week before clinical manifestations of GVHD occured (Fig.4). However, elevated neopterin levels have not only been described in malignant (15), rheumatic (16) and renal (17) diseases, but also in viral infections (9). An influence of pharmacological factors on neopterin levels cannot be excluded in our patient.

Graft rejection could also cause an elevation in neopterin,even though the involvement of immunological factors is unlikely in Patient K.K. (Fig.5), who received bone marrow from his identical twin. If those various factors which additionally influence neopterin levels are taken into account, its determination might be helpful in early diagnosis of GVHD. The lymphocytic biopterin levels also exhibited highly elevated levels in a patient with a localized dermal GVHD and preceded the clinical course (Fig.2). Therefore, the lymphocytic biopterin could also be of clinical value for the prediction of GVHD, but since a highly elevated blood biopterin in an early phase of a hepatitis B has been reported (Fig.6 in [4]), an influence of viral infections has to be considered, too.

Acknowledgement

We thank Mrs. Demtröder, Mrs. Schuri, Mrs. Wachholz, Mrs. Doischer and Mrs. Högner for their skillful technical assistance.
We also thank the nursing staff of our transplant unit F12 for cooperation.
This study was supported by grants of the Deutsche Forschungsgemeinschaft.

References

1. Storb, R.: Transplant. Proc. 13, 221-225 (1981).
2. Ziegler, I., Fink, M., Wilmanns, W. : Blut 44, 231-24o (1982).
3. Ziegler, I., Kolb, H.J., Bodenberger, U., Wilmanns, W.: Blut 44, 261-27o (1982).
4. Fink, M., Jehn, U., Wilmanns, W., Ziegler, I..In: Biochemical and Clinical Aspects of Pteridines (Eds. H. Wachter, H.Ch. Curtius and W. Pfleiderer), Vol.1. W. de Gruyter, Berlin, 241-25o (1982).
5. Margreiter, R., Spielberger, M., Fuchs, D., Hausen, A., Reibnegger, G., Wachter, H., Huber, Ch.. In: Biochemical and Clinical Aspects of Pteridines (Eds. H. Wachter, H.Ch, Curtius and W. Pfleiderer), Vol.1. W. de Gruyter, Berlin, 233-24o (1982),
6. Fuchs, D., Hausen, A., Huber, Ch., Margreiter, R., Reibnegger, G., Spielberger, M. Wachter, H. : Hoppe Seyler's Z. Physiol. Chem. 363, 661-664 (1982).

7. Baker, H., Frank, O., Shapiro, A., Hutner, S.H.. In : Methods in Enzymology (Eds. Mc. Cormick, D.B., Wright, L.D.), Vol.66. Academic Press, New. York, 49o-5oo (198o).

8. Rokos, H., Rokos, K.. In : Chemistry and Biology of Pteridines (Ed. J. Blair), W. de Gruyter, Berlin (in press)

9. Wachter, H. Hausen, A., Grassmayr, K. : Hoppe Seyler's Z. Physiol. Chem. 36o, 1957-196o (1979).

1o. Fefer, A., Freemann, H., Storb, R., Hill, J., Singer, J., Edwards, A., Thomas, D. : Ann. Int. Med. 84, 692-695 (1976).

11. Weiden, P., Sullivan, K., Flournoy, N., Storb, R., Thomas, D. : New Engl. J. Med. 3o4, 1529-1533 (1981).

12. Huber, Ch., Schönitzer, D., Fuchs, D., Hausen, A., Reibnegger, G., Wachter, H. : This Symposium.

13. Ziegler, I., Hamm, U., Berndt, J. : This Symposium.

14. Margreiter, R., Spielberger, M., Fuchs, D., Hausen, A., Reibnegger, G., Wachter, H., Huber, Ch. : This Symposium.

15. Rokos, K., Rokos, H., Frisius, H.. In : Biochemical and Clinical Aspects of Pteridines (Eds. H. Wachter, H.Ch. Curtius and W. Pfleiderer), Vol.1. W. de Gruyter, Berlin, 117-13o (1982).

16. Hausen, A., Fuchs, D., Reibnegger, G., Wachter, H., Egg, D. : This Symposium.

17. Dhont, J., Vanhille, Ph.. In : Biochemical and Clinical Aspects of Pteridines (Eds. H. Wachter, H.CH. Curtius and W. Pfleiderer), Vol.1. W. de Gruyter, Berlin, 328-336 (1982).

18. Rokos, K., Rokos, H., Frisius, H., Kirstaedter, H.J. : Clin. Chim. Acta 1o5, 275-286 (198o).

CHEMICAL DIVERSITY AND LYMPHOCYTE STIMULATING PROPERTIES OF A PTERIDINE-BINDING VARIANT OF ALPHA$_1$-ACID GLYCOPROTEIN AND ITS DIAGNOSTIC APPLICATION

Irmgard Ziegler

Institut für Toxikologie und Biochemie der Gesellschaft für Strahlen-und Umweltforschung, Abteilung Zellchemie; D-8000 München 2, Germany

Michael Fink

III.Medizinische Klinik der Universität, Klinikum Großhadern; D-8000 München 70, Germany

Introduction

In previous investigations a pteridine-binding variant of alpha$_1$-acid glycoprotein (AGP$_M$) had been isolated from pooled blood samples of cancer patients (1,2). Chemical and clinical aspects had indicated common features with the abnormal orosomucoid described by Rudman (1,3,4). In control blood it makes up less than 1% of the total AGP. The main differences as compared to normal AGP are:(i) it possesses a chromophoric group whose characteristics indicated a pteridine; (ii) some of its microheterogenic bands show less negative charge even though its sialic acid content is not reduced; (iii) it exists partially in a polymeric form, possibly brought about by the cross-linking effect of the chromophoric group; (iv) its extinction values and ORD data indicate differences in the secondary structure.

An indirect method of AGP$_M$ quantitation had determined its glycoprotein moiety; in addition to malignancies, increased amounts also had been found during inflammatory diseases (3). In the following experiments the chemical characteristics of AGP$_M$ preparations from individual clinical cases such as chro-

Biochemical and Clinical Aspects of Pteridines, Vol. 2

mophore content, microheterogeneity and polymeric forms are compared.
Some pteridines recently have been described as functioning constituents of the lymphokine system (5). These findings encouraged an examination of AGP_M for a possible lymphocyte stimulating property; especially since reports on immunosuppressive (6,7) and growth stimulating activities (8) of $alpha_1$-acid glycoprotein isolations are most contradictory at present.

Materials and methods

The methods of isolation and chemical characterization of AGP_M were the same as described (1).
Mouse spleen lymphocytes were obtained as described (5). Human peripheral blood lymphocytes were obtained by the Ficoll-Hypaque density gradient centrifugation. The Microtiter assay system and the growth determination by incorporation of ^{3}H-thymidine is described in (5). Aliquots of the AGP_M preparations were lyophilized and redissolved in the culture medium at the concentrations indicated.

Results

1) The chromophore content of AGP_M. If AGP_M is isolated from individual clinical cases the chromophore contents of the preparations can vary greatly. The range of variation is illustrated in fig.1. The Con A reactive and the Con A-non reactive portion respectively (1) had about the same chromophore load. Fig.2 summarizes the chromophore content of all isolations by giving the ratio of optical density at 400 nm/280 nm. The clinical status of the patients is described in table 1.Even though the chromophore content may vary among the samples obtained from blood of malignant

cases, it clearly differentiates the AGP_M samples from non-malignant diseases.

Table 1 Sources of AGP_M and clinical status of the patients

Sample number	Diagnosis	Therapy
5	myelomonocytic leukemia	cytostatic since 4 days
6	malignant lymphoma (centrocytic-centroblastic), Stad.IVb; Pseudomonas sepsis	resistant to therapy
7	M.Hodgkin, lymphocyte depleted	before therapy
9	CGL, Phil.Chrom.negativ; WBC 67 000	cytostatic
10	diabetes mellitus with vascular and infectious complications	antibiotic
12	pleural effusion of unknown etiology	antiphlogistic
13	testicular teratoma, disseminated; in short partial remission	cytostatic
14	AML; WBC 80 000	before therapy
15	M.Hodgkin, Stad.IIb, lymphocyte depleted	cytostatic 23-9 days before
16	T-ALL; WBC 2 000, 71% blasts	before therapy
17	testicular teratoma, undifferen - tiated, disseminated; terminal	cytostatic 4 months before
20	c/T-ALL, relapse; leukopenia, sepsis	cytostatic
22	M.Boeck, erythema nodosum	before therapy
23	large cell bronchial carcinoma; disseminated	dexamethasone, CNS rad.
24	cancer of the uterine cervix, Stad.IV	after radiation
25	signet-ring cell carcinoma of the stomach, disseminated	before therapy
26	CGL; blastic crisis; WBC 120 000	before therapy
27	as 26, 11 days later; WBC 20 800; (patient died 9 days later)	cytostatic since 11 days
28	pancytopenia with hypercellular marrow; preleukemia? (status 1 year later: leukemia not yet apparent)	transfusions
29	diabetes mellitus with vascular and infectious complications	antibiotics
30	sepsis	antibiotics
32	probably autoimmune disease following influenza vaccination	none

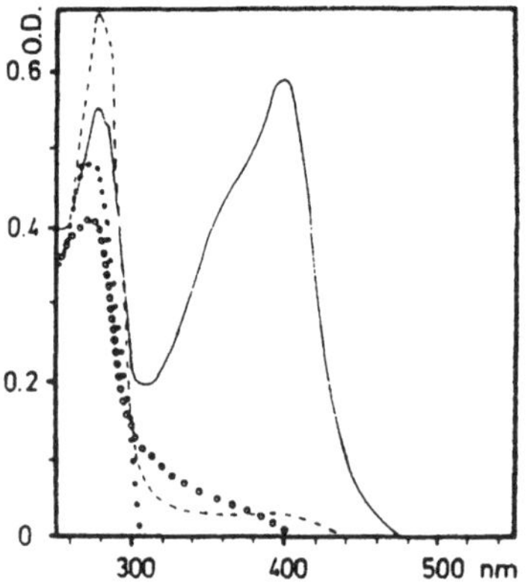

Fig.1 Absorption spectrum of Con A - reactive AGP_M in H_2O (qualitatively)

——— AGP_M from case 23 (bronchial carcinoma)

- - - AGP_M from case 30 (sepsis)

o o o AGP_C (AGP variant from controls; see ref.1,3)

.... total AGP

This differentiation had not been possible by quantitation of the glycoprotein moiety which had been found in increased amounts in either group (3).

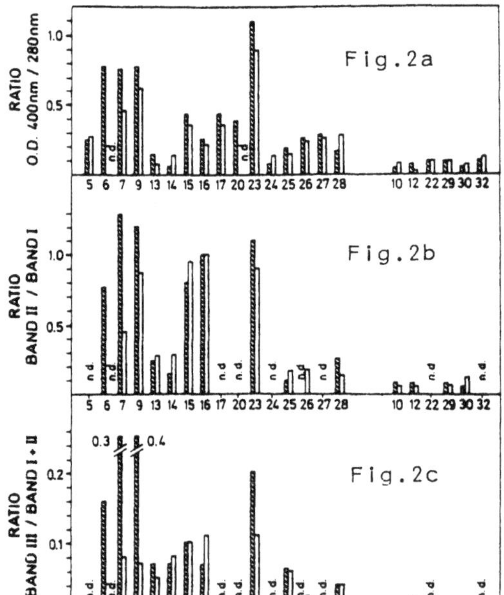

Fig.2a Chromophore content of AGP_M from individual clinical cases, indicated by the ratio in O.D. at 400 nm/280 nm. For the description of cases see table 1.

▧ Con A-reactive fraction

□ Con A-non reactive fraction

Left group: AGP_M from malignant cases

Right group: AGP_M from non-malignant cases

Fig.2b Portions of decreased negative charge (Band II) in AGP_M from individual cases. n.d.=not determined. For further legend see fig.2a

Fig.2c Portions of polymeric form (Band III) in AGP_M from individual cases. For further legend see fig.2a

It is noteworthy that case 28 had been classified by conventional clinical criteria as a presumably preleukemic stage. The quantitation of the glycoprotein moiety (28 µg/ml blood; normal values: 75 $\pm$ 20 µg/ml) had given extremely low amounts but the chromophore content grouped it with the malignancies. Final cases (e.g. samples 6,17 and 20) were characterized not only by high amounts of the glycoprotein moiety in the blood (255, 178 and 179 µg AGP_M/ml blood, respectively) but also by a high chromophore content.

2) Microheterogeneity. Previous experiments had indicated that the pteridine chromophore is protonated by the sialic acid residues of the glycoprotein (2). This anticipates higher p_I values of the molecule with increasing chromophore content. The portion of decreased negative charge can be either determined by polyacrylamide gel electrophoresis in the absence of denaturating conditions or by isoelectric focusing (1). The range of variability which may occur in charge distribution of AGP_M preparations is illustrated in fig.3. As seen in fig.2b indeed, the chromophore content of the various AGP_M preparations is closely correlated with the portions of higher p_I. Thus the charge distribution

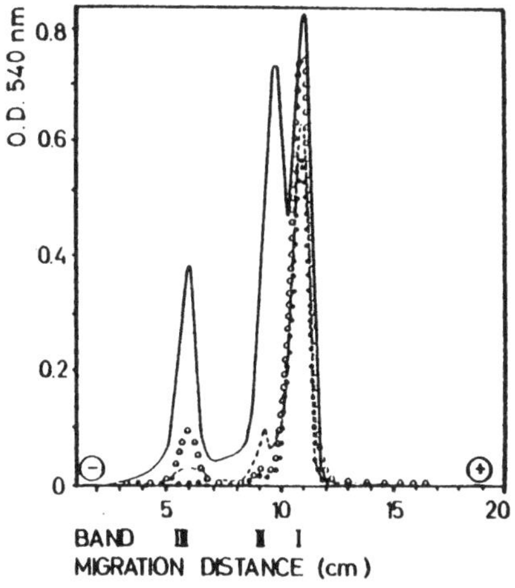

Fig.3 Densitometric scan after polyacrylamide gel electrophoresis under non-denaturating conditions of Con A-reactive AGP_M. Band II represents the portion of reduced negative charge (1)

—— AGP_M from case 23 (bronchial carcinoma)

- - -AGP_M from case 30 (sepsis)

o o o AGP_c (AGP variant from controls; see ref.1,3)

.... total AGP

For chromophore content see fig.1. For description of cases see table 1 clearly differentiates AGP_M from malignant cases from that obtained during inflammatory diseases.

3) Polymeric form. In the absence of denaturating conditions, a band with slow mobility is found after gel electrophoresis. This Band III was attributed to a polymeric form (1). Fig.3 illustrates the range of variation which may occur in the portion of the polymeric form within the AGP_M

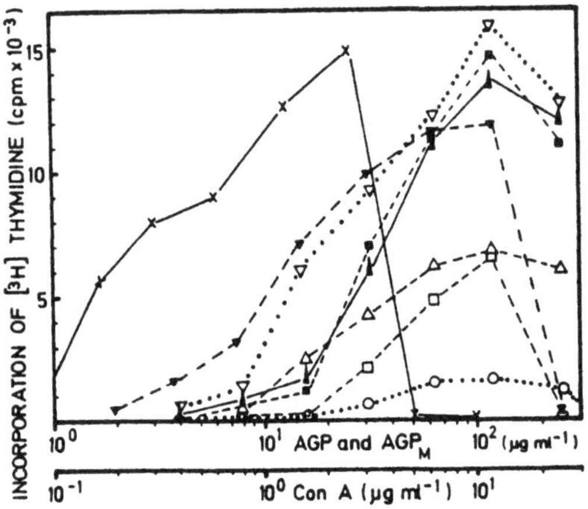

Fig.4a Dose response curve for the stimulation of ^{3}H-thymidine incorporation into the DNA of mouse spleen lympho - cytes by AGP_M ; assay in RPMI 1640 medium + 5% fetal calf serum.

■- -■ AGP_M from case 20
▼- -▼ AGP_M from case 25
▲——▲ AGP_M from case 26
▽···▽ AGP_M from case 22
△- -△ AGP_M from case 29
□···□ AGP_M from case 30
o····o AGP
X——X concanavalin A

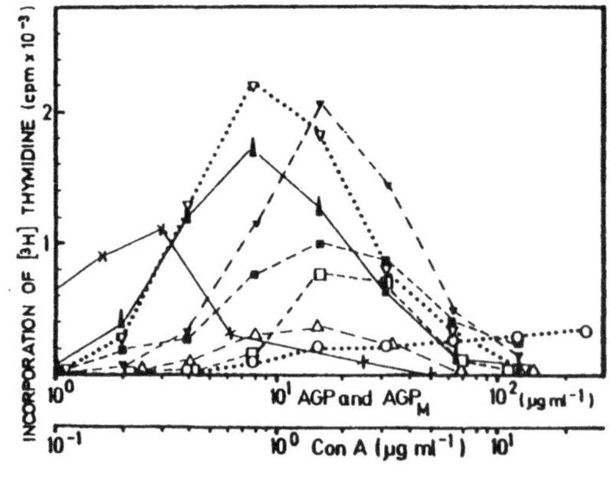

Fig.4b For legends see Fig.4a. Assay in serum free medium

preparations investigated. Fig. 2c shows that this portion in turn, is correlated with the chromophore content of the sample. Thus the molecular weight distribution also differentiates AGP_M from malignant cases from that obtained during inflammatory diseases. Moreover, the analysis corroborates the view that the pteridine chromophore exerts a cross-linking effect between the glycoprotein moieties of AGP_M.

4) The stimulation of lymphocytes by AGP_M. In contrast to AGP, all AGP_M preparations activated lymphocytes from mouse spleen and from peripheral human blood as well. The activation followed an optimum curve with respect to concentration. It is shown for 6 AGP_M samples in figs.4a and 4b. In serum free medium, the concentration needed for optimum activation is shifted to much lower values as compared to 5% FCS medium. Moreover, the stimulatory capacity of AGP_M, which may attain levels as high as concanavalin A in 5% FCS medium, is about double that of the lectin in serum free medium.

Discussion

The analysis of AGP_M preparations from individual clinical cases with respect to chromophore content, charge distribution and molecular weight corroborates the view that both the higher p_I values of AGP_M and the occurrence of higher molecular weight forms are secondary effects and are caused by its pteridine chromophore. All 3 characteristics thus provide a tool for the better discrimination between malignant cases on the one hand and inflammatory diseases on the other. The quantitative determination of the glycoprotein moiety did not enable us to distinguish this, but rather indicated increased amounts of AGP_M also in the second group.

The most prominent character of AGP_M is its ability to activate resting lymphocytes. It may be safely assumed that the two characteristics both participate in the elaboration of this property. As compared to AGP these are the difference in secondary structure (2) which seems to be an essential prerequisite for its signal character and the presence of the pteridine chromophore. The co-stimulating effect of some intermediates of pterin metabolism (5) may indicate that the pteridine chromophore additionally contributes to the cascade of amplifying events, known to occur during lymphocyte activation (cf. 9). This results in a marked superiority of AGP_M in stimulation capacity as compared to the lectin, if the fetal calf serum is withdrawn, which usually co-supplies such amplification factors. The AGP_M concentration needed for optimal stimulation of lymphocytes in serum containing medium are usually found at initial stages of malignant diseases but largely increasedat final stages (3). Further investigations have to clarify the biological significance of the lymphocyte stimulating ability of AGP_M.

References

1. Ziegler,I., Maier,K., Fink,M.: Cancer Res.42, 1567 - 1573 (1982).

2. Ziegler,I., Armarego,W.L.F., Fink,M.: In Chemistry and Biology of Pteridines. (Ed.J.Blair). W.de Gruyter, Berlin (in press).

3. Fink,M., Ziegler,I., Maier,K., Wilmanns,W.: Cancer Res.42, 1574 - 1578 (1982).

4. Rudman,D., Treadwell,P.E., Vogler,W.R., Howart,C.H., Hollins,B.: Cancer Res.32, 1951 - 1959 (1971).

5. Ziegler,I., Hamm,U., Berndt,J.: In Biochemical and Clinical Aspects of Pteridines. W.de Gruyter,Berlin (this symposium).

6. Chiu,K.M., Mortensen,R.F., Osmand,A.P., Gewürz,H.: Immunology 32, 997 - 1005 (1977).

7. Tamura,K., Shibata,Y., Matsuda,Y., Ishida,N.: Cancer Res. 41, 3244 - 3252 (1981).

8. Maeda,H., Murakami,O., Kann,M., Yamane,I.: Proc.Soc.Exp. Biol.Med. 163, 223 - 227 (1980).

9. Waksman, B.H., Namba,Y.: Cell.Immunol.21, 161 - 176 (1976).

The work was supported by the Deutsche Forschungsgemeinschaft.

NEOPTERIN AS INDEX FOR ACTIVITY OF DISEASE IN PATIENTS WITH RHEUMATOID ARTHRITIS

Arno Hausen, Dietmar Fuchs, Gilbert Reibnegger, Helmut Wachter
Institute of Medical Chemistry and Biochemistry, University of Innsbruck

Dietmar Egg, Robert Günther
Department of Physical Medicine, University of Innsbruck

Neopterin evaluation has been found useful as a diagnostic parameter in allograft recipients (1), in patients with certain malignancies (2,3,4) and in patients with pulmonary tuberculosis (5). Recently, correlations have been shown between neopterin excretion and immunopathological features. Antigen specific stimulation of T-lymphocytes in vitro and in vivo is accompanied strongly by secretion of neopterin (6,7).

In various inflammatory rheumatic diseases, e.g. rheumatoid arthritis and connective tissue diseases a striking variety of immunological phenomena have been observed (8,9,10). Therefore we were interested to study associations between excretion of neopterin and activity or development of rheumatic diseases. In the present study urinary neopterin levels were measured in 30 patients with definite or classical form of rheumatoid arthritis and compared with clinical, radiographic and currently used laboratory data.

We have chosen rheumatoid arthritis because it has many diagnostic problems especially on early forms of disease. Diagnosis of rheumatoid arthritis is difficult in the early course of the disease due to the failure of pathognomonic laboratory and clinical criterias (11).

Biochemical and Clinical Aspects of Pteridines, Vol. 2

Rheumatoid arthritis is a chronic and often progressive disease associated with autoimmune phenomena. The process affects predominantly synovial joints, bone, muscle, fascia, ligaments and tendons. In the early course of the disease the clinical picture is marked by tenderness, pain and swelling of the joints. Progression of disease is characterized by cartilage erosion and bone destruction. Limitation of motion, joint stiffness and deformity are common.

The classification of patients with rheumatoid arthritis is best done by the functional analysis (Table 1) and by the anatomic-radio-graphic staging system (Table 2) according to Steinbrocker (12). In the differential diagnosis of inflammatory rheumatic diseases laboratory tests are regarded helpful (13). The most widely employed indicators in rheumatology are measurements of the erythrocyte sedimentation rate and of the C-reactive protein. These parameters have been used to examine the in-

Table 1 FUNCTIONAL CAPACITY IN RA*

GRADE	DEFINITION	REMARKS
I	Capable of all activites	
II	Moderate restriction	Adequate for normal activities despite handicap of discomfort or limited motion at one or more joints
III	Marked restriction	Activity limited to self-care and few or no duties of a usual occupation
IV	Bed and/or chair	Capable of little or no self-care

* From Steinbrocker, O., et al.: JAMA 140:659, 1949.

Table 2 RADIOGRAPHIC STAGING SYSTEM FOR RA*

GRADE	DEFINITION
1	No destructive changes; periarticular osteopenia may be present
2	Osteoporosis is definite, with or without slight subchondral bone destruction; slight joint space narrowing may be present
3	Evidence for cartilage and bone destruction; joint deformity (e.g. subluxation, ulnar deviation); no bony or fibrous ankylosis
4	Stage 3 criteria plus fibrous or bony ankylosis

* From Steinbrocker, O., et al.: JAMA 140:659, 1949.

flammatory process initially as well as for monitoring the course of disease in a given patient. This is regarded as critical because the course of the disease process is generally not reflected by the rise and fall of the levels of these two parameters. Remission produced by drugs such as gold, azathioprime and penicillamine is often accompanied by reduction of erythrocyte sedimentation rate and C-reactive protein to normal titers. In contrast, successful treatment with non-steroidal anti-inflammatory drugs is not generally associated with a return to normal titers (13).

Similar to erythrocyte sedimentation rate and C-reactive protein, rheumatoid factors are not specific for rheumatoid arthritis although during the course of disease 70% or more patients with rheumatoid arthritis show titers falling into the pathological range (14). At the onset of disease the titers of the rheumatoid factor are often correlated with clinical symptoms, but they will not always fall during apparently successful management of disease,particularly with non-steroidal anti-inflammatory drugs.

As a result of the poor correlation between rheumatoid factor and activity of disease, response to therapy in rheumatoid arthritis is still assessed by evaluation of clinical symptoms and signs.

In the present pilot study, urinary neopterin levels were measured in 30 patients with definite or classical form of rheumatoid arthritis. All patients received antirheumatic therapy, partly by long acting remission inducing agents and/or by nonsteroidal anti-inflammatory drugs. The neopterin levels were correlated with clinical and radiographic assessment of patients given as functional capacity and radiographic staging according to the criterias of the American Rheumatism Association (Tables 1 and 2) and with the clinical activity of patients assessed by physical examination and by evaluation of

Table 3

URINARY NEOPTERIN LEVELS OF PATIENTS WITH RHEUMATOID ARTHRITIS IN COMPARISON WITH FUNCTIONAL AND RADIOGRAPHIC STAGE AND CLINICAL ACTIVITY OF DISEASE.

	Functional analysis (I-IV) and radiographic stage (1-4)			Clinical activity a)		
	I/1	II/2	III/3	remission or low	moderate	high
Number of patients	8	12	8	9	8	12
Mean neopterin levels b)	240	519	593	222	342	668
SE	97	263	402	106	187	345
P<	0.0005 c)	0.01 d)	0.30 d)	0.0005 c)	0.10 d)	0.025 d)
percentage with elevated levels e)	38	100	100	44	75	100

a) The clinical activity was assessed by physical examination and evaluation of clinical symptoms and signs

b) Neopterin, μmol/mol creatinine

c) The p-values were obtained by the student's t-test when mean neopterin levels were compared with the control group

d) The p-values were obtained by the student's t-test when mean neopterin levels were compared with the stage presenting lower stage or activity of disease , respectively

e) With levels above the previously published upper confidence limits

clinical symptoms and signs. On the other hand, neopterin levels were compared with currently used laboratory parameters, erythrocyte sedimentation rate, C-reactive protein and rheumatoid factor.

Neopterin and creatinine measurements were performed simultaneously by HPLC on reversed phase (15). Neopterin levels were related to urinary creatinine levels. Healthy subjects (n = 417) served as controls. There was a slight dependence of mean values and upper confidence limits on age and sex. These limits include, with 95% probability, 97.5% of healthy subjects.

As shown in table 3 and figure 1, the mean neopterin level of patients with functional and radiographic stage I/1 was highly significantly elevated compared to the sex and age-matched healthy control group. The mean neopterin level of functional and radiographic stage II/2 was raised with probability over that of stage I/1. The mean level of stage III/3 was higher than that of stage II/2, but without significance.

Figure 1 Urinary neopterin levels of patients with rheumatoid arthritis placed according to stage and activity of disease

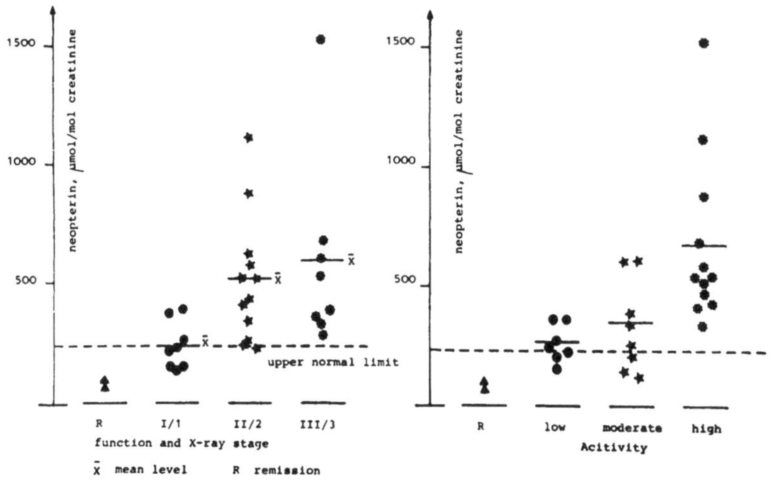

Similarly, different mean neopterin levels were observed when the patients were placed according to the activity of disease. The mean levels of patients with low activity were compared to those of the healthy control group. Those of patients with moderate activity were compared with those of patients with low activity and the mean level of patients with high activity with that of patients with moderate activity. Except for the first comparison, the differences were statistically not significant by the student's t-test.

In a similar manner, the incidence of neopterin levels higher than the upper normal limits depended on stage and activity of disease.

The individual values grouped according to the activity and stage of disease are plotted in figure 1. Based on this plot we feel that the lack of significance is due to the small number of patients studied. However, it should be noted that the classification of activity was done at the time of urine sampling by all means without knowledge of neopterin levels.

Table 4 URINARY NEOPTERIN LEVELS OF PATIENTS WITH RHEUMATOID ARTHRITIS IN COMPARISON WITH ERYTHROCYTE SEDIMENTATION RATE AND WITH C-REACTIVE PROTEIN

	Erythrocyte sedimentation rate (mm/1 h)			C-reactive protein	
	20	20-40	40	neg.	pos.
Number of patients	9	10	11	7	23
Mean neopterin levels[a)]	356	399	532	448	430
SE	320	222	369	352	305
$P<$	0.0005[b)]	0.40[c)]	0.20[c)]	0.0005[b)]	0.45[c)]
Percentage with elevated levels[d)]	56	80	91	71	78

a) Neopterin, μmol/mol creatinine

b) The p-values were obtained by the student's t-test when mean neopterin levels were compared with the control group

c) The p-values were obtained by the student's t-test when mean neopterin levels were compared with the group presenting lower levels of the studied parameter

d) With levels above the previously published upper confidence limits

The results of the comparison of neopterin levels with erythrocyte sedimentation rate, with C-reactive protein and with rheumatoid factor are demonstrated in tables 4 and 5. According to the erythrocyte sedimentation rate, patients were divided into three groups. A modest increase of mean neopterin was

Table 5 URINARY NEOPTERIN LEVELS OF PATIENTS WITH RHEUMATOID ARTHRITIS IN COMPARISON WITH RHEUMATOID FACTOR (DETERMINED BY LATEX TEST OR BY ROSE-WAALER TEST)

	Rheumatoid factor			
	Latex test		Rose-Waaler test	
	neg.	pos.	neg.	pos.
Number of patients	7	23	8	22
Mean neopterin levels[a)]	330	467	308	481
SE	267	321	254	321
$P<$	0.0005[b)]	0.20[c)]	0.0005[b)]	0.10[c)]
Percentage with elevated levels[d)]	57	83	50	86

a) Neopterin, μmol/mol creatinine

b) The p-values were obtained by the student's t-test when mean neopterin levels were compared with the control group

c) The p-values were obtained by the student's t-test when mean neopterin levels were compared with the group presenting lower levels of the studied parameter

d) With levels above the previously published upper confidence limits found for the groups presenting higher erythrocyte sedimentation rate. There was no difference of mean neopterin levels between patients without and with presence of C-reactive protein. The percentage of patients with elevated levels exhibited an increase dependent on the degree of the compared parameter.

A comparison of neopterin levels with absence or presence of rheumatoid factors is shown in table 5. The mean neopterin levels and the frequency of elevated levels were slightly, but not significantly higher in patients with positive Latex test as well as positive Rose-Waaler test (two tests for the detection of rheumatoid factor).

The sensitivity of neopterin evaluation in rheumatoid arthritis is compared with that of the currently used laboratory parameters, erythrocyte sedimentation rate, C-reactive protein and rheumatoid factor in table 6. There was no difference in correlation with clinical status between neopterin, erythrocyte sedimentation rate and C-reactive protein evaluation. The sen-

Table 6 SENSITIVITY OF NEOPTERIN EVALUATION IN RHEUMATOID ARTHRITIS COMPARED WITH ERYTHROCYTE SEDIMENTATION RATE (ESR), WITH C-REACTIVE PROTEIN (CRP) AND WITH RHEUMATOID FACTOR (RF, DETERMINED BY LATEX-TEST OR ROSE-WAALER-TEST) WHEN ESTIMATIONS WERE PERFORMED CONCURRENTLY

				R F	
	Neopterin[a)]	ESR	CRP	Latex-test	Rose-Waaler-test
Number of patients	30	30	30	30	30
Patients with elevated levels[b)]	25	25	25	23	22
Corresponce with clinical status (%)	83	83	83	77	73

a) Neopterin, µmol/mol creatinine

b) With levels above the previously published upper normal limits. These limits tested according to Liebermann contained with 95% probability 97.5% of healthy subjects sitivity of these parameters was slightly superior to that of the rheumatoid factor.

It should be noted that this pilot study certainly comprises an insufficient number of patients and that the influence of treatment remains unclear. It is known that steroidal anti-inflammatory drugs depress neopterin excretion. However, the results of this preliminary report indicate a possible correlation between urinary neopterin levels and stage and, particularly, with the activity of rheumatoid arthritis. We observed a weak association between neopterin levels and currently used laboratory parameters. It is well known that the development of positivity of rheumatoid factor and erythrocyte sedimentation rate lags behind clinical manifestations.

Thus, neopterin measurement may reflect changes in activity of rheumatoid arthritis more rapidly. Studies to determine the reliability of neopterin evaluation in monitoring response to treatment and in the differentiation between inflammatory and non-inflammatory rheumatic processes are in progress.

Conclusions

- The results of this pilot study suggest a correlation between neopterin levels and activity of rheumatoid arthritis.

- The sensitivity of neopterin evaluation is comparable to that of the erythrocyte sedimentation rate and the C-reactive protein.

- We feel that neopterin levels as an index of T-cell activity may be useful as a criterion for the management of therapy in patients with rheumatoid arthritis.

References

1. Margreiter, R., Fuchs, D., Hausen, A., Huber, Ch., Reibnegger., Spielberger, M., Wachter, H.: Transplantation, in press.

2. Bichler, A., Fuchs, D., Hausen, A., Hetzel, H., Reibnegger, G., Wachter, H.: Arch. Gynecol. in press.

3. Hausen, A., Wachter, H.: J. Clin. Chem. Clin. Biochem. 20, 593-602 (1982).

4. Kostron-Krainz, Ch., Fuchs, D., Hausen, A., Reibnegger, G., Wachter, H.: In: Biochemical and Clinical Aspects of Pteridines, Vol. 1; Wachter, H., Curtius, H.Ch., Pfleiderer, W. (Hrsg.), Walter de Gruyter: Berlin, New York (1982), pp 157-171.

5. Fuchs, D., Hausen, A., Knosp, O., Reibnegger, G., Wachter, H., Kofler, M., Kosanowsky, H., Huber, Ch., Niederwieser, D.: In: Biochemical and Clinical Aspects of Pteridines, Vol. 2; Curtius, H.Ch., Pfleiderer, W., Wachter, H.(Hrsg.) Walter de Gruyter: Berlin, New York (in press).

6. Fuchs, D., Hausen, A., Huber, Ch., Margreiter, R., Reibnegger, Spielberger, M., Wachter, H.: Hoppe-Seyler's Z. Physiol. Chem. 363, 661-664 (1982).

7. Huber, Ch., Fuchs, D., Hausen, A., Margreiter, R., Reibnegger, G., Spielberger, M., Wachter, H.: J. Immunol. 130, 1047-1050 (1983).

8. Janossy, G., Panayi, G., Duke, O., Bofill, M., Poulter, L.W., Goldstein, G.: Lancet, 839-842 (1981).

9. Fox, R.J., Fong, S., Sabharwal, N., Carstens, S.A., Kung, P.C., Vaughan, J.H.: J. Immunol. 128, 351-354 (1982).

10 Rosenthal, M., Sendelova, S.: Schweiz. med. Wschr. 113, 394-398 (1983).

11. Harris, E.D.: In: Kelley, W.N., Harris, E.D. jr., Ruddy, S., Sledge, C.B. (eds), Textbook of Rheumatology, W.B. Saunders Comp., Philadelphia, London, Toronto, 1981, pp 928-963.

12. Steinbrocker, O., Traeger, C.H., Battermann, R.C.: JAMA 140, 659 (1949).

13. Kushner, I.: IN: Kelley, W.N., Harris, E.D.jr., Ruddy, S., Sledge, C.B. (eds), Textbook of Rheumatology, W.B. Saunders Comp., Philadelphia, London, Toronto, 1981, pp 669-676.

14. Carson, D.A.: In: Kelley, W.N., Harris, E.D.jr., Ruddy, S., Sledge, C.B. (eds), Textbook of Rheumatology, W.B. Saunders Comp., Philadelphia, London, Toronto, 1981, pp 677-690.

15. Hausen, A., Fuchs, D., König, K, Wachter, H.: J. Chromatogr. 227, 61-70 (1982).

Acknowledgement

This study was financially supported by the Austrian Research Funds "Zur Förderung der wissenschaftlichen Forschung", project P 4905.

4

PTERIDINES IN CANCER AND OTHER DISEASES

URINARY NEOPTERIN LEVELS IN PATIENTS WITH MALIGNANT TUMORS OF THE GENITOURINARY TRACT

Wolfgang Aulitzky, Julian Frick
Urological Department, General Hospital Salzburg
A-5020 Salzburg

Dietmar Fuchs, Arno Hausen, Gilbert Reibnegger, Helmut Wachter
Department of Medical Chemistry, University of Innsbruck
A-6020 Innsbruck

Introduction

Previous studies by Wachter and coworkers (1) have demonstrated elevated urinary neopterin levels in malignancy. In the meantime it became evident by other studies that there is a correlation between neopterin excretion and the clinical course of the disease in patients suffering from gynecological (2), hematological (3) and some other malignant tumors (4). From recent results it may be concluded that neopterin release in vitro and in vivo is closely related to stimulation of T-lymphocytes (5). Also rejection of allografts is strongly accompanied by increased neopterin secretion (6). In this presentation we shall show our results of urinary neopterin excretion in patients suffering from malignant tumors of the genitourinary tract.

Biochemical and Clinical Aspects of Pteridines, Vol. 2

Material and method

Urinary neopterin and creatinine levels were measured in four groups.
Group I: Patients with bladder tumors
Group II: Patients with carcinoma of the prostate
Group III: Patients with renal cell carcinoma and
Group IV: Patients with a malignant tumor of the testis
The samples were taken from the first portion of the early morning urine and stored by -20°C until examination. Neopterin excretion was correlated to urinary creatinine levels.
Neopterin and creatinine levels were simultaneously determined by high-performance liquid Chromatography on reversed phase according to Hausen and coworkers (7).

Results and Discussion

1) Bladder tumor

Our results in patients with bladder tumor are shown in Fig. 1. 31 patients took part in this study. The neopterin levels were related to the stage of the tumor.
All patients with a tumor stage I had neopterin levels within the normal range; only one patient was above the upper confidence limit.
All patients with stage II tumors had neopterin levels above the upper confidence limit. In the stage III group we had two patients with normal neopterin values - although both patients exhibited extensive tumor growth. All the other patients had high urinary neopterin levels. The urine of the 3 patients in stage 4 was sampled preterminal and presented very high neopterin levels.

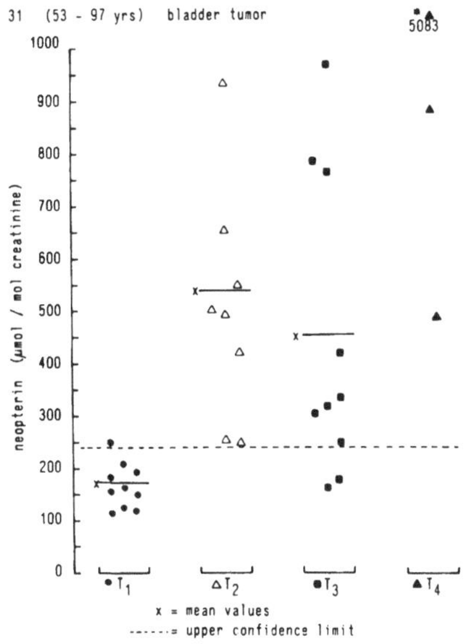

Fig. 1: **results bladder tumor**

Inspite of the progress in diagnosis and therapy of the bladder cancer during the last decade, this disease has still a poor prognosis. Even when the patients were treated most extensively by cystectomy and radiation therapy the five years survival rate ranges between 36% - 53%. Staging and grading techniques in the diagnostic management of the bladder tumor have improved, the biological behaviour of the tumor however is still unpredictable and an appropriate therapeutic timing is very difficult. It is generally agreed that bladder cancer begins with superficial lesions like atypical epithelium proliferation with flat or papillary noninvasive growth. This stage is represented by the group T_1. There is no clear answer why and when the tumor starts its invasive growth deeper into the bladder wall, a fact which requires clarification by new parameters or methods.

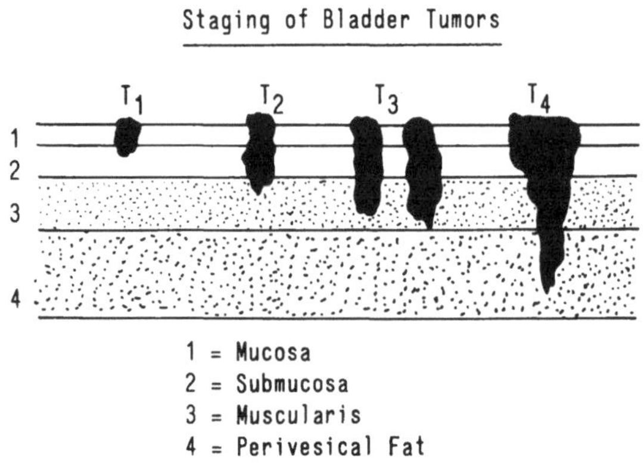

Fig. 2: Staging of Bladder Tumor

Neopterin may be such a parameter since we have found low neopterin levels in almost all T_1 cases and levels above the upper confidence limit in the higher stages. Of course the number of patients examined is still too small for any final conclusions.
However these results indicate that there must be a difference in neopterin secretion between stage T_1 and stages T_2 - T_4. Probably there is an activation of T-lymphocytes when the tumor penetrates the lamina propria and when invasive growth is initiated. On the other hand however the activation of T-lymphocytes could be caused by transformation of tumor cell surface antigens or when dedifferentiation proceeds.

2) Carcinoma of the prostate

Fig. 3 shows the urinary neopterin levels from 34 patients who are suffering from carcinoma of the prostate, the levels are plotted according to tumor stage. In stage A/B we usually intend to perform more radical surgery because it may be curative at this stage of the disease, when the tumor is entirely restricted to the prostate. Our patients in this group except one had neopterin levels within the normal range.

The patients in group C - D presented high neopterin levels with two exceptions.

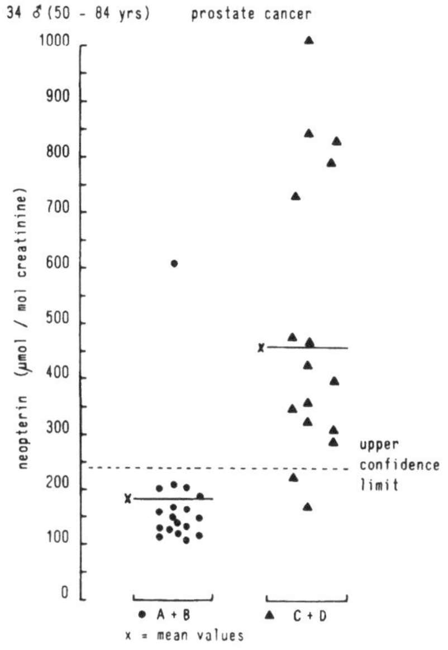

Fig. 3: results prostate cancer

The carcinoma of the prostate is a very heterogenic tumor even if the histopathologist finds in a fairly high percentage an adenocarcinoma. The histopathological classification in one and the same organ or even in small biopsies may vary from well differentiated tumor cells to highly dedifferentiated cells. The unpredictable clinical course of the disease in many instances may be due to these findings. Therefore the tumor stage is the main parameter today for choosing the therapeutical procedure.

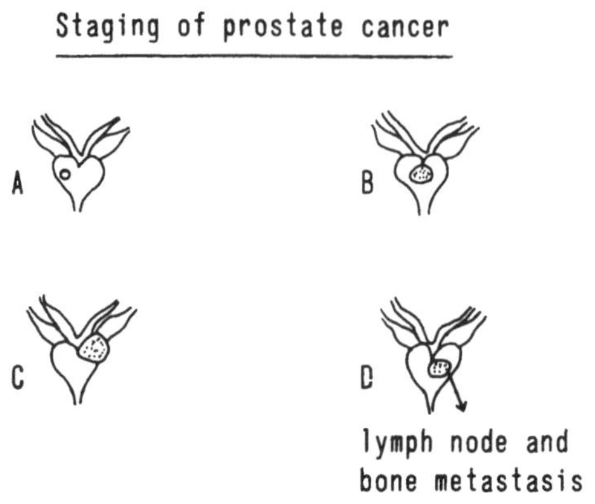

Fig. 4: Staging of prostate cancer

Stage A: lesion completely limited to prostate;
Stage B: lesion involving most of the prostate;
Stage C: lesion locally extended beyond prostate to regions of the base of the bladder and seminal vesicles;
Stage D: lesions clinically disseminated to lymph nodes and to bones.

Nevertheless a new parameter which tells something about tumor behaviour would be of great help for the physician. Our results showed a good correlation between low tumor grade and low neopterin levels. It may be that the very early tumor stage and the relatively low histopathological grade are responsible for these results, but this is a speculative assumption only. Our results however indicate that there might be also some influence on neopterin secretion or T-lymphocyte activation by tumor progression, or by the development of distant metastases or dedifferentiation of the tumor cells.

3) Renal cell carcinoma

In Fig. 5 we demonstrate the urinary neopterin values in 11 patients with renal cell carcinoma. We could not find any correlation between tumor stage and neopterin excretion.

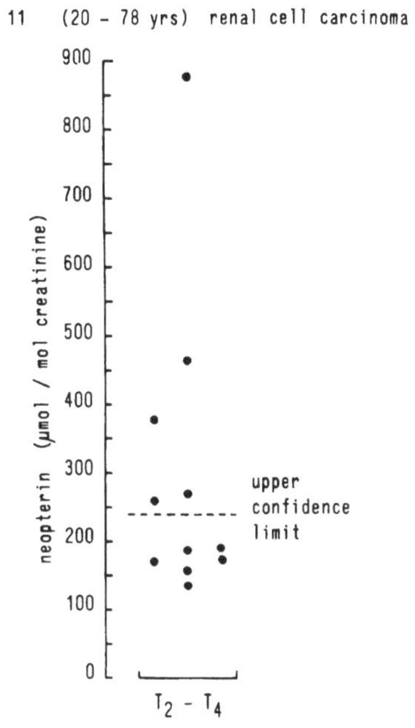

Fig. 5: results renal cell carcinoma

More patients need to be evaluated for this purpose. The negative correlation may be due to the fact that most renal cell carcinoma have already reached a higher tumor stage and large tumor size when first diagnosed. We know from the literature that there is a negative correlation between activation of the immune system and tumor size (8). We also know from the literature that the cellular immune response is impaired in patients with renal cell carcinoma, and T-lymphocyte levels in tumorous venous blood are significantly decreased (9).

4) Testicular tumors

In Fig. 6 the neopterin levels from 6 patients with testicular tumors are presented. Because of the small number of patients no conclusions should be drawn from these results. It is however still striking that the two patients with normal neopterin excretion had a small lesion entirely confined to the testis which means stage I, and the histology revealed a seminoma. The remaining 4 patients had metastases and a more malignant histopathological classification, and the neopterin levels are above the upper confidence limit.

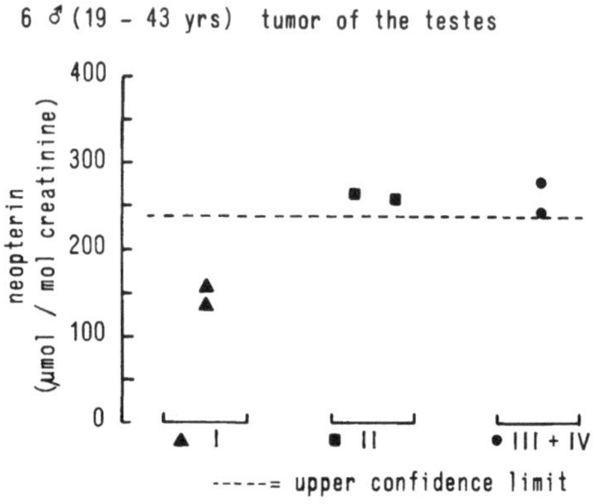

Fig. 6: tumor of the testes

Conclusion

As conclusion we believe that neopterin is not a tumor-marker in the common sense. It might however give some evidence for a better understanding of the biological behaviour of malignant diseases of the genitourinary tract. Neopterin might also be of great help in case selection for the most convenient therapeutic management and follow up of malignant diseases. Since neopterin seems to be one of the first biochemical immunological parameters, it may again focus the interest of oncologists more to the immunology of malignancy.

References

1. Hausen, A., Wachter, H.: J.Clin.Chem.Clin.Biochem. 2o,593-6o2 (1982)

2. Bichler, A., Hetzel, H., Marth, Ch., Fuchs, D., Hausen, A., Reibnegger, G., Wachter, H.: Biochem.and Clin.Aspects of Pteridines 1, 141-156 (1982) Walter de Gruyter

3. Grünewald, K., Huber, H., Fuchs, D., Hausen, A., Reibnegger, G., Wachter, H.: Biochem. and Clin.Aspects of Pteridines 1, 173-206 (1982) Walter de Gruyter

4. Kostron-Krainz, Ch., Fuchs, D., Hausen, A., Reibnegger, G., Wachter,H.: Biochem.and Clin.Aspects of Pteridines 1, 157-171 (1982) Walter de Gruyter

5. Huber, Ch., Fuchs, D., Hausen, A., Margreiter, R., Reibnegger, G., Spielberger,M., Wachter, H.: J.Immunology 130 (3) (1983)

6. Margreiter, R., Spielberger, M., Fuchs, D., Hausen, A., Reibnegger, G., Wachter, H.: Biochem. and Clin.Aspects of Pteridines 1, 233-240 (1982) Walter de Gruyter

7. Hausen, A., Fuchs, D., König, K., Wachter, H.: J.Chromatography 227, 61-70 (1982)

8. Klippel, K.F.: Urologie in Klinik und Praxis, Band I, 464-481, Georg Thieme Verlag Stuttgart - New York (1982)

9. Carmignani, G., Belgrano, E., et al.: J.Urol. 120, 673-675 (1978).

Acknowledgement

This study was financially supported by the Austrian Research Funds "Zur Förderung der wissenschaftlichen Forschung", project P 4905.

THE CLINICAL SIGNIFICANCE OF NEOPTERIN SAMPLING IN THE FOLLOW-UP OF PATIENTS WITH GENITAL CANCER

Heinz Hetzel, Alfred Bichler, Christian Marth
Universitätsklinik für Frauenheilkunde Innsbruck
A-6020 Innsbruck, Austria

Dietmar Fuchs, Arno Hausen, Gilbert Reibnegger, Helmut Wachter
Institut für Medizinische Chemie und Biochemie der Universität Innsbruck
A-6020 Innsbruck, Austria

Introduction

In previous publications our group (1,2) presented our initial results concerning neopterin values in patients with genital cancer. Thereafter, these findings could be confirmed by analysis of a greater number of patients. The purpose of the present paper was to go a step further in our investigations concerning the value of neopterin determination in the follow-up of patients with genital cancer. In other words, we wanted to answer the question: Does the monitoring of urinary neopterin levels have value as a tumor marker?; the outstanding feature of which consists of an early indication of a tumor recurrence.

Patients and methods

Urinary neopterin determinations were performed in the following patients:

1) On 224 patients with different types of genital as well as breast cancer.
2) As serial determinations on 125 patients with genital can-

Biochemical and Clinical Aspects of Pteridines, Vol. 2

cer within a follow-up period of 6 to 30 months.

3) On 7 patients, who were requested to come for a clinical check-up because of elevated neopterin values in the follow-up for detection of a tumor recurrence.

	n	correspondence
cervical cancer	127	72 (57 %)
endometrial cancer	50	35 (70 %)
ovarian cancer	25	23 (92 %)
cancer of the vulva	9	7 (78 %)
breast cancer	13	9 (69 %)
total	224	146 (65 %)

Table 1

Significance of urinary neopterin values (μmol / mol creatinine) prior to the beginning of therapy in patients with genital cancer and breast cancer.

Results

Since our last publication (2) we were able to double the number of patients in our study group. The percentage of patients with elevated neopterin values before treatment has not changed very much:
elevated neopterin values were found in 65% of all patients with genital and breast cancer (table 1).
In table 2 patients with either normal or elevated neopterin values prior to treatment were placed into one group in the follow-up of these cancer patients. A correlation between neopterin values and the clinicoroentgenological findings was found in 85% of the cases (table 2).
In tables 3 and 4 the patients with elevated and normal neopterin values before treatment are shown separately. The corresponding and non-corresponding neopterin values in the follow-up of the cancer patients did not change very much.

	BEFORE THERAPY	FOLLOW - UP (between 6 - 30 months)			
		correspondence		false	
	normal and elevated neopterin values n	no recurrence - normal neopterin values n	recurrence - elevated neopterin values n	negative n	positive n
cervical cancer	77	51	14	1	11
endometrial cancer	31	19	6	-	6
ovarian cancer	14	4	10	-	-
cancer of the vulva	3	-	3	-	-
total	125	74	33	1	17
		107 (85 %)		18 (15 %)	

Table 2

Urinary neopterin concentration (μmol / mol creatinine) and clinicoroentgenological findings in the follow - up of those patients with elevated and normal neopterin values before treatment.

	BEFORE THERAPY	FOLLOW - UP (between 6 - 30 months)			
		correspondence		false	
	elevated neopterin values n	no recurrence - normal neopterin values n	recurrence - elevated neopterin values n	negative n	positive n
cervical cancer	47	28	12	-	7
endometrial cancer	22	12	5	-	5
ovarian cancer	12	3	9	-	-
cancer of the vulva	2	-	2	-	-
total	83	43	28	-	12
		71 (86 %)		12 (14 %)	

Table 3

Urinary neopterin concentration (μmol / mol creatinine) and clinicoroentgenological findings in the follow - up of those patients with pretherapeutic elevated neopterin values.

Summarizing the results from the tables 2-4 we can conclude that there is no difference in the recognition of a tumor recurrence in patients who had either normal or elevated neopterin values before treatment.
The frequency of tumor recurrence in patients with normal and elevated neopterin välues before therapy is listed in table 5. Patients with elevated neopterin values prior to treatment displayed a significantly higher frequence of tumor recurrence ($p < 0.025$).
To elucidate the value of neopterin sampling in the follow-up of patients with genital cancer some characteristic neopterin curves are demonstrated. The first example concerns patients with cervical cancer. In figure 1 the neopterin values during combined radium x-ray therapy are shown. A 69 year old patient with cervical cancer stage III and elevated neopterin values before treatment received three radium applications within an intervall of 14 days with concurrent gamma irradiation. The neopterin values dropped to within the normal range after the completion of therapy. 7 months later the patient was still clinically free of disease and the neopterin values were normal.
Figure 2 shows the neopterin curve of a 57 year old patient with cervical cancer stage III who developed a tumor progression shortly after the completion of radiation therapy. The neopterin values remained elevated during the radiotherapy and the patient died from cancer 4 months later.
The elevated neopterin values might have been the first evidence for a tumor recurrence in this patient.
Figures 3 and 4 deal with patients having endometrial cancer. In figure 3 the neopterin values remained high and the patient died from cancer ten months after the commencement of therapy. In figure 4 the neopterin curve of a 66 year old patient with endometrial cancer stage III can be seen. Primary treatment for this patient was abdominal hysterectomy and bilateral adnexectomy followed by gamma irradiation 6 weeks later. The neopterin values increased within four months and they were far

above the normal level. The clinicoroentgenological findings

	BEFORE THERAPY	FOLLOW - UP (between 6 - 30 months)			
		correspondence		false	
	normal neopterin values n	no recurrence - normal neopterin values n	recurrence - elevated neopterin values n	negative n	positive n
cervical cancer	30	23	2	1	4
endometrial cancer	9	7	1	-	1
ovarian cancer	2	1	1	-	-
cancer of the vulva	1	-	1	-	-
total	42	31	5	1	5
		36 (86 %)		6 (14 %)	

Table 4

Urinary neopterin concentration (μmol / mol creatinine) and clinicoroentgenological findings in the follow - up of those patients with pretherapeutic normal neopterin values.

	total number of patients with genital cancer n	tumor recurrence n	
normal neopterin values before therapy	42	6 (14 %)	$p < 0,025$
elevated neopterin values before therapy	83	28 (34 %)	

Table 5

Tumor recurrence in cancer patients with normal and elevated neopterin values prior to treatment.

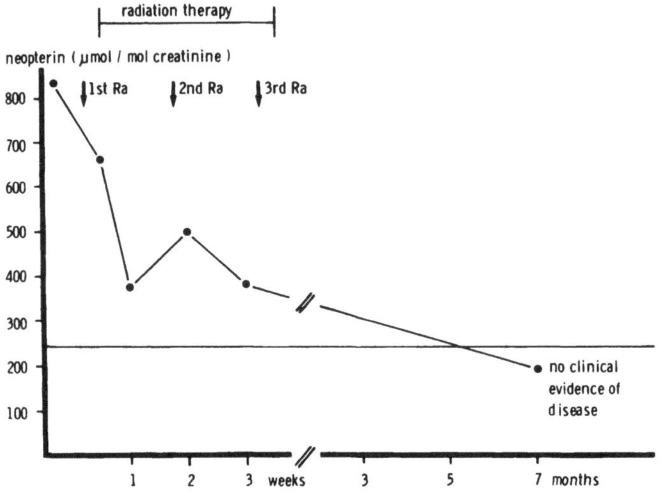

Figure 1
N. P. 69 years, cervical cancer, stage III.

Horizontal line : Upper limit of tolerance of normal neopterin values.

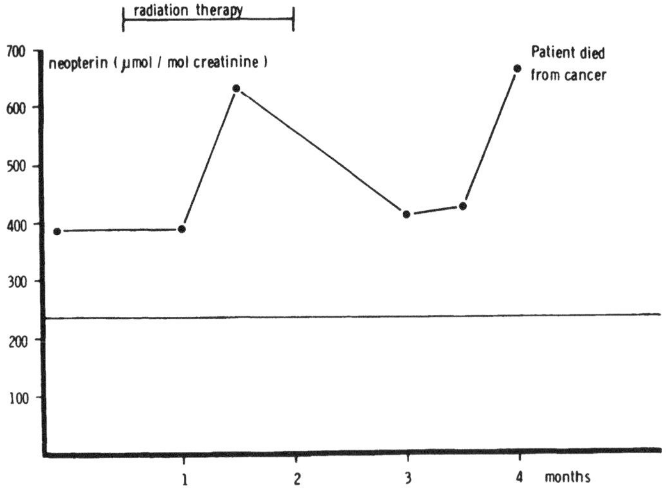

Figure 2
O. M. 57 years, cervical cancer, stage III.

Horizontal line : Upper limit of tolerance of normal neopterin values.

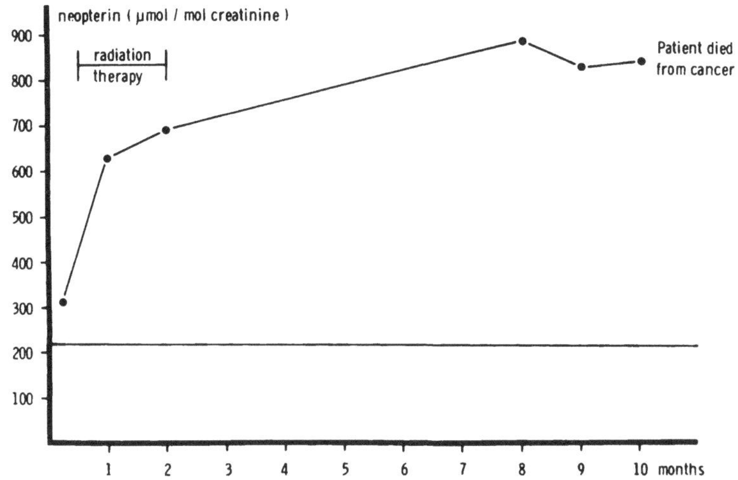

Figure 3
B. M. 68 years, endometrial cancer, stage III.

Horizontal line : Upper limit of tolerance of normal neopterin values.

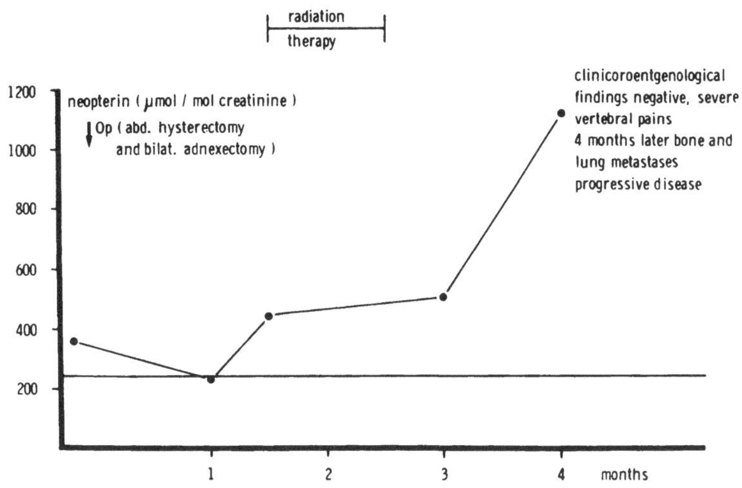

Figure 4
K. A. 66 years, endometrial cancer, stage III.

Horizontal line : Upper limit of tolerance of normal neopterin values.

and gynecological palpation were negative although the patient suffered from severe vertebral pain. Four months later bone metastases were visible on x-rays.
Figure 5 and 6 show the neopterin curves of patients with ova-

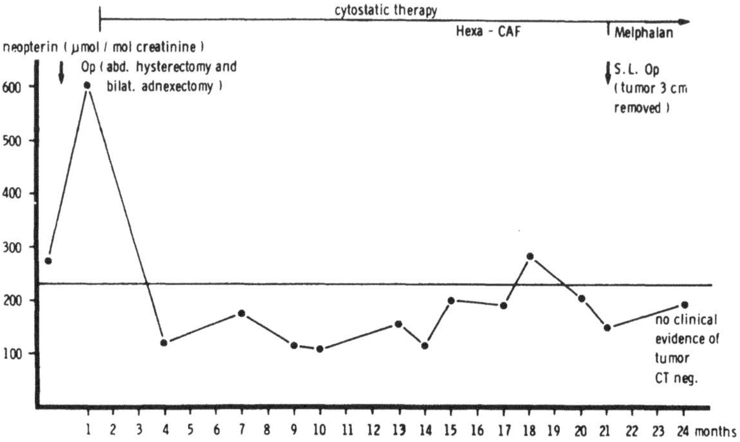

Figure 5
E. E. 42 years, ovarian cancer, stage II b.

Horizontal line : Upper limit of tolerance of normal neopterin values.
S.L. Op : Second look operation
CT : Computer tomography

rian cancer. Figure 5 demonstrates the neopterin values of a 42 year old patient with an ovarian cancer stage IIb who experienced a partial tumor remission lasting about 2 years. Two months after the surgery neopterin values dropped into the normal range and remained there. The patient received cytostatic treatment until "second look" surgery which was performed about 21 months after the first operation. During this "second look" operation a tumor with a diameter of about 3cm was removed. This patient remained without any clinical evidence of tumor recurrence 24 months after primary therapy and the neopterin values were normal. Since the patient is free of tumor for more than 6 months after this "second look" operation, we may prosume that the removed tumor cells were inactivated by the cytostatic therapy. Therefore the normal neopterin values at this time were correct. In figure 6 one sees the neopterin values of a 65 years old patient with an ovarian cancer stage IV, who developed a tumor progression following an 11 months period of partial remission. This patient suffered from pulmonary metastases before treatment. Following surgery a cyto-

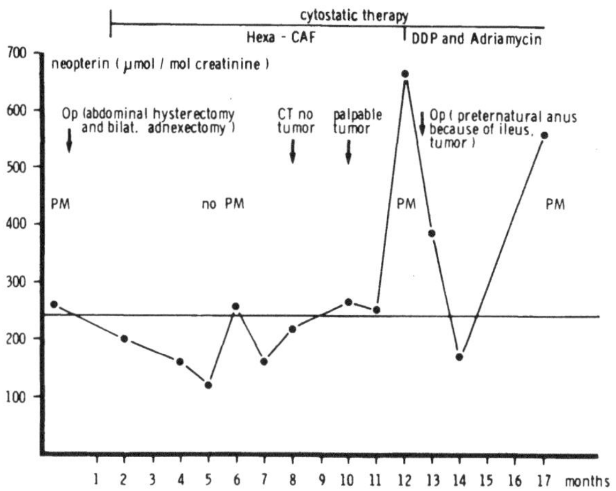

Figure 6
T. B. 65 years, ovarian cancer, stage IV.

Horizontal line: Upper limit of tolerance of normal neopterin values.
PM: Pulmonary metastases
CT: Computer tomography

static treatment regimen was performed; the pulmonary metastases vanished and the computer-tomography and the gynecological palpation were negative. Two months later a tumor was palpable and the pulmonary metastases were detected again. A colostomy had to be performed because of an ileus caused by a bowel obstruction. The cytostatic treatment was changed to more aggressive chemotherapeutic agents. The neopterin values displayed a transient decrease into the normal range although the patient suffered from progressive disease. In the further course the neopterin values were again elevated and the patient was in poor physical condition.

The next two cases will deal with patients with choriocarcinoma. This tumor appears very rarely. In Europe this type of cancer is seen with a frequency of about 1 : 12.000 - 40.000 births. Choriocarcinoma is classified in the group of trophoplastic tumors. This malignant tumor is diagnosed by curettage, human chorinic gonadotrophin (hCG) and ß-human chorionic gonadotrophin (ß-hCG) hormone measurement in urine and serum. This hormone is produced by chorionic tissue. This malignant disease state allows one the opportunity to compare neopterin values with these two well established tumor markers. Figure 7 shows the follow-up of a 38 year old patient with choriocarcinoma who was treated with cytostatic drugs. hCG and ß-hCG dropped to normal values whereas the neopterin levels always remained high. The patient died from cancer 15 months after the first cytostatic treatment. The course of another 54 year old patient with choriocarcinoma is demonstrated on figure 8. After the second cytostatic treatment hCG and ß-hCG dropped to normal values, whereas the neopterin values remained high. With exception of a few changes the pulmonary metastases remained during therapy. We conclude from these two choriocarcinoma patients that the neopterin determinations are more tumor sensitive than the hormonal parameters.

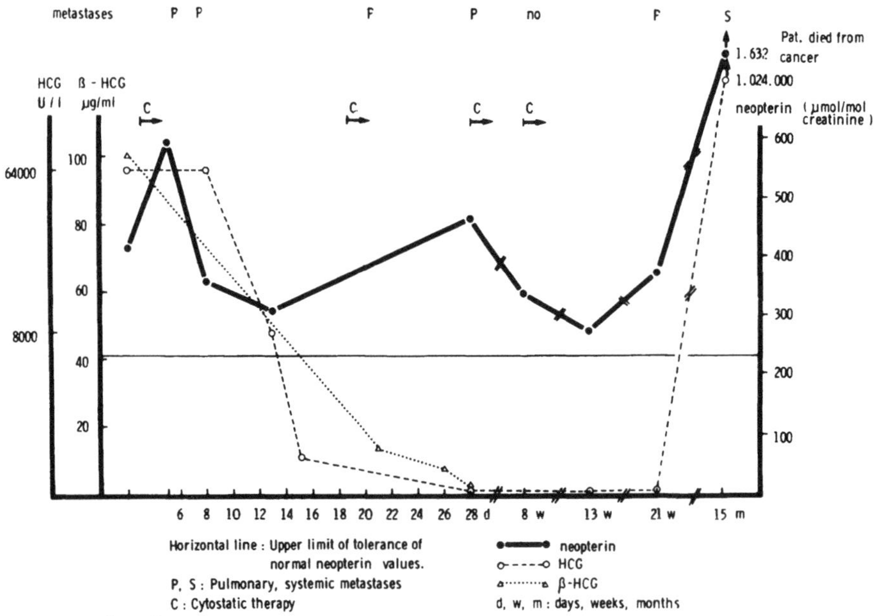

Figure 7
N. M., 38 years. Chorionic carcinoma of the uterus with formation of metastases. Surgical treatment and serial cytostatic therapy preceded this period of observation.

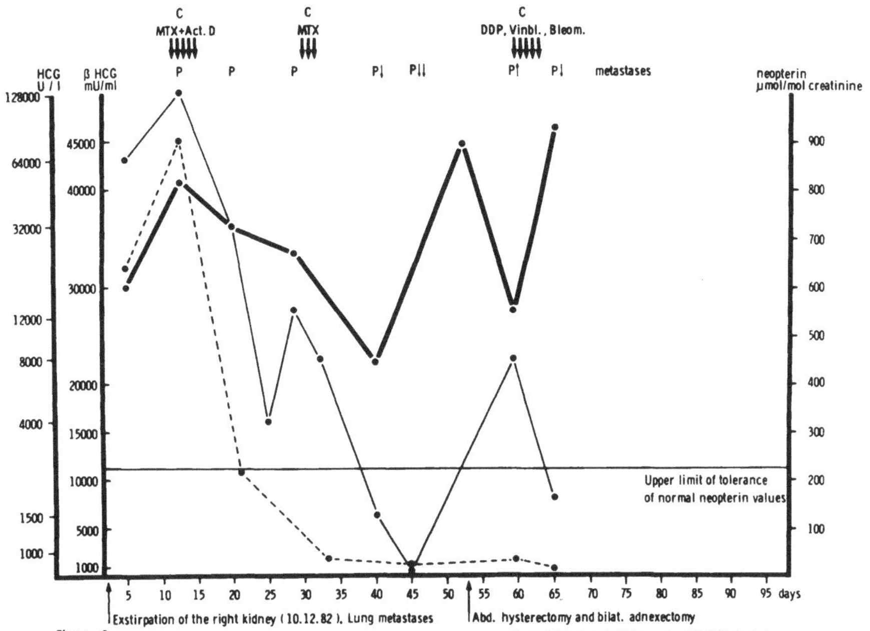

Figure 8
O. F. 45 years, choriocarcinoma of the uterus with formation of metastases.

We were able to demonstrate a good correspondence between neopterin values and the clinical follow-up in patients with all types of genital cancer. It would be of great clinical utility to diagnose a tumor recurrence as early as possible. To prove if neopterin might be an early parameter for detecting a tumor recurrence 7 patients with elevated neopterin values but with negative gynecological palpation in the follow-up were requested to come to a control examination. They were examined again by computertomography, urography, erythrocyte sedimentation, and biopsy. The results are listed on table 6. In 3 patients (Nr. 1,2,4) a tumor recurrence was found. Three other elderly patients (Nr. 3,6,7), two of them suffering from endometrial and one from cervical cancer were obviously without any clinical evidence of tumor; however, in one of these patients, (Nr. 3), a control neopterin value was within the normal range. The other two patients (Nr. 6,7) will be re-examined at short intervalls; perhaps they will develop a tumor recurrence in the up-coming months. One patient (Nr.5) suffered from influenza which is known to cause elevated neopterin levels (3).

Table 6

Nr.	Pat.	age (years)	type of cancer	palp.	neopterin µmol/mol cr.	erythrocyte sedimentation	CT	biopsy	urography	
1	G. M.	74	cervical cancer III	neg.	366	30 / 48	tumor	neg. (?)	hydronephrosis left side (already before therapy)	tumor recurrence
2	K. H.	65	cervical cancer III	neg.	505	4 / 9	tumor	pos.	hydronephrosis left side (already before therapy)	tumor recurrence
3	M. H.	75	cervical cancer III	neg.	367 171	6 / 10	no tumor	-	normal	no recurrence
4	G. H.	72	cervical cancer III	neg.	250	110 / 130	tumor	pos.	hydronephrosis left side (already before therapy)	tumor recurrence
5	E. R.	60	endometrial cancer I a	neg.	265	33 / 56	no tumor	-	normal	no recurr. influenca
6	St. M.	74	endometrial cancer I	neg.	416	19 / 34	-	neg.	normal	no recurrence
7	K. B.	78	endometrial cancer I a	neg.	316	5 / 9	no tumor	-	normal	no recurrence

Conclusions

Summarizing our findings we can draw the following conclusions:

Neopterin sampling appears to be of great help in monitoring patients with genital cancer. The rational basis for elevated neopterin excretion in some patients with cancer might be a consequence of T-cell-stimulation (4). A single elevated neopterin value has to be confirmed by control values. Increased levels may be caused by non-tumor associated T-cell-immunity (4). Sustained elevation of neopterin values after treatment is a sign of poor prognosis.

Neopterin sampling is also useful in detecting tumor recurrence in patients with normal neopterin values before treatment. Therefore patients treated elsewhere with missing neopterin values before treatment can also be monitored by neopterin sampling.

Our results point to the conclusion that elevated urinary neopterin values observed during the follow-up of genital cancer patients may be an early sign of tumor recurrence.

References

1. Bichler, A., Fuchs, D,, Hausen, A., Hetzel, H., Reibnegger, G., Wachter, H.: Arch. Gynecol., in press
2. Bichler, A., Hetzel, H., Marth, Ch., Fuchs, D., Hausen, A., Reibnegger, G., Wachter, H.: In: Biochemical and Clinical Aspects of Pteridines, Volume 1, Cancer - Immunology - Metabolic Diseases, Walter de Gruyter & Co., Berlin - New York, 141-156 (1982)
3. Wachter, H., Hausen, A., Grassmayr, K.: Hoppe-Seyler's Z. Physiol. Chem. 360, 1957-1960 (1979)
4. Fuchs, D., Hausen, A., Huber, Ch., Margreiter, R., Reibnegger, G., Spielberger, M., Wachter, H.: Hoppe-Seyler's Z. Physiol. Chem. 363, 661-664 (1982)

Achnowledgement
This study was financially supported by the Austrian Research Funds "Zur Förderung der wissenschaftlichen Forschung", project P 4905.

NEOPTERIN EVALUATION IN PATIENTS SUFFERING FROM PULMONARY TUBERCULOSIS

Dietmar Fuchs, Arno Hausen, Olga Knosp, Gilbert Reibnegger, Helmut Wachter
Department of Medical Chemistry and Biochemistry, University of Innsbruck, Austria

Manfred Kofler, Hans Kosanowsky
Natters Hospital, Austria

Christoph Huber, Dietger Niederwieser
Department of Internal Medicine, University of Innsbruck, Austria

Summary

Neopterin excretion as marker of activated T-lymphocytes is tested in pulmonary tuberculosis. The clinical value of neopterin estimation in diagnosis as well as in the follow-up and therapy control of pulmonary tuberculosis is pointed out. There exists a high correlation between clinical course and neopterin excretion. Fast amelioration of the disease status is predicted by rapid normalization of neopterin values; slow improvement in response to therapy is characterized by relatively unchanged elevated neopterin excretion.

Introduction

Pteridines undergo characteristic changes during development and growth of various cells (1). The strong connection between activation of T-lymphocytes by histoincompatible HLA

Biochemical and Clinical Aspects of Pteridines, Vol. 2

antigens and secretion of pteridines (2,3) represents a new and exciting finding. Based on these results we discussed the use of urinary neopterin excretion as a marker for cell mediated immune response at the 1st Winterworkshop 1982 (4). We concluded that neopterin estimation should be useful in the clinical monitoring of diseases which are associated with or caused by activation of T-lymphocytes. Disorders fulfilling these prerequisites are:

- allograft rejections
- virus infections
- infections with intracellular microorganisms
- cutaneous delayed hypersensitivity
- graft versus host disease (bone marrow transplantation)
- autoimmune diseases
- certain neoplasias

Whereas tuberculosis is one of the oldest life-threatening infectious diseases it is still affecting humans throughout the world. Pulmonary tuberculosis continues to pose a major health problem especially for many third world countries although an effective vaccine and assorted chemotherapeutic agents are at hand (5). Due to closer economic communication with these countries tuberculosis seems to be increasing again slightly in the rest of the world. Tuberculosis is a disease caused by infection with mycobacterium tuberculosis. This pathogen is preferentially eliminated by cellular immune reaction. The present study addresses the question of the clinical usefulness of the neopterin marker for monitoring of patients suffering from pulmonary tuberculosis.

Patients and Methods

Urinary neopterin determinations were performed in early

morning urine of 52 patients with acute pulmonary tuberculosis, aged from 16 to 94 years; 417 healthy volunteers served as controls (6). The chosen upper confidence limits were slightly dependent on age and sex and included, with 95% probability, 97.5% of healthy subjects (table 1).

Table 1: Upper confidence limits of urinary neopterin for different sex and age groups (6)

AGE years	NEOPTERIN (µMOL/MOL CREATININE) limit of tolerance* ♂	♀
18-25	195	208
26-35	182	209
36-45	176	239
46-55	197	229
56-65	218	249
>65	229	251

* tested according to Lieberman

Neopterin was measured by high performance liquid chromatography (HPLC) on reversed phase; neopterin levels were related to creatinine concentrations measured simultaneously. It should be mentioned that there is no oxidation step during sample preparation. The data shown thus represent "native" or "true" neopterin (7). In parallel, the extent of disease was determined by physical examination and chest Xray. Additional tests included evaluation of total and differential white blood cell counts of erythrocyte sedimentation rate, serum electrophoresis, chemical and/or microscopic investigation of urine, sputum and serum. Patients were treated for pulmonary tuberculosis with standardized therapy (8) using Isoniazid, Ethambutol and Rifampicin.

Table 2: Neopterin excretion of patients suffering from pulmonary tuberculosis

number of patients	number of observations	above the upper normal limit	NEOPTERIN (µmol/mol creatinine) $\bar{x}$	∓ S.D.	sensitivity
52	109	94	365	216	86%

52 patients were evaluated repeatedly during the active stages of their disease; results are composed in table 2. Patients with acute pulmonary tuberculosis show highly significantly elevated neopterin levels if compared with normal controls ($p < 0.0005$). In the following chapters neopterin levels were correlated with clinical parameters.

Correlation between neopterin excretion and radiological findings

Follow up studies of chest Xray (CR_x) represent a reliable means to assess disease activity in patients suffering from pulmonary tuberculosis. In figure 1 neopterin levels in relation to Xray findings are demonstrated. Two different schemes of classification are applied:

1. The scheme of the American Sanatorium Association of National Tuberculosis as follows:

   stage 1 indicating extent less than the equivalent of one lung top

   stage 2 an extent less than the equivalent of one lung

   stage 3 involvement of more than one lung or cavities with more than 4 cm diameter

2. Activity of disease was grouped according to the qualitative description of the Xray findings.

Highly significantly increased mean neopterin levels ($p < 0.0005$) and high percentage of levels above the upper normal limits were seen in each group. Moreover, a significant and positive correlation was observed between neopterin levels and the extent or activity of the disease. The mean neopterin level of activity stage A (exudative) is highly significantly raised over that of activity stage B ($p < 0.0005$). Statistically a difference between extent 1 and 2 and between extent 2 and 3 is probable

Figure 1: Urinary neopterin levels of patients with active pulmonary tuberculosis in relation to the extent (left side) and the activity (right side) of disease.

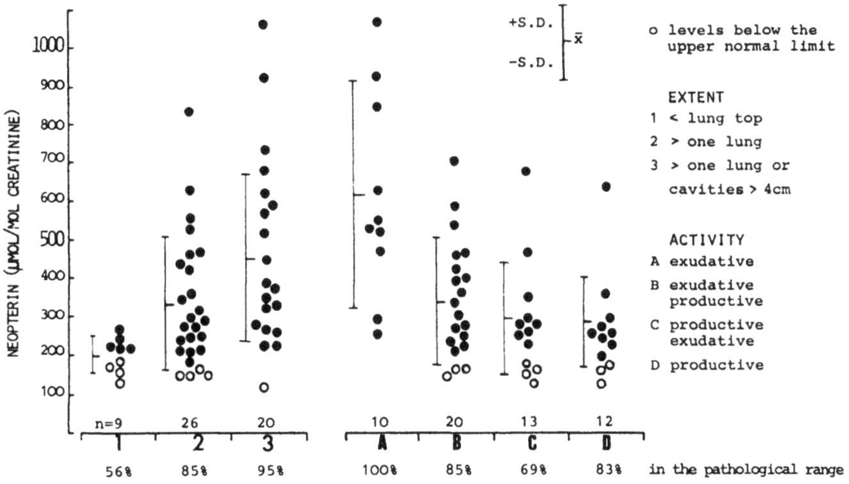

($p < 0.0125$ and $p < 0.125$). No difference is manifest between activity stages B:C and C:D ($p < 0.3$ and $P < 0.495$).

Relationship between neopterin excretion and erythrocyte sedimentation rate and between serum electrophoresis

Evaluation of blood sedimentation rate and of serum electrophoresis patterns represent other useful tools in the clinical monitoring of these diseases.

In table 3 the correlations between urinary neopterin levels and erythrocyte sedimentation rate (ESR) or serum electrophoresis are demonstrated. The mean neopterin levels are found to be elevated in all groups. In addition a positive correlation between neopterin levels and the extent of elevation of ESR and of α_2 and γ globulins was observed. In the case of ESR, patients were placed into three groups: those with normal,

Table 3: Urinary neopterin levels of patients with active pulmonary tuberculosis in relation to erythrocyte sedimentation rate and to serum electrophoresis

	ERYTHROCYTE SEDIMENTATION RATE (mm/hour)			SERUM ELECTROPHORESIS (rel.%)			
				α_2 [a]		γ [b]	
	≤10	11-30	>30	<13	>13	<25	>25
NUMBER OF PATIENTS	33	21	33	11	9	9	11
MEAN NEOPTERIN LEVELS (µmol/mol creatinine)	247	346	483	354	548	366	507
STANDARD DEVIATION (µmol/mol creatinine)	37	76	222	107	172	120	181
p <	0.0005[c]	0.0005[d]	0.005[d]	0.0005[c]	0.0005[d]	0.0005[c]	0.0005[d]
SENSITIVITY % [e]	76	91	97	100	100	100	100

a) normal range: 5-9%

b) normal range: 10-20%

c) the p-values were obtained by the student's t-test when mean neopterin values were compared with the control group

d) the p-values were obtained by the student's t-test, mean neopterin values were compared with the group presenting lower levels of the compared parameter

e) levels above the upper confidence limits

those with moderately high and those with the highest rate. The neopterin mean value of the group with a normal ESR (≤ 10 mm/h) is highly significantly lower than that of the group with a moderately elevated ESR. The group with the highest ESR (> 30 mm/h) differs significantly from the middle group. Neopterin levels above the upper limit of normal were seen in 25 out of 33 patients with acute tuberculosis but with normal ESR. Sensitivity of the neopterin assay is even better in patients with abnormal ESR; in the group of ESR above 30 mm/h only one patient out of 33 showed a normal neopterin value.

A similar relationship was also observed to exist between neopterin levels and extent of α_2- and γ-globulin elevation. All patients showed neopterin excretion above the confidence limits. All groups were highly significantly elevated compared with normal controls. In both cases (α_2 and γ) the group with the higher content of the protein fraction is highly significantly different from the lower group.

Relationship of neopterin excretion to leukocyte count and staff cell granulocyte percentage (SCG)

Evaluation of total and differential white blood cell counts represent further means to assess disease activity. Best results were also obtained when neopterin levels were compared to these parameters (SCG). Patients with a normal leukocyte count show a significantly lower neopterin level than patients with more than 10 x 10^9 leukocytes per liter. Both groups, however, are high significantly elevated when compared with the normal controls (table 4). All patients with abnormally elevated leukocyte counts also showed elevated neopterin levels.

Patients with acute tuberculosis but with a SCG within the normal range (≤ 5%) show significantly lower neopterin levels than those with an abnormal differential blood picture. Both groups were significantly elevated compared with the controls. Two out of 35 patients with abnormal and 6 out of 30 with normal SCG, but with active disease status, had normal neopterin excretion.

Table 4: Urinary neopterin levels of patients with active pulmonary tuberculosis in comparison with leukocyte count and percentage of staff cell granulocytes

	LEUKOCYTES/LITER		STAFF CELL GRANULOCYTES (percentage)	
	$\leq 10\cdot10^9$	$>10\cdot10^9$	0–5	>5
NUMBER OF PATIENTS	53	13	30	35
MEAN NEOPTERIN LEVELS (µmol/mol creatinine)	302	521	292	404
STANDARD DEVIATION (µmol/mol creatinine)	161	280	157	224
p <	0.0005[a)]	0.0005[b)]	0.0005[b)]	0.0125[b)]
SENSITIVITY % [c)]	85	100	80	95

[a)] p-values were obtained by the student's t-test when mean neopterin levels were compared with the control group

[b)] p-values were obtained by the student's t-test when mean neopterin levels were compared with the group presenting lower levels of the studied parameter

[c)] levels above the upper confidence limits

Table 5: Sensitivity of neopterin evaluation compared with erythrocyte sedimentation rate (ESR), leukocyte count (LC) and percentage of staff cell granulocytes (SCG)

	ESR	NEOPTERIN	LC	NEOPTERIN	SCG	NEOPTERIN
NUMBER OF PATIENTS	87	87	66	66	65	65
NUMBER OF PATIENTS WITH ELEVATED LEVELS	54	76[a]	13	58[a]	35	58[a]
SENSITIVITY %	70	88[a]	20	88[a]	54	89[a]

[a] above the upper confidence limits

Sensitivity of the neopterin marker in comparison with the well established laboratory parameters

Simultaneous estimations of urinary neopterin, of erythrocyte sedimentation rate and of staff cell granulocyte percentage were used to assess the sensitivities of these parameters to define clinically active disease. Data are summarized in table 5. The highest sensitivity was observed for neopterin with 88% elevated levels in patients with acute pulmonary tuberculosis.

These studies demonstrate that neopterin levels in the majority of patients with active pulmonary tuberculosis are elevated above the upper normal limit. Neopterin values are positively correlated to the extent and to the activity of disease.

Neopterin follow-up of patients suffering from tuberculosis

Detailed long term follow-up of neopterin levels was performed in twelve patients. Two representative courses from these patients are discussed in detail.

Patient 1: A 50-year-old man presented with extensive exudative productive tuberculosis of both upper lobes, bilateral cavita-

Figure 2: Urinary neopterin levels compared with the clinical course of patients with pulmonary tuberculosis

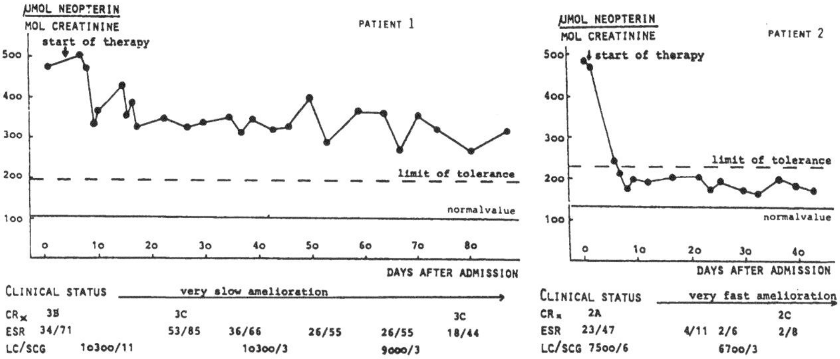

tion and numerous spreads in both lungs. During the last six months he had lost 7 kg in weight, had had night sweats, cough and sputum, but no fever. Following treatment his clinical condition improved very slowly. His weight increased within ten weeks by 4 kg. In accordance with the neopterin levels chest Xray, blood count, erythrocyte sedimentation rate and serum electrophoresis still indicated active disease refractory to treatment. No neopterin value was found below the upper normal limit.

Patient 2: A 16-year-old man exhibited exudative tuberculosis of the right upper lobe with cavitation and fresh spreads in the left central field. Three weeks before admission he had had fever of up to 39°C, 3 kg weight loss, chest pain and cough with yellow sputum. Following therapy a rapid improvement of clinical symptoms was observed. The patient soon became afebrile, put on 2 kg in weight in two weeks and the cough was distinctly reduced. Similarly, chest Xray and laboratory parameters exhibited fast amelioration during the 40 day observation period. It can be seen that neopterin levels decrease even faster to normal than the Xray and the other laboratory data, suggesting an even better correlation of the neopterin marker with the clinical course. Effectiveness of treatment and the normalization of laboratory parameters could be predicted by urinary neopterin levels below the upper normal limit within the first week after therapy was started.

Similar results were obtained from further ten patients studied under treatment; when disease activity improved slowly neopterin levels fluctuated in the higher range. Conversely, fast response to therapy was indicated by sharp drops in urinary neopterin levels. It is of note that a marked decrease of urinary neopterin occurs before changes of other laboratory parameters, particularly Xray findings, are detectable.

Conclusion

Our results indicate that the neopterin evaluation in patients suffering from tuberculosis is sufficiently sensitive to detect active disease. Moreover, changes in disease activity are indicated earlier by this marker than by currently employed radiological or laboratory methods. From our experience it appears that neopterin levels might reflect the disease state rapidly and accurately. This was particularly seen in long term follow-up studies where a good correlation between neopterin levels and clinical course was obtained. It is noteworthy to mention that so far no specific biochemical marker for staging and classifying pulmonary tuberculosis exists. Chest Xray, as a quick method and a reliable parameter for diagnosing and monitoring of disease activity, is usually employed only once a month to keep Xray-burden low. Neopterin, a cheap and safe biochemical parameter, seems to reflect changes of the disease status earlier and more accurately than currently applied parameters.

Although useful in the monitoring of pulmonary tuberculosis, elevation of neopterin levels is not specific for tuberculosis. Disorders associated with or caused by a cell-mediated immune response are invariably accompanied by enhanced urinary neopterin levels.

References

1. Wachter, H., Fuchs, D., Hausen, A., Reibnegger, G.: Chemistry and Biology of Pteridines, ed. J.A. Blair, W. de Gruyter, Berlin-New York(1983),in press.

2. Fuchs, D., Hausen, A., Huber, Ch., Margreiter, R., Reibnegger, G., Spielberger, M., Wachter, H.: Hoppe Seyler's Z. Physiol. Chem. 363, 661-664 (1982).

3. Huber, Ch., Fuchs, D., Hausen, A., Margreiter, R., Reibnegger, G., Spielberger, M., Wachter, H.: J. Immunol. 130, 1047-1050 (1983).

4. Wachter, H., Hausen, A., Fuchs, D., Reibnegger, G.: Biochemical and Clinical Aspects of Pteridines, eds: H. Wachter, H.Ch. Curtius, W. Pfleiderer, W. de Gruyter Berlin-New York 277-281 (1982).

5. Collins, F.M.: Am. Rev. Respir. Dis. 125/3, 42-49 (1982).

6. Hausen, A., Wachter, H.: J. Clin. Chem. Clin. Biochem. 20, 593-6o2 (1982).

7. Fuchs, D., Hausen, A., Reibnegger, G., Wachter, H.: Biochemical and Clinical Aspects of Pteridines, eds: H. Wachter, H.Ch. Curtius, W. Pfleiderer, W. de Gruyter Berlin-New York 67-79 (1982).

8. Urbanczik, R.: Lungentuberkulose: Pathogenese, Klinik, Therapie, Epidemiologie, Herausgeber: J. Hein, R. Ferlinz, Georg Thieme Verlag, Stuttgart-New York 8-19 (1982).

Acknowledgement

This study was financially supported by the Austrian Research Funds "Zur Förderung der wissenschaftlichen Forschung", project P 4905.

CONFIGURATION OF BIOPTERIN AND NEOPTERIN EXCRETED BY PATIENTS WITH VARIOUS DISEASES

Hartmut Rokos, Katharina Rokos
Forschungslabor, Henning Berlin GmbH
D-1000 Berlin 42

Irmgard Ziegler
Institut für Toxikologie und Biochemie der Gesellschaft für Strahlen- und Umweltforschung
D-8000 München 2

Introduction

Many recent reports have shown alterations in the excretion of pteridines, e.g. biopterin and neopterin, in patients with a variety of diseases. However, little is known of the configuration of the side chain in the 6-position of the pterin ring, because the method which was mostly used for their determination - HPLC - does not discriminate between the enantiomers. D-*erythro*- and L-*erythro*-neopterin as well as L-*erythro*- and D-*erythro*-biopterin (Figure 1) are not separated on HPLC. The *threo* forms which have the hydroxy groups in trans configuration are diastereomers of the *erythro*-neopterins and are well separated from these in HPLC. It has been previously shown (1) that healthy persons excrete L-*erythro*-biopterin, D-*erythro*-neopterin and L-*threo*-neopterin. Besides "*erythro*"-neopterin - no differentiation between the D- and L-form could be made by HPLC - also "*threo*"-neopterin was excreted in patients with various diseases to an extent of about 10-15 % of "*erythro*"-neopterin irrespective of the absolute quantities which varied greatly (2).
In the biosynthetic pathway of biopterin changes of the configuration of the side chain occur: GTP has a D-*erythro*-ribose

Biochemical and Clinical Aspects of Pteridines, Vol. 2

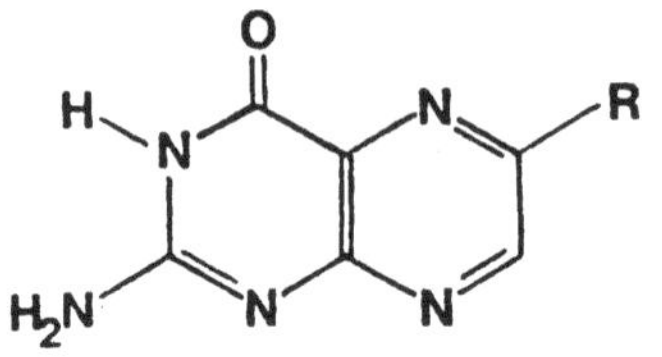

	R	Configuration
L-e-Biopterin	H H —C — C — CH_3 OH OH	L-*erythro*
D-e-Neopterin	OH OH —C — C — CH_2OH H H	D-*erythro*
L-e-Neopterin	H H —C — C — CH_2OH OH OH	L-*erythro*
D-e-Biopterin	OH OH —C — C — CH_3 H H	D-*erythro*
L-t-Neopterin	OH H —C — C — CH_2OH H OH	L-*threo*
D-t-Neopterin	H OH —C — C — CH_2OH OH H	D-*threo*

Figure 1. Stereoisomers of biopterin and neopterin

moiety, which is retained in D-erythro-dihydroneopterin triphosphate. The primary hydroxy group in 3'-position and the 5.6-double bond are reduced, and both chiral centres of the side chain are inverted, yielding tetrahydrobiopterin, the cofactor of the aromatic hydroxylases, probably via several intermediates. How L-threo-neopterin is formed, remains unknown.

This pathway was established for normal persons; it is not known whether the same mechanism applies in diseases with pathological pteridine excretion. The biochemical pathways may be partly altered in cancer cells or in cells transformed by viruses, in which the biosynthesis of pteridines has not yet been investigated. Therefore, it might be premature to assume without further evidence the same configuration of the neopterin and biopterin excreted in pathological quantities.

Our aim was to develop a radioimmunoassay for neopterin, as this seemed to have the best diagnostic potential of the pteridines examined. For this purpose we had to be certain which isomer was elevated in various diseases, as we expected from our experience in immunoassays to be able to obtain antibodies which could discriminate between all the possible isomers.

Results and Discussion

For discrimination of the isomers of biopterin and neopterin, the bioassay with Crithidia fasciculata was used (3). These flagellata require biopterin or related pterins for growth. The concentration of some pteridines needed for half-maximal stimulation of growth vary greatly according to the configuration of the side chain (4,5): L-erythro-compounds are needed in the ng/l range, whereas D-erythro-compounds are required in the µg/l range (Table 1). Therefore, the L/D-configuration of the side chain of biopterin and neopterin isolated from urine of patients can be determined with the bioassay.

	effective concentration (ng/l)	
Biopterin (L-e-B)	30	
D-*erythro*-Biopterin	5000	
Neopterin (D-e-N)	10000	10000[a]
L-*erythro*-Neopterin	50	30[a]
L-*threo*-Neopterin		2000[a]
D-*threo*-Neopterin		10000[a]
6-Hydroxymethylpterin	400	

[a] reference (4)

Table 1. Effective concentration needed for half-maximal stimulation of the growth of *Crithidia fasciculata*

Urinary pteridines were separated and purified by preparative HPLC (2,6,7). Repeated purifications were run until the absorption spectra showed maximum to minimum ratios comparable to pure authentic standards. Then they were quantified by absorption and fluorescence spectroscopy and tested in the bioassay. Later the radioimmunoassay (8) was used, too, which was developed after the first results with the bioassay.

As previously shown (2), there are two distinct patterns of abnormal pteridine excretion in cancer patients: 1.) high neopterin/low biopterin excretion with high N/B-ratios and 2.) normal neopterin/high biopterin excretion with low N/B-ratios. Results of the bioassay of one typical patient each of these two groups (osteomyelofibrosis and breast cancer, respectively) are given in Table 2. The *Crithidia* test shows unambigiously that the forms of biopterin and neopterin excreted by these cancer patients are the same as in controls, that is L-*erythro*-biopterin and D-*erythro*-neopterin.

	Absorption[a]	Fluorescence[a]	Crithidia[b]
Biopterin			
sample 1*	(100)	92	90
sample 2**	(100)	98	100
Neopterin			
sample 1	(100)	96	100
sample 2	(100)	102	100

a arbitrary units, the absorption value is taken as 100 %

b determination of half-maximal stimulation of growth of Crithidia fasciculata (ATCC 12857) in comparison to the standard L-e-B or D-e-N, respectively

* breast cancer (N 340, B 1332, µmol/mol creatinine)

** osteomyeolofibrosis (N 2101, B 398, µmol/mol creatinine)

Table 2. Quantification of urinary biopterin and neopterin by optical data and bioassay

Therefore, a radioimmunoassay for D-*erythro*-neopterin was developed (8). N^2-Carboxypropylneopterin was used as actual hapten, its radioiodinated tyramine derivative as the tracer (Figure 2). We choose the carboxypropyl group as spacer to the immunogenic protein as this - known from our studies with hormones - should give antibodies which would not bind the tracer too tightly and would consequently give a sensitive assay.

RIA-Incubation: 100 µl sample (in buffer pH 6.5)
100 µl N-cp-t(^{125}I)
100 µl pre-precipitated antibody complex
2 h at 37 °C

As shown by the cross-reactivities with various structurally related pteridines (Figure 3), the radioimmunoassay is highly specific for D-*erythro*-neopterin.

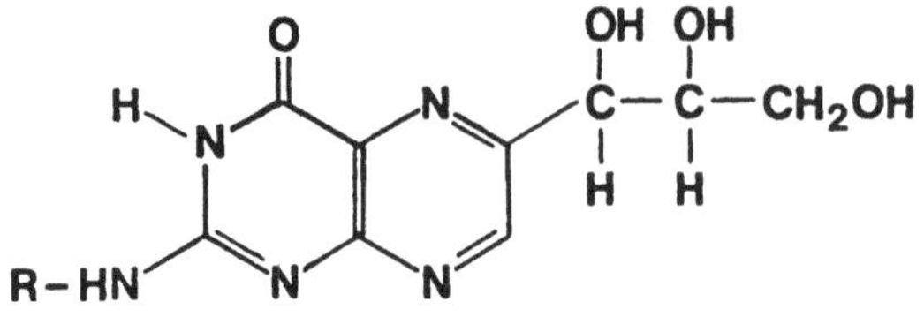

	R
N-cp	$-(CH_2)_3-COOH$
N-cp-t	$-(CH_2)_3-CO-NH-(CH_2)_2-$ 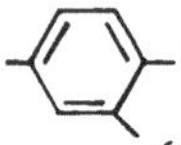 $-OH$
N-cp-t-(^{125}I)	(^{125}I)
N-cp-Protein	$-(CH_2)_3-CO-NH-$ Protein

Fig. 2. Neopterin derivatives for the radioimmunoassay

	Neopterin HPLC µg/l	fluorescence (100 %)	RIA %
Cancer			
Breast ca	1440	100	107
Bronchus ca	2350	100	101
Plasmocytoma	5000	100	96
AML	10700	100	98
Various diseases			
Renal insufficiency	880	100	106
Pyelonephritis	2070	100	101
Viral infection	2450	100	103
PKU	1080	100	97

Table 3. Determination of erythro-neopterin from urine of patients, isolated by preparative HPLC, by fluorescence spectroscopy (set at 100 %) and radioimmunoassay.

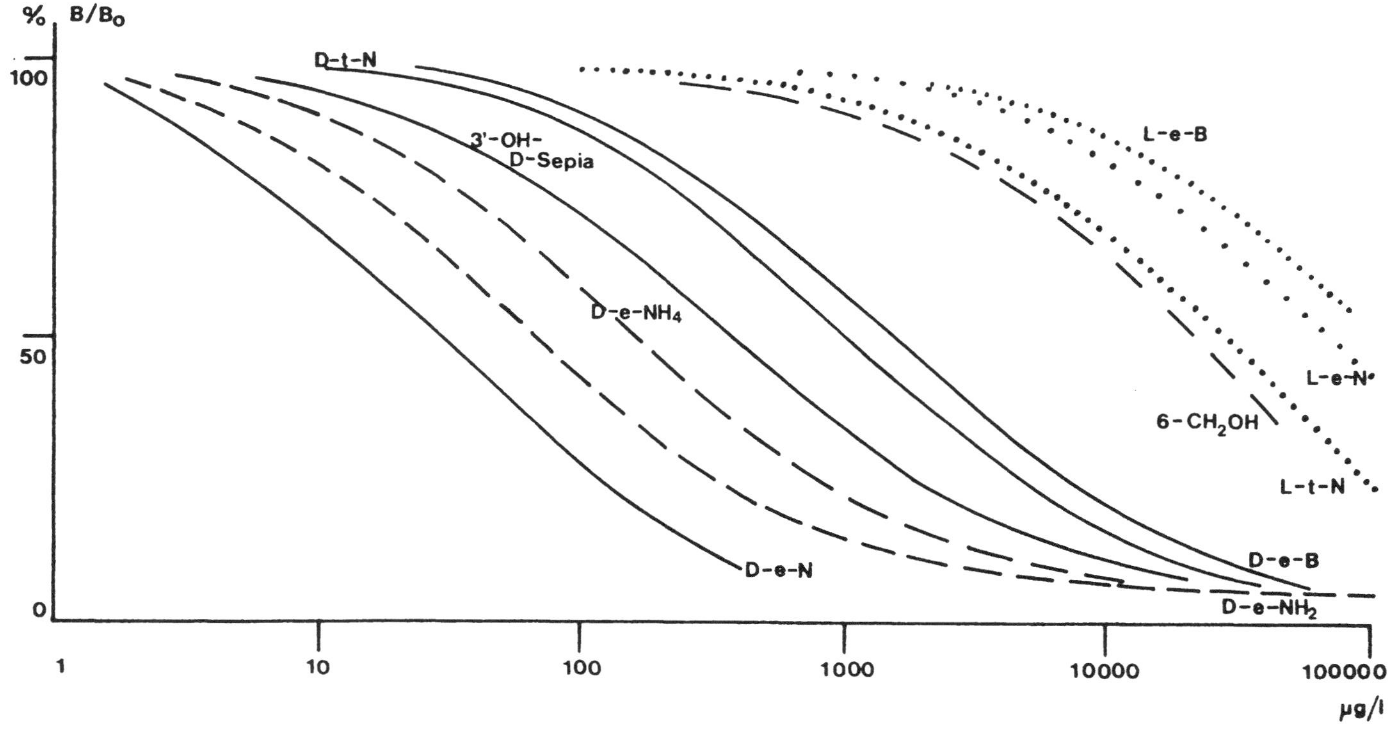

Figure 3. Cross-reactivities of various 6-substituted pterins with the D-*erythro*-neopterin antiserum used in the radioimmunoassay

In Table 3 data of patients with various diseases are given, who excrete quite different amounts of neopterin. After HPLC isolation of the *erythro*-neopterin, the radioimmunoassay showed that most, if not all, *erythro*-neopterin is excreted as the D-enantiomer by these patients.

Conclusion

It was shown by *Crithidia fasciculata* bioassay and radioimmunoassay for D-*erythro*-neopterin that patients with various diseases excrete the same isomers of neopterin and biopterin as healthy controls, L-*erythro*-biopterin and D-*erythro*-neopterin. Therefore, the pathway of biopterin synthesis seems to be not principally altered in these patients, despite abnormal levels of some pteridines in urine and serum.

References

1. Fukushima, T., Shiota, T.: J. Biol. Chem. 247, 4549-4556 (1972).
2. Rokos, H., Rokos, K., Frisius, H., Kirstaedter, H.-J.: Clin. Chim. Acta 105, 275-286 (1980).
3. Baker, H., Frank, O., Shapiro, A., Hutner, S.H.: In Methods of Enzymology Vol. 66, Eds.: McCormick, D.B. and Wright, L.D.: 490-500, Academic Press, New York (1980).
4. Rembold, H., Buschmann, L.: Chem. Ber. 96, 1406-1410 (1963).
5. Ziegler, I., Böhme, M., Pfleiderer, W., Elstner, E.F.: J. Protozool. 28, 354-357 (1981).
6. Rokos, K., Rokos, H., Frisius, H.: In Biochemical and Clinical Aspects of Pteridines Vol. 1, Eds.: Wachter, H., Curtius, H.-Ch., Pfleiderer, W.: 117-130, de Gruyter, Berlin (1982).
7. Fukushima, T., Nixon, J.C.: Anal. Biochemistry 102, 176-188 (1980).
8. Rokos, H., Rokos, K.: In Chemistry and Biology of Pteridines, Ed.: Blair, J.A., de Gruyter, Berlin (in press).

5

TETRAHYDROBIOPTERIN DEFICIENCIES, DIAGNOSIS AND THERAPY

TETRAHYDROBIOPTERIN METABOLISM AND MENTAL DISEASE.

P.A. Barford, J.A.Blair, C. Eggar and C. Hamon.
Department of Chemistry, University of Aston in Birmingham

and

R.J. Leeming
General Hospital, Steelhouse Lane, Birmingham.

Introduction

Tetrahydrobiopterin is an essential cofactor for the hydroxylation of phenylalanine to tyrosine, tyrosine to dopa and tryptophan to 5-hydroxytryptophan. Its essential role in neurotransmitter biosynthesis is clearly demonstrated in a rare genetic disorder, malignant hyperphenylalaninaemia, in which there is a severe deficiency of tetrahydrobioptern (1). Affected infants fail to develop normally, have hyperphenylalaninaemia, and usually die within a few years of birth, although several cases are now known in which patients have survived to adolescence, but are severely mentally retarded. The parents of affected children are normal. In addition several cases of a partial failure to synthesise tetrahydrobiopterin are known (2), some of whom are mentally retarded. It is, therefore, apparent that although a complete lack of tetrahydrobiopterin always results in mental retardation, it does not always result in early death, and that a partial lack of tetrahydrobiopterin may result in mental retardation.

Previous work has shown abnormalities in tetrahydrobiopterin metabolism in patients with certain types of mental retardation. Thus in a group of patients with Down's syndrome serum biopterin concentrations were significantly higher than normal (3). A group of non specific mentally retarded patients from the same hospital had serum biopterin concentrations significantly lower than normal (3).

Biochemical and Clinical Aspects of Pteridines, Vol. 2

Patients with untreated classical phenylketonuria due to phenylalanine hydroxylase deficiency have significantly increased serum biopterin levels and Blair and Leeming (4) have suggested that a disturbance in tetrahydrobiopterin metabolism is responsible for the mental retardation seen in untreated phenylketonuria.

A disturbance in tetrahydrobiopterin metabolism is also seen in lead poisoning. Purdy et al. (5) have shown that lead significantly inhibits both tetrahydrobiopterin synthesis and dihydropteridine reductase (DHPR) in preparations from the rat. Inhibition of both systems by lead would result in lowered cellular levels of tetrahydrobiopterin and could contribute to the neurotoxicity of lead. Although serum biopterin levels rise with age (6), they are significantly lower in the serum of patients with senile dementia of the Alzheimer type (SDAT) than in age matched controls (7, 8). Lowered biopterin concentrations are also seen in CSF from SDAT patients (9). There is therefore evidence of an abnormality in tetrahydrobiopterin metabolism in SDAT.

It is, therefore, apparent that cellular tetrahydrobiopterin concentrations play a critical role in mental health in man. Any abnormality resulting in decreased cellular tetrahydrobiopterin concentrations would lead to mental disease or mental retardation. Cellular tetrahydrobiopterin concentrations could be reduced in two ways, either by decreasing the synthesis of tetrahydrobiopterin or by inhibiting the salvage of quinonoid dihydrobiopterin by DHPR. Purdy and Blair (10) have shown that phenylpyruvate, produced in large amounts in phenylketonuria, inhibits dihydropteridine reductase. Plasma phenylalanine levels correlate with plasma biopterin concentrations, and inversely with I.Q. score in phenylketonuria (11). These findings provide a model for mental retardation caused by depletion of cellular tetrahydrobiopterin concentrations by inhibition of dihydropteridine reductase.

The results presented in this paper represent a study of tetrahydrobiopterin metabolism in normal man and man with various conditions which result in mental impairment. In addition we report the effect of oestrogens on dihydropteridine reductase activity in man. Leeming and Blair (6) reported that serum biopterin concentrations fall when oestrogen levels rise, thus small changes in performance which occur in the menstrual cycle (12) may be related to changes in tetrahydrobiopterin metabolism.

Materials and Methods

1) Assay of dihydropteridine reductase activity in whole blood and from Guthrie card spots. DHPR activity was assayed by the method of Marisawa et al (13). Each incubation contained 10^{-4}M NADH, 5 x 10^{-5}M ferricytochrome C, 5 x 10^{-2}M tris buffer pH 7.4, 10^{-5}M 6-methyltetrahydropterin $\pm$ 0.1ml of the enzyme solution in a total volume of 1ml. For controls assays were run without tetrahydropterin. Whole blood samples were diluted x 10 with distilled water for assay. Under these conditions no DHPR activity could be detected in white cells, or platelets. For Guthrie card spots 2 discs of 8mm diameter were punched from the spots and eluted for 45 minutes at room temperature in 0.5ml distilled water. At the end of this time the filter paper discs were removed and the eluate stored on ice until required. 0.1ml of the eluate was used in the assay as described above. Haemoglobin concentrations in diluted blood samples were measured by the method of Drabkin (14).

2) Assay of dihydropteridine reductase activity in rat livers and human and rat brains. DHPR activity was measured by the method of Craine et al (15). Each incubation contained 10^{-4}M NADH, 10^{-3}M H_2O_2, 8 µg horseradish peroxidase, 2.5 x 10^{-4}M sodium azide $\pm$ 10^{-4}M 6.7 - dimethyltetrahydropterin in 0.05M tris buffer pH 7.6 and 0.02ml of the appropriate tissue extract in a total volume of 1ml. For controls assays were run without 6,7 - dimethyltetrahydropterin.

3) Measurement of tetrahydrobiopterin biosynthesis. Tetrahydrobiopterin synthesis in tissue preparations was measured by the method of Brown (16). Each incubation contained 6mM GTP, 3mM NADPH, 1mM $MgCl_2$ and 0.1ml of the appropriate tissue extract in 0.75mM Tris buffer pH 8.0 in a total volume of 1ml. For controls incubations were carried out in the absence of GTP and/or NADPH. Incubation was for 3h at 37°C. The reaction was stopped with 2ml of 0.1M.HCl and reduced biopterins oxidised with I_2 (Fukushima and Nixon, 17). Total biopterin was measured by HPLC.

4) Chronic lead toxicity in rats. Immediately after mating female rats were kept in cages and supplied with drinking water containing lead (300ppm, and 1000ppm). Female rats were given only leaded water until the offspring were 21 days old, at which stage the young were removed and killed (18). Brain extracts were prepared as described below and used for assay of DHPR activity and measurement of tetrahydrobiopterin biosynthesis.

5) Oestrogen effects in rats. Weanling female rats (100g body wt) were dosed with diethylstilboestrol in corn oil (50 mg/kg body wt), for 3 days. On day 5 the animals were killed and livers removed and tissue extracts prepared as described. For controls animals were dosed with corn oil alone.

6) Preparation of tissue extracts. Brains from rats drinking leaded water were obtained from Carmichael et al (18) and stored at -70^{0}C until required. Human brain samples were supplied by the MRC brain bank and stored at −70^{0}C until required. Livers from rats dosed with diethylstilboestrol were removed into ice cold beakers. 20% homogenates in 0.5M Tris buffer pH 7.6 were prepared from all tissues and centrifuged at 40,000 Xg for 40 mins. Supernatants were stored frozen until required. The supernatants were used for all assays.

7) Protein estimations. Proteins were estimated by the biuret method.

Results

1) Assay of dihydropteridine reductase activity from Guthrie card blood spots. Guthrie cards spotted with blood samples from normal adults, DHPR deficient children, heterozygotes for DHPR deficiency and neonates with hyperphenylalaninaemia were eluted and assayed for DHPR activity as described. (Table 1). Sixteen normal adults had mean DHPR activity of 0.96 units. In children aged 1 to 12 years with classical phenylketonuria the level of activity was slightly lower, but not significantly so. Neonates with hyperphenylalaninaemia had substantially more reductase activity, but this

was most probably due to the increased red cell content in blood of neonates. Children who lack DHPR had no detectable activity on Guthrie cards. Parents of reductase dificient children had reductase activity on Guthrie cards which was less than half that of normal adults ($P < 0.1\%$).

Table 1. Dihydropteridine reductase activity assayed from Guthrie card blood spots. For details see text. Number of subjects in brackets.

	NADH used ηmol/min mean ± standard deviation
Normal adults (16)	0.96 ± 0.20
Children 1-12 years with classical phenylketonuria (9)	0.84 ± 0.14
Neonates with hyperphenylalaninaemia (23)	1.19 ± 0.38
Known adult heterozygotes (3)	0.34 ± 0.04
Reductase deficient (3)	not detectable

Table 2. Dihydropteridine reductase activity in whole blood. For details seetext.

	rate NADH used η mol/min x mg haemoglobin. mean ± standard deviation
Normal males (18)	0.84 ± 0.17
Normal females (13)	0.93 ± 0.23
Male heterozygotes (1)	0.30
Female heterozygotes (2)	0.45 ± 0.05
Dihydropteridine reductase deficient (2)	not detectable

Table 3. Dihydropteridine reductase activity in whole blood from patients with mental retardation. For details see text.

	rate NADH used ηmol/min x mg haemoglobin. mean ± standard deviation
Untreated adults with phenylketonuria	
males (5)	0.96 ± 0.18
females (5)	0.99 ± 0.15
Adults with fragile X chromosome	
males (8)	0.85 ± 0.16
Adults with Down's syndrome	
males (6)	0.98 ± 0.15
females (5)	0.99 ± 0.15
Children with mental retardation of unknown cause (5)	1.08 ± 0.14

2) Assay of dihydropteridine reductase activity in whole blood. DHPR activity was assayed in whole blood of normal adults, reductase deficient children, parents of reductase deficient children and in adults with mental retardation from different causes (tables 2 and 3). Normal adult males have a mean DHPR activity of 0.84 NADH usedηmol/min x mg haemoglobin. The activity in the blood of normal adult females is higher but not significantly so. Heterozygotes have whole blood DHPR activity less than half that of normal adults ($P < 1\%$), and no activity was detected in blood from children with dihydropteridine reductase deficiency. DHPR activity in the blood of untreated adults with classical phenylketonuria was higher than that found in normal adults, but not significantly so. In adults with fragile X chromosome, and with Down's syndrome DHPR levels were not significantly different from normal. Children with mental retardation of unknown cause had higher reductase levels in blood than normal adults, but this is most probably due to the lower haemoglobin content of red cells in children.

Table 4. Dihydropteridine reductase activity in whole blood from workers in the lead industry. For details see text. Number of subjects in brackets.

Blood lead level µg/l	rate NADH used ηmol/min x mg haemoglobin. mean ± standard deviation
A. < 300 (6)	0.93 ± 0.11
B. 300 - 500 (33)	0.93 ± 0.20
C. > 500 (18)	1.02 ± 0.18

A. V/S normal males not significant

B. V/S normal males not significant

C. V/S normal males $p < 1\%$

Free erythrocyte porphyrin µ g/l	
C. 600 (16)	0.94 ± 0.15
D. 600 - 1000 (8)	1.00 ± 0.17
E. 1000 (14)	1.00 ± 0.17

C. V/S normal males not significant

D. V/S normal males $p < 5\%$

E. V/S normal males $p < 1\%$

3) <u>The effect of lead on tetrahydrobiopterin metabolism</u>. DHPR activity was measured in whole blood samples from workers in the lead industry, together with blood lead levels and free erythrocyte porphyrin (table 4). DHPR activity is significantly elevated in the blood of workers in the lead industry when blood lead levels are greater than 500 µg/l ($p < 1\%$). DHPR levels are significantly elevated at free erythrocyte porphyrin levels (a better indicator of lead toxicity) between 600-1000 µg/l ($p < 5\%$) and at free erythrocyte porphyrin levels greater than 1000 µg/l ($p < 1\%$). When pregnant female rats are maintained on water containing lead (300ppm and 1000ppm) during pregnancy and until the offspring are 21 days old, then DHPR activity in the brains of the offspring is significantly decreased at both lead levels. ($p < 1\%$) (table 5).

Table 5. Dihydropteridine reductase activity in the brains of rats born to mothers drinking leaded water. Number of animals in brackets.

[Lead] in water ppm.	Dihydropteridine reductase activity NADH ηmol/min x mg protein. mean ± standard deviation.
0 (6)	224.8 ± 35.0
300 (6)	177.9 ± 21.6
1000 (6)	177.4 ± 43.1

4) <u>The effect of oestrogens on dihydropteridine reductase activity in vivo.</u> DHPR is significantly elevated in the livers of rats dosed with diethylstilboestrol compared to control rats ($p < 0.1\%$) (table 6). DHPR activity was measured in blood samples of males, females, females taking the oral contraceptive pill and pregnant females (table 7).

Table 6. The effect of oestrogen on dihydropteridine reductase activity in rat livers. For details see text. Number of animals in brackets.

	Dihydropteridine reductase activity rate NADH ηmol/min x mg protein. mean ± standard deviation.
controls (5)	368.4 ± 72.82
Diethylstilboestrol treated (5)	637.3 ± 141.0

Table 7. Dihydropteridine reductase activity and total biopterin concentrations (by Crithidia assay) in blood of males, females, females on the oral contraceptive pill and pregnant females. Number of subjects in brackets.

	DHPR activity rate NADH ηmol/min x mg haemoglobin mean ± standard deviation			Total biopterin * μg/l		
Males (18)	0.84	±	0.17	1.75	±	0.03
Females (14)	0.90	±	0.25			
Females on "pill" (7)	1.10	±	0.24	1.53	±	0.04
Pregnant females (13)	1.33	±	0.23	1.20	±	0.30

* (6)

There is no significant elevation in DHPR activity between men and women, DHPR activity is significantly elevated in women on the "pill" compared to men ($p < 5\%$) but not compared to normal women. DHPR levels in pregnant women are significantly elevated compared to men and normal women ($p < 1\%$ in both cases). The rise in DHPR activity is matched by a fall in serum total biopterin (Leeming).

5) <u>Tetrahydrobiopterin metabolism in human brain.</u> Tetrahydrobiopterin biosynthesis and DHPR activity were measured in human temporal lobes obtained at post mortem. (Table 8) All samples had considerable DHPR activity. The patient with Down's syndrome had severe senile dementia of the Alzheimer type, and apparently very little ability to synthesise tetrahydrobiopterin.

Table 8. Tetrahydrobiopterin synthesis and Dihydropteridine reductase activity in human temporal lobe. For details see text.

	Age in years	Tetrahydrobiopterin synthesis total biopterin ηmol/hr x mg protein	Dihydropteridine reductase activity NADH ηmol/min x mg protein
Patient La (right)	32	1.86	217
Patient La (left)	32	1.86	165
Patient Ha	78	0.43	245
Patient Th	66	0.28	246
Down's	50	0.1	395

Discussion

The results presented here demonstrate clearly that assay of DHPR activity from Guthrie card blood spots can provide an effective screen for DHPR deficiency. No special storage of Guthrie cards is required and they can be posted to laboratories where screening can be carried out. To date we have screened only the cards from neonates with hyperphenylalaninaemia, but since reductase deficient children with only mild hyperphenylalaninaemia are known (Clayton personal communication) it may be appropriate to screen all neonates.

Assay of DHPR activity in blood of normal adults has given a normal range for men and women. There is no alteration in DHPR activity in the blood of adults with phenylketonuria, Down's syndrome and fragile X-chromosome compared to normal adults. Any apparent alteration in tetrahydrobiopterin metabolism in these patients is, therefore, not caused by an alteration in DHPR activity in blood.

In workers from the lead industry DHPR activity is elevated compared to normal males and this is not due to a decrease in haemoglobin concentrations. Purdy et al. (5) reported that lead was an irreversible inhibitor of DHPR, and of tetrahydrobiopterin synthesis. The increase in DHPR activity may be a response to a decrease in tetrahydrobiopterin synthesis. In the brains of rats born to mothers drinking leaded water DHPR activity is significantly decreased compared to controls. These animals have developed in a high lead environment, and have not been exposed to high lead levels in adult life. This fact may explain the difference in results between rats born to mothers drinking leaded water and workers from the lead industry.

That DHPR activity can increase in response to physiological stimuli is shown clearly in women. DHPR activity is increased in women compared to men. In women taking the oral contraceptive pill, and pregnant women there is a further increase in activity. Thus DHPR activity increases as oestrogens increase. This finding agrees with the work of Dhondt et al. (19) who showed that DHPR levels correlate positively with oestrogen receptor levels in human breast tumours. We have shown (20) that DHPR activity is significantly increased in human breast tumours compared to apparently normal tissue from the same breast, but not in human large intestine tumours compared to normal tissue. It is, therefore, apparent that high oestrogen levels increase DHPR activity in sensitive tissues. This is confirmed by administration of a synthetic oestrogen, diethylstilboestrol, to female rats. This results in a large increase in DHPR activity in rat liver. The reason for this increase in DHPR activity is not clear. Oestrogens increase proliferation of certain tissues. Dihydropteridine reductase may play a role in folate metabolism (21) in maintaining the cellular levels of tetrahydrofolate. Oestrogens have been shown to have a pronounced effect on folate metabolism in the rat, (20) increasing the amount of 10-formylfolates and decreasing folate catabolism by scission, changes which would facilitate cellular proliferation. Thus, DHPR may be increased as part of the system which alters the folate metabolism of the cell into a form which is more suited to cellular proliferation.

The results obtained from human brains show a decrease in the ability to synthesise tetrahydrobiopterin with increasing age and activity. The Down's patient had severe SDAT and had almost lost the ability to synthesise tetrahydrobiopterin.

Total biopterin and neopterin concentrations in the cerebrospinal fluid of Down's patients are significantly below normal (9). This, together with the lack of tetrahydrobiopterin synthesis in Down's brain, suggests that the mental retardation seen in Down's patients could be related to a failure to synthesise tetrahydrobiopterin. Aziz et al (7) have shown that total biopterin concentration in the serum of patients with SDAT is significantly lower than that of age-matched controls. The lack of synthesis of tetrahydrobiopterin in the brain of the patient with severe SDAT suggests that this disease may be caused by a failure to synthesise tetrahydrobiopterin.

The results presented here do not demonstrate a role for dihydropteridine reductase in mental disease except in reductase deficient children. In the normal population decreased levels of dihydropteridine reductase in blood are not seen. The increase in dihydropteridine reductase activity in response to oestrogens and in workers from the lead industry clearly demonstrates that the activity of this enzyme can be increased in response to physiological stimuli. Studies of tetrahydrobiopterin synthesis in human brain samples strongly suggest that the failure to make tetrahydrobiopterin may contribute to the mental retardation seen in Down's syndrome, and may be involved in the aetiology of Alzheimer's disease.

Acknowledgements

The authors are grateful to Shell U.K. Ltd and Billition Ltd for financial support and to all the clinicians who have supplied samples.

References

1. Danks, D.M., Bartolome, K., Clayton, B.E., Curtius, H., Grobe, H., Kaufman, S., Leeming, R.J., Pfleiderer, W., Rembold, H. Rey, F. J. Inher. Metab. Dis. 1, 49-53 (1978).
2. Rey, F., Leeming, R.J., Blair, J.A. Rey, J. Arch. Dis. Childhood. 55, 637-639 (1980).
3. Aziz, A., Blair, J.A., Leeming, R.J., Sylvester, P.E., J. Ment. Defic. Res. 26, 67-71 (1982).
4. Blair, J.A., Leeming, R.J. In First European Symposium on Scientific Studies in Mental Retardation. (1983) in Press.
5. Purdy, S.E., Blair, J.A., Leeming, R.J., Hilburn, M.E. Intern. J. Environmental Studies, 17, 141-145 (1981).
6. Leeming, R.J., Blair, J.A. Clin. Chim. Acta. 108, 103-111 (1980).
7. Aziz, A., Leeming, R.J., Blair, J.A. J. Neurol. Neurosurgery & Psych. (1983) in Press.
8. Young, J.H., Kelly, B., Clayton, B.E. J.Clin. Expt. Gerontol (1982) in Press.
9. Morar, C., Whitburn, S.W., Blair, J.A., Leeming, R.J., Wilcock, G.K. J. Ment. Defic. Res. (1983) in Press.
10. Purdy, S.E., Blair, J.A. Biochem. Soc. Trans. 8, 565 (1980).
11. D.H.S.S. Phenylketonuria newsletter No. 8 (1982).
12. d'Orban, P.T., & Dalton, J. Psychol. Med. 10, 353-359 (1980).
13. Marisawa, K., Arai, N., Hayakawa, H., Tada, K. International Symposium on Neonatal Screening for Inborn Errors of Metabolism. p.87 (1982).
14. Drabkin, D.L., Austin, J.H. J. Biol. Chem. 98, 719-723 (1932).
15. Craine, J.E., Hall, K.S., Kaufman, S. J. Biol. Chem. 247, 6082-6091 (1972).
16. Brown, S.E. Ph.D. Thesis. University of Aston (1981).
17. Fukushima, J., Nixon, J.C. Anal. Biochem. 102, 176-188 (1980).
18. Carmichael, N.G., Winder, C., Lewis, P.D. Toxicol 21, 117-128 (1981).
19. Dhondt, J.L., Bonneterre, J., Farriaux, J.P., Lefebvre, J., Demaille, A. Biomedicine 35, 58-60 (1981).
20. Eggar, C., Barford, P.A., Blair, J.A., Pheasant, A.E., Guest, A.E., Oates, G.D. In Seventh International Symposium on Pteridines and Folic Acid Derivatives (1983) in Press.
21. Pollock, R.J., Kaufman, S. : J. Neurochem. 31, 115-123 (1978).

DIHYDROBIOPTERIN-SYNTHETASE DEFICIENCY

Diagnosis problems of two atypical cases.

Jean-Louis Dhondt, Jean-Pierre Farriaux
Institut Pasteur, 59000 Lille, France

B. Leroux
Service de Pédiatrie, Centre Hospitalier Regional,51100 Reims,France

Ian Lyon
Dept.of Community Health,School of Medicine, University of Auckland, New Zealand

Introduction

Published cases of hyperphenylalaninemia due to tetrahydrobiopterin (BH_4) deficiency have been generally diagnosed after the first three months of life, following the onset of atypical neurological signs, even if a phenylalanine-restricted diet had been instituted. The metabolic error may be a defective dihydrobiopterin synthesis (dihydrobiopterin- synthetase or GTP-cyclohydrolase deficiency) or a defective regeneration of BH_4 (DHPR deficiency).

The lack of BH_4 is responsible for a deficient synthesis of catecholamine and serotonin neurotransmitters, resulting in neurological abnormalities. This defect could be corrected by a neurotransmitter replacement therapy and, in some cases, by a BH_4 monotherapy (1).

Biochemical and Clinical Aspects of Pteridines, Vol. 2

To recognize such atypical forms of hyperphenylalaninemia, diagnostic tests have been developed. Consequently, the screened babies could be tested for BH_4 deficiency; however, new variant forms were discovered, as it is illustrated in the present report.

Case reports

Case 1. TH, a boy, born on November 29,1980 after an uncomplicated pregnancy and delivery (2450 g, Apgar score 7) was the first child of non consanguinous and healthy parents. Routine screening tests were performed at 5 days of age and revealed a blood phenylalanine of 20 mg/dl, while the child was breast fed. In a control sample collected at 19 days of age, blood phenylalanine was 16 mg/dl.

At the age of 1 1/4 month, a BH_4 loading test (10 mg) was performed and a significant decrease of blood phenylalanine occured. Leukocyte dihydropteridine-reductase (D.M.Danks, Melbourne) was normal and analysis of urinary pterins showed an increase of neopterin and a decrease of biopterin concentrations.

Case 2. TS, a girl, born on October 26,1981 after an uncomplicated pregnancy and delivery, was the first child of non-consanguinous and healthy parents. Routine screening tests were performed in the neonatal period and revealed a blood phenylalanine of 7 mg/dl. At 3 weeks of age, a repeated determination yielded a serum phenylalanine of 25.4 mg/dl. A systematic analysis of urinary pterins showed high neopterin and low biopterin concentrations. Enzymatic studies in a liver biopsy confirmed normal phenylalanine-hydroxylase and dihydropteridine-reductase activities.

Methods

Pterin concentrations in urine were determined according to the method previously described (2) and blood biopterin measured by a C.fasciculata assay (R.J.Leeming,Birmingham and Nicol, Perth). CSF neurotransmitters were analyzed by I.Butler (Houston, Texas).

Results

1) Urinary profiles of pteridines: In both patients, the longitudinal survey of urinary excretion of pterins showed a reduction of neopterin, whereas biopterin excretion did not significantly vary (Table I).

Table I: Urinary pterins (µmol/mmol creatinine) in cases TH and TS.

patient	age	neopterin	biopterin
TH	2 months	11.05	0.35
	6 months	4.84	0.30
	2 years	3.61	0.40
TS	1 month	16.84	0.41
	2 months	3.88	0.16
	1 year	4.36	0.43
controls	1-2 months	2.83 ± 1.09	1.91 ± 0.41
	2-4 years	1.24 ± 0.59	1.61 ± 0.92

2) Phenylalanine loading tests: In both patients, changes of blood phenylalanine following an oral phenylalanine load (100 mg/kg) were consistent with a mild impairement of the phenylalanine hydroxylating system (Table II). A deuterated phenylalanine load confirmed in TS that the <u>in vivo</u> capacity to hydroxylate phenylalanine was 10% of that of controls (H.Ch.Curtius).

Table II: Results of oral phenylalanine loading tests.

phenylalaninemia (mg/dl)	hrs 0	1	2	3	4	8	12	24
TS	2	7.5	22	23.5	26	22	-	7.5
TH	2	6.3	8.5	-	12	-	8.9	3.4

In both cases, it was possible to rapidly increase the diet phenylalanine intake and the children are presently on normal proteic diet with low blood phenylalanine levels.

3) BH_4 loading tests:

Patient TH: Three loading tests have been performed, with different preload phenylalaninemia values (Table III), each time a significant decrease of blood phenylalanine was observed.

Table III: Phenylalaninemia (mg/dl) after oral BH_4 loads.

BH_4 dose	hrs 0	4	6
2.5 mg/kg	4.1	0.9	1.0
2.5 mg/kg	18.0	2.7	1.6
7 mg/kg	7.6	1.8	1.2

Patient TS: Since fasting blood phenyalanine concentration was normal, two loads have been performed: a phenylalanine load (100 mg/kg), and 48 hrs later a phenylalanine + BH_4 load (BH_4: 7.5 mg/kg).

Table IV: Phenylalaninemia (mg/dl) following Phe and Phe+BH_4 loading test.

patient	test	hrs 0	1	2	3	4	24
TS	Phe	2	7.5	21	22	23	7.5
	Phe+BH_4	2	7.5	7.5	2.5	2	2
mild HPA	Phe	2	18.5	19.5	16.5	7.5	2
	Phe+BH_4	2	15.5	19.5	16.5	11.5	2

The results, shown in Table IV and compared to that of similar loads performed in a patient with a "mild hyperphenylalaninemia" due to a partial defect of phenylalanine-hydroxylase, demonstrated a significant responseto BH_4 in TS, whereas no change was noted in the "mild HPA" patient.

4) Pterins levels following phenylalanine load: Blood and urine biopterin levels are known to increase following a phenylalanine load in normal controls and, at a lesser extent, in PKU patients (3,4). In both the present patients, a significant increase of biopterin excretion occured (Table V) following an oral phenylalanine load (100 mg/kg), whereas blood C.fasciculata positive material did not significantly change (results not shown).

Table V: Variations of pterin excretion following phenylalanine load.

urinary excretion (µmol/mmol creatinine)		neopterin	biopterin
patient:	hrs		
TS	0	9.03	0.08
	2	9.06	0.17
TH	0	3.28	0.19
	2	10.12	0.33
mild HPA	0	3.65	3.05
	2	3.93	6.37
typical PKU	0	3.99	3.27
	2	5.98	3.90
controls (n=22)	2/0	1.0 ± 0.3	1.7 ± 0.4

Question mark

In these two observations an impaired synthesis of BH_4 is clearly demonstrated and evidence was risen that the defect is not complete. The remaining question concerned the in vivo efficiency of the residual BH_4 synthesis. In liver, the BH_4 supply appeared to be enough to ensure a normal clearance of phenylalanine from a normal diet.

However, was this residual BH_4 synthesis sufficient for a normal neurotransmitter synthesis ?

In case TH, metabolites of monoamine neurotransmitters have been measured in cerebro-spinal fluid (CSF).

Table VI: CSF neurotransmitter metabolite levels in case TH.

(ug/l)	HVA	5 HIA	MHPG
controls	132.4 ± 10.3	39.6 ± 4.9	12.9 ± 0.8
untreated PKU	87.4 ± 9.1	16.0 ± 2.8	8.9 ± 2.6
TH *	55.6	11.2	4.9
**	43.2	14.6	4.9

data from I.Butler (5 and personal communication)
* 1 week off neurotransmitter therapy and ** 2 weeks off.
HVA: homovanillic acid; 5HIA: 5 hydroxyindole-acetic acid; MHPG: 3 methoxy-4-hydroxyphenylethylene glycol.

The results (Table VI) showed a decreased synthesis of neurotransmitters. The values were below those found in untreated PKU patients (5), a situation known to reduce neurotransmitter synthesis to a level incompatible with a normal brain function.

A specific neurotransmitter therapy was initiated in both cases. Respectively at age 2 1/2 and 1 1/2 years, physical and neurological developments were normal.

Discussion

Deficient BH_4 syntheses have been initially reported in patients suffering from a PKU syndrome but abnormal neurological signs even if a low phenylalanine diet has been early initiated. Since the development of methods for the early recognition of these "atypical" variants of hyperphenylalaninemia, screening programs for the BH_4 deficiency in hyperphenylalaninemic babies have been initiated by many groups and, consequently, new metabolic situations have been and will be discovered. Concerning "dihydrobiopterin-synthetase deficiency", the occurence of different variant forms recently emerged:

- typical forms: with hyperphenylalaninemic syndrom and neurological signs; most of these cases have been detected only rather late because of the unexpected appearance of these neurological signs.
- partial forms: the two present cases could be called "partial form" with regard to the results previously discussed. Four additional cases have been reported (6,7) with similar presentations.
- peripheral defects: the observation reported during this meeting by S.Milstien and one previously published (8) of patients having adequate CSF BH_4 and neurotransmitter levels despite a persistant peripheral defect and having consequently a normal development without specific therapy.
- transient form: such a case had been reported by Matalon (9); however sufficient data are not available to discuss this possibility, in regard to the large physiological variations in the developmental change of pteridine metabolism in childhood (4).

These various presentations reflect a wide variety in the intensity of the metabolic blockage. However, no direct test is yet able to discriminate these different forms except the CSF study of pterins and neurotransmitters in the latter form (peripheral form).

References

1. Niederwieser,A., Curtius,H-Ch., Wang,M., Leupold,D.: Eur.J.Pediatr. 138, 110-112 (1982)

2. Dhondt,J.L., Largilliere,C., Ardouin,P., Farriaux,J.P., Dautrevaux,M.: Clin.Chim.Acta 110, 205-214 (1981)

3. Leeming, R.J., Blair,J.A., Green,A., Raine, D.N.: Arch.Dis.Child. 51, 771-777 (1976)

4. Dhondt,J.L., Ardouin,P., Hayte,J.M., Farriaux,J.P.: Clin.Chim.Acta 116, 143-152 (1981)

5. Butler,I., O'Flynn,M.E., Seifert, W.E., Howell,R.R.: J.Pediat. 98, 729-733 (1981)

6. Rey,F., Leeming,R.J., Blair,J.A., Rey,J.: Arch.Dis.Child. 55,637-639 (1980)

7. Rey,F. PKU workshop, Paris (1982)

8. Hreidarsson,S., Valle,D., Holtzman,N., Coyle,J., Singer,H., Kapatos,G., Kaufman,S.: Pediat.Res. 16, 192 A (1982)

9. Matalon,R., Kimberlee,M., Lee,C-L., Nixon,J.C.: Pediat.Res. 15, 565 (1981)

RECENT ADVANCES IN TETRAHYDROBIOPTERIN BIOSYNTHESIS AND THE TREATMENT OF HUMAN DISEASE

Robert A. Levine*, Dorothee Heintel, Walter Leimbacher, Alois Niederwieser, and Hans-Christoph Curtius

Division of Clinical Chemistry, Department of Pediatrics, University of Zurich, CH-8032 Zurich, Switzerland

and

Sandro Ghisla

Department of Biology, University of Constance, Postfach 5560, D-7750 Constance, FRG

Introduction

The treatment of a human disease with tetrahydrobiopterin (BH_4) was initiated when it was discovered that certain cases of hyperphenylalaninemia in newborns were caused by a metabolic defect in BH_4 biosynthesis (for review, see 1). In contrast to patients with a liver defective phenylalanine hydroxylase (classical phenylketonuria), the patients with BH_4-deficient hyperphenylalaninemia, [also referred to as atypical phenylketonuria (PKU)], were unresponsive to the traditional low phenylalanine diet and also exhibited neurological symptoms that were indicative of central nervous system (CNS) involvement in the disease. BH_4 alone or in combination with biogenic amine neurotransmitter precursors has proved to be a successful therapy in certain BH_4-deficient patients. The theoretical

*Robert A. Levine was a visiting scientist in the Department of Pediatrics, Children's Hospital, Zurich. His current address is the Laboratory of Cell Biology and Genetics, National Institutes of Health, Bldg. 4, Room 312, Bethesda, MD 20205,USA.

Biochemical and Clinical Aspects of Pteridines, Vol. 2

mechanism of therapeutic action of BH_4 in the treatment of human diseases (2) is the elevation of the activities of phenylalanine, tyrosine, and tryptophan hydroxylases, for which BH_4 serves as the obligatory cofactor (3,4). Within the last two years, BH_4 therapy has been attempted with some successes in Parkinson's disease (5,6,7), familial dystonia(8) and endogenous depression (9,10). There have also been reports of new variants of BH_4- deficient hyperphenylalaninemia. Studying BH_4-deficient hyperphenylalaninemia, in general, has aided in the further understanding of human BH_4 biosynthesis, which has been a very difficult and controversial area of research. Regarding BH_4 biosynthesis, several new developments have helped clarify some issues. Continued studies on BH_4 biosynthesis could define the role that altered BH_4 metabolism may have in the etiology of certain diseases. The rapid advances that have been accomplished recently in the areas of BH_4 biosynthesis and the treatment of human disease will be evaluated.

BH_4 BIOSYNTHESIS

One major problem in studying BH_4 biosynthesis is that the intermediates on the pathway have been difficult to isolate. In addition, many variations of the biosynthetic pathway to BH_4 that have been proposed have contained certain internal inconsistencies, either from a chemical or biological standpoint, or both. Gál and coworkers earlier proposed that the initial reaction in BH_4 biosynthesis involved two enzymes to catalyze the conversion of guanosine triphosphate (GTP) to dihydroneopterin triphosphate (NH_2P_3), which they suggested was in the quinoid form (11). However, there is general agreement amongst other researchers that the first reaction in BH_4 biosynthesis is the conversion of GTP to NH_3P_3 (nonquinoid by a single enzyme by GTP-cyclohydrolase I (12).

Figure 1 outlines our current concepts on BH_4 biosynthesis in the context of past work and also the most recent information on the existence of two possible pathways for BH_4 biosynthesis from NH_2P_3 (13). The experiments in Zurich during the last year attempted to put in perspective some of the controversial aspects of BH_4 biosynthesis. It was originally proposed by Goto (14) and demonstrated by Tanaka and coworkers (15) that NH_2P_3 might be converted through an intermediate compound "X" having a diketo-dihydropterin structure to sepiapterin, which is also a dihydropterin. Sepiapterin can be converted to dihydrobiopterin (BH_2) by sepiapterin reductase in the presence of NAD(P)H. The conversion of BH_2 to BH_4 has been proposed by Kaufman (16) to be catalyzed by dihydrofolate reductase (DHFR). On the other hand, Nichol and coworkers (13) have demonstrated that BH_4 formation can occur from NH_2 P_3 (but not from sepiapterin) in the presence of methotrexate (MTX), which is an inhibitor of DHFR. Their studies cast some doubt as to the involvement of sepiapterin, BH_2 and DHFR in the formation of BH_4 in vivo. Therefore, it appears that sepiapterin may result from a side reaction of an intermediate that is normally on the BH_4 biosynthetic pathway as suggested by Smith at the meeting and discussed in the volume.

During the many discussions at the St. Christoph meeting, it seemed likely that NH_2P_3 could be converted to some intermediate with a tetrahydropterin ring, which could explain how BH_4 is formed without involving sepiapterin, BH_2 and DHFR. One possible mechanism that was suggested by our group involves an internal oxidoreduction reaction during the enzymatic transformation of NH_2P_3. It seems chemically possible that the elimination of the triphosphate moiety in the 6 position side chain could occur concomitantly with the transfer of reducing equivalents from the side chain to the ring. As a result, the most likely compound that might be formed would have a keto function in the 1' and 2' position of the side chain (referred to as diketo-tetrahydropterin in figure 1). It is possible that Tanaka's proposed compound "X" (15) could be formed from a chemical oxidation of diketo-tetrahydropterin to the diketo-dihydropterin. If a diketo-tetrahydropterin was formed, two successive side chain reductase reactions could lead to BH_4 formation. It is possible that sepiapterin reductase plus NAD(P)H could first form 5,6-dihydrosepiapterin,which actually has a tetrahydropterin ring (the correct chemical name is 6-lactoyl-tetrahydropterin). Presumably, 5,6-dihydrosepiapterin could be reduced directly to BH_4 by sepiapterin reductase in the presence of NAD(P)H. Thus, diketo-tetrahydropterin and 5,6-dihydrosepiapterin may be intermediates on the BH_4 biosynthetic pathway that does not involve sepiapterin or BH_2 as intermediates. Regarding the proposed internal oxidoreduction reaction, there are examples in sugar chemistry of this type of reaction, for instance, the Lobry de Bruyn-Alberda van Ekenstein transformation (17). This type of oxidoreduction reaction was originally proposed by Pfleiderer (18) as a mechanism for the chemical formation of sepiapterin from BH_4.

Another possible explanation that cannot be ruled out is that NH_2P_3 could be converted to a quinoid intermediate(figure 1). If a quinoid-dihydropterin intermediate was formed, there could be two possible ways to achieve BH_4 that would also not necessarily involve sepiapterin, BH_2, or DHFR. A quinoid compound could first be transformed by sepiapterin reductase and NAD(P)H into quinoid-dihydrobiopterin (q-BH_2) (11), which could then be converted by DHPR and NAD(P)H to BH_4. Alternatively, a quinoid intermediate could first be converted by DHPR and NAD(P)H to 5,6-dihydrosepiapterin, which could be converted to BH_4 by sepiapterin reductase. We feel that the more likely possibility is the formation of a diketo-tetrahydropterin intermediate, and we have some preliminary information on this point, described below.

It is interesting that our group in Zurich has recently isolated a nonfluorescent, reduced intermediate (as detected by

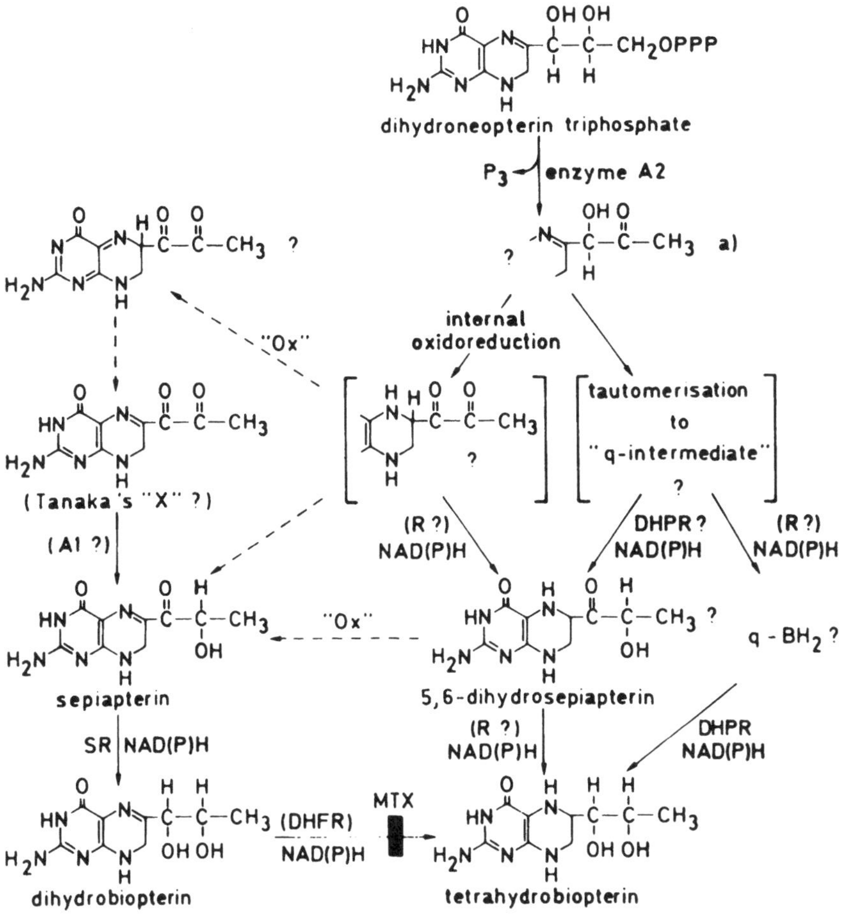

Fig. 1 Proposed pathway for tetrahydrobiopterin biosynthesis. The chemical structures of NH_2P_3, BH_4, BH_2 and sepiapterin have been established; those of the intermediates diketo-tetrahydropterin and 5,6-dihydrosepiapterin are hypothetical. Accordingly sepiapterin and Tanaka's compound "X" might be formed from these tetrahydrobiopterin intermediates by non-enzymatic side reactions denoted by dashed arrows.

electrochemical detection after high-pressure liquid chromatography) from a human liver biosynthetic system. Preliminary chemical analyses suggest that this compound may have undergone some of the rearrangements described in this section. A conclusive structural identification of this compound is currently underway. The possibility of forming a diketo-tetrahydropterin, or 5,6-dihydrosepiapterin (and possibly a quinoid compound) as intermediates in BH_4 biosynthesis remains to be proven experimentally. However, these theoretical considerations can serve as a framework for investigating these questions.

ATYPICAL PHENYLKETONURIA

Research on BH_4 biosynthesis and BH_4-deficient hyperphenylalaninemia (atypical PKU) have paralleled each other because this disease in newborns is caused by defects in certain of the enzymes responsible for normal BH_4 metabolism. Until recently, the known causes of BH_4-deficient hyperphenylalaninemia involved either defects in one of the BH_4 biosynthetic enzymes after NH_2P_3 or a defect in dihydropteridine reductase (DHPR). However, the first case of an enzyme defect in GTP-cyclohydrolase I was recently reported (19), where urinary excretion and CSF content of all pterins were markedly reduced. Further study of this cyclohydrolase deficient patient could yield more information on the possible role(s) of pterins in the response of the immune system as is discussed in this volume.

Another exciting development regarding organ system involvement in these BH_4 deficiency conditions is the report by Kaufman and coworkers (28) of a patient having BH_4 -deficient hyperphenylalaninemia but showing no neurological symptoms and a normal profile of pterins and biogenic amine metabolites in the CSF. The majority of reported cases prior to this one involved enzyme defects in both the liver and brain (and possibly other systems). It seems that in this patient, the liver has been a selected target and other organ systems, particularly the brain, have been spared. Thus, it is possible that the converse situation exists where certain patients could have normal pterin metabolism in the liver, but not in the brain. This hypothesis is disturbing in the sense that such patients would not be detected by current PKU screening methods because any patient with a "CNS-selective BH_4 deficiency" would not exhibit hyperphenylalaninemia, since liver BH_4 metabolism would be normal. Relating to the potential existence of patients with a selective CNS abnormality in pterin metabolism, Dhondt presented some very interesting results at this meeting, which are also contained in this volume. Two patients were detected at birth with hyperphenylalaninemia that were shown to be BH_4 deficient in both the periphery (liver) and the brain. However, the peripheral defect seemed to be transient because after some time, the serum phenylalanine levels in these patients remained normal on a regular diet without BH_4 administration. However, administration of the biogenic amine neurotransmitter precursors, L-dihydroxyphenylalanine (L-DOPA) and 5-hydroxytryptophan (5-HTP), remained necessary to alleviate the neurological symptoms of these patients. These may be the first examples of a pterin metabolic disease that we suggest could be referred to as "CNS-selective BH_4 deficiency". Future efforts should focus on ways in which newborns with this condition might be detected. It seems likely that some mentally disturbed adults may actually have a CNS-selective BH_4 deficiency that was not detected at birth by typical PKU screening methods. The analysis of CSF from a broad cross-section of mentally disturbed patients might shed light on this hypothesis.

Another involvement that is difficult to explain is the variability in patient responsiveness to both BH_4 and the synthetic cofactor, 6-methyltetrahydropterin (6-MPH_4). With regard to BH_4-deficient hyperphenylalaninemia, Kaufman and coworkers have successfully treated a patient with 6-MPH_4 (20) and they have suggested that this compound may represent a significant improvement over BH_4 for several reasons including that 6-MPH_4 has been shown to cross the blood-brain barrier much more effectively than BH_4 in rats (21,22). However, results were presented at this meeting by two pediatricians concerning their experiences with 6-MPH_4 administration. Dr. D. Leupold, Ulm, FRG has been treating a BH_4-deficient patient (central and peripheral) for several years with BH_4-monotherapy (23). When this patient was treated with 6-MPH_4 alone for almost one week, the blood phenylalanine rose continuously to very high levels and the patient's neurological symptoms reappeared, which precluded further testing of 6-MPH_4 in this patient. The other case, presented by Dr. S. Scheibenreiter, Wien, Austria was also a BH_4-deficient patient that was not responsive to BH_4 alone and required neurotransmitter precursors to alleviate clinical symptoms. However, low doses of BH_4 were able to control serum phenylalanine levels when the patient was on a normal diet. When this patient was switched to high-dose 6-MPH_4 alone, the clinical symptoms worsened and during the following 4 days on 6-MPH_4, serum phenylalanine levels rose steadily to unacceptable levels. While these results are preliminary, they indicate that BH_4-deficient patients may have to be evaluated on a case by case basis to determine which cofactor is optimal for counteracting the cofactor deficiency condition.

The question of why only certain BH_4-deficient hyperphenylalaninemic patients respond to BH_4 monotherapy is perplexing. One explanation could be that there is some variability in the population with respect to the ability of exogenously administered BH_4 to enter the brain. Another possibility is that unresponsive patients do not have an adequate circulating level of precursor amino acids (tyrosine and tryptophan). Trials with unresponsive patients should include the combination treatment of BH_4, tyrosine and tryptophan. It also cannot be ruled out whether these might be a multiple enzyme block involving not only BH_4 metabolizing enzymes, but possibly also tyrosine and/or tryptophan hydroxylase. However, unlikely this might explain why L-DOPA and 5-HTP, which bypass the hydroxylation steps, are effective in alleviating clinical symptoms when BH_4 alone is not.

The idea of using synthetic analogues of BH_4 for treatment of various diseases resulted primarily from early reports that BH_4 crossed the blood-brain barrier very poorly after peripheral administration (24,25). Later studies demonstrated that peripherally administered 6-MPH_4 entered the brains of rats (21,22) much more effectively than BH_4, which penetrated only slightly. Ayling and Bailey presented data (see this volume) on another synthetic cofactor, 6,6-dimethyltetrahydropterin (6,6-$DMPH_4$), suggesting that the enhanced test tube stability of this compound in the quinoid form compared to either BH_4 or 6-MPH_4 would make 6,6-$DMPH_4$ a more favorable cofactor than BH_4 or 6-MPH_4 for the treatment of BH_4-deficiency conditions. They suggest that 6,6-$DMPH_4$ might be especially useful in a DHPR-deficient patient, since there might be a greater possibility for nonenzymatic reduction (possibly by ascorbic acid) back to the active cofactor. While this argument may be valid for patients with a DHPR deficiency, it is not clear that the above logic can be extrapolated to other BH_4 deficiency conditions based on all the currently available evidence. For instance, prior experience with BH_4 and 6-MPH_4 administration to both humans (10) and rats (21,22) indicates that BH_4 and 6-MPH_4 are quite stable once in the body and remains predominantly in the fully reduced, tetrahydro form. Therefore, for cofactors that are stabile in the body in the reduced form, such as BH_4 and 6-MPH_4, the most relevant question with regard to treating cofactor deficiency conditions (possible exception is DHPR deficiency) may be their ability to penetrate the blood-brain barrier and elevate hydroxylase enzyme activity, rather than their test tube stability characteristics. Of course, any 6,6-disubstituted tetrahydropterin exhibiting favorable blood-brain barrier penetration could be considered a potential alternative to BH_4 therapy in a variety of diseases. However, there still is not a clear understanding at present of the full therapeutic potential of the natural cofactor BH_4.

PARKINSON'S DISEASE, DYSTONIA, AND ENDOGENOUS DEPRESSION

The initial association of altered BH_4 metabolism and a neurological disease was the report (26) that the CSF cofactor content was reduced by approximately 50% in patients with Parkinson's disease. This indicated the potential for a primary role of BH_4 in the disease process, although this decrease could simply be reflective of a loss of nigrostriatal dopamine neurons that is known to occur in Parkinson's disease. It was later shown that BH_4 content was decreased by about 80 % in postmortem striatal samples from Parkinson's disease patients (27). The decrease of BH_4 in the CNS of Parkinson's disease patients together with the suggestion that BH_4 may limit dopamine synthesis under normal conditions (2) has led to recent clinical trials in Parkinson's disease. Reports from Switzerland (5) and Japan (6) indicated some positive responses in mild cases of parkinsonism, whereas a limited study with 4 patients at the National Institutes of Health (7) obtained no positive responses. However, in a related trial with BH_4 in familial dystonia, where some patients also exhibit a decrease of BH_4 in the CSF (7), several patients have been treated successfully (8). At this time, it can tentatively be concluded that BH_4 therapy may be of value in some cases of the above mentioned neurological diseases, although larger studies are necessary to evaluate further the efficacy of BH_4 therapy. Some of the same reasons that were suggested in the prior section to explain variability in patient responsiveness in atypical PKU might also help explain the variable response to

BH_4 in the neurological diseases.

Since endogenous depression was very early on classified as a biogenic amine deficiency condition (lack of serotonin and/or norepinephrine), it was of interest to attempt the treatment of depressed patients with BH_4. The initial report (9) of success with BH_4 therapy in two out of three depressed patients described a short-term (several hours) improvement as judged by qualitative physician assessment in an open study. Although these results of subjective improvement are somewhat tenuous, the urinary excretion of dopamine and serotonin were only increased by BH_4 in the two responsive patients (9).

In an attempt to check the responsiveness of depressed patients to BH_4 in a more analytical fashion, another trial was initiated in Zurich that involved the measurement of CSF content of pterins and homovanillic acid (HVA) and 5-hydroxyindoleacetic acid (5-HIAA) as well as commonly accepted psychological rating scales. In this study, we detected a dramatic clinical improvement in one out of three patients tested. Further testing with this patient revealed that CSF content of BH_4, HVA, and 5-HIAA was elevated 4-5 fold when the mood improvement occurred (10). However, the temporal relationship between quality of mood and CSF content of BH_4 and metabolites was not completely clear due to a limitation on the number of lumbar punctures that could be performed. Larger studies with BH_4 in endogenous depression should help clarify the therapeutic potential of BH_4 either alone or in combination with traditional antidepressive therapy.

References

1. Niederwieser, A., Matasovic, A., Staudenmann, W., Wang, M., Curtius, H.-Ch.: In: Biochemical and Clinical Aspects of Pteridines, Vol. 1, CAncer-Immunology-Metabolic Diseases ed. by Wachter, Curtius, Pfleiderer, de Gruyter, Berlin pp. 293 -306 (1982).
2. Levine, R.A., Lovenberg, W., Curtius, H.-Ch., Niederwieser, A.: In: Chemistry and Biology of Pteridines, ed. by Blair, de Gruyter, Berlin (1983).
3. Kaufman, S.: In: Aromatic Amino Acids in the Brain, ed.by Wolstenholme and Fitzsimons, Elsevier, New York, pp. 85-115 (1974).
4. Lovenberg, W., Ames, M., Lerner, P.: In: Psychopharmacology: A Generation of Progress, ed. by Lipton, DiMascio, Killam, Raven Press, New York, pp.247-259 (1978).
5. Curtius, H.-Ch., Niederwieser, A., Levine R.A., Muldner, H.: Ad. Neurol. (in press).
6. Narabayashi, H., Kondo, T., Nagatsu, T., Matsuura, S.: Proc. Japan Acad. 58, 283-287 (1982).
7. LeWitt, P.A., Miller, L.P., Newman, R.P. et al.: Adv. Neurol. (in press).
8. LeWitt, P., Newman, R., Miller, L., Lovenberg, W., Eldridge, R.: New England Journal of Medicine 308, no. 3, 157-158 (1983).
9. Curtius, H.-Ch., Muldner, H., Niederwieser, A.: J. Neural Transmission 55, 301-318 (1982).
10. Curtius, H.-Ch., Niederwieser, A., Levine, R.A., Lovenberg, W., Woggon, B., Angst, J.: The Lancet, March 19, 657-658 (1983).
11. Gál, E.M., Nelson, J.M., Sherman, A.D.: Neurochem. Res.3, 69-88 (1978).
12. Blau, N., Niederwieser, A.: Analytical Biochemistry 128, 446-452 (1983).
13. Nichol, C.A., Viveros, O.H., Duch, D.S., Abou-Donia, M.M., Smith, G.K.: In: Chemistry and Biology of Pteridines, ed. by Blair, de Gruyter, Berlin (1983).
14. Suzuki, A. and Goto, M.: Biochimica et Biophysica Acta 304, 222-224 (1973).
15. Tanaka, K., Akino, M., Hagi, Y., Doi, M., Shiota, T.: J. Biol. Chem. 256, 2963-2972 (1981).

16. Kaufman, S.: J. Biol. Chem. 242, 3934-3943 (1967).

17. Speck, J.C.: Adv. Carboh. Chem. 13, 63 (1958).

18. Pfleiderer, W.: In: Chemistry and Biology of Pteridines, ed. by Pfleiderer, de Gruyter, Berlin, pp. 941-949 (1975).

19. Niederwieser, A., Staudenmann, W., Wang, M., Curtius, H.-Ch., Atares, M., Cardesa-Garcia, J.: In: Chemistry and Biology of Pteridines, ed. by Blair, de Gruyter, Berlin, pp. 183-187 (1983)

20. Kaufman, S., Kapatos, G., McInnes, R.R., Schulmann, J.D., Rizzo, W.B.: Pediatrics 70, 376-380 (1982).

21. Kapatos, G., Kaufman, S.: Science 212, 955-956 (1980).

22. Levine, R.A., Lovenberg, W., Niederwieser, A., Leimbacher, W., Redweik, U., Staudenmann, W., Curtius, H.-Ch.: In: Chemistry and Biology of Pteridines, ed. by Blair, de Gruyter, Berlin (1983).

23. Niederwieser, A., Curtius, H.-Ch., Wang, M., Leupold, D.: Eur. J. Pediatr. 138, 110-112 (1982).

24. Kettler, R., Bartholini, G., Pletscher, A.: Nature 249, 476-478 (1974).

25. Gál, E.M., Hanson, G., Sherman, A.: Neurochemical Research (1) 511-523 (1976).

26. Lovenberg, W., Levine, R.A., Robinson, D.R., Ebert, M., Williams, A.C., Calne, D.B.: Science 204, 624-626 (1979).

27. Nagatsu, T., Yamaguchi, T., Kaio, T., Sugimoto, T., Matsuura, S., Akino, M., Nagatsu, I., Iisuka, R., Narabayashi, H.: Clin. Chim. Acta 109, 305-311 (1981).

28. Kaufman, S. et al.: Case presentation at the Phenylketonuria Workshop in Paris, France, Sept. 1982.

6

FOLATE QUEUOSINE AND ACRASIN

ROLE OF THYROXINE IN REGULATION OF FOLIC ACID METABOLISM

E.L. Robert Stokstad, Margaret M-S Chan, Joe E. Watson, Gabriela Urban
Department of Nutritional Sciences
University of California, Berkeley
Berkeley, CA 94720 U.S.A.

Summary

This report deals with the effect of thyroxine on folate function as measured by the oxidation of histidine to carbon dioxide in normal and vitamin B_{12} deficient rats. The effects of hypo and hyperthyroidism in promoting and decreasing, respectively, the oxidation of (2 ^{14}C) histidine to $^{14}CO_2$ via the folate dependent formiminotransferase, is consistent with their effects in decreasing and increasing, respectively, the levels of methylenetetrahydrofolate reductase in liver. While hyperthyroidism increases the level of methylenetetrahydrofolate reductase in liver it does not increase this enzyme in kidney. Hypothyroidism increases methionine synthetase in both liver and kidney. The production of a severe vitamin B_{12} deficiency by the feeding of pectin results in a marked decrease in the oxidation of histidine which is consistent with the "methyl trap" theory that vitamin B_{12} deficiency increases the proportion of methyltetrahydrofolate which is inactive in most folate dependent reactions. The administration of the thyroid inhibitor, thiouracil, to vitamin B_{12} deficient animals increases histidine oxidation to normal levels indicating that the "methyl trap" produced by vitamin B_{12} deficiency has been relieved. All these observations are consistent with the concept that the primary factor in the regulation of folate activity is methylenetetrahydrofolate reductase.

Biochemical and Clinical Aspects of Pteridines, Vol. 2

Indroduction

The effects of thyroxine and vitamin B_{12} deficiency on histidine metabolism via formiminoglutamic acid (FlGlu) have provided useful information in the study of factors which regulate folate metabolism. Vitamin B_{12} deficiency increases the excretion of FlGlu by reason of the impairment of methionine synthesis resulting in the accumulation of methyletrahydrofolate with a corresponding deficiency of the non-methyltetrahydrofolate coenzymes which facilitate the further metablism of FlGlu. The feeding of thyroid powder was early shown to produce a growth depression and pseudo-vitamin B_{12} deficiency which could be partially alleviated by administration of vitamin B_{12} (1,2,3). This also resulted in an increase in FlGlu excretion (4,5,6) which could be normalized by addition of dietary methionine. This effect is similar to that produced by administration of methionine to a vitamin B_{12} deficient diet which reduces FlGlu excretion to normal (7).

Previously reported studies of hypothyroidism produced by either thyroidectomy (8) or the feeding of a thyroid inhibitor (9) have shown that it reduces FlGlu excretion, and increases the proportion of non-methyltetrahydrofolates similar to that produced by the feeding of methionine. Hypothyroidism also decreases the level of hepatic methylenetetrahydrofolate reductase. The overall effect on the activity of this enzyme would thus be similar to that produced by methionine which after conversion to S-adenosyl-methionine decreases the activity of this enzyme by allosteric inhibition (10,11). A reduced methylenetetrahydrofolate reductase activity would be consistent with a decreased level of methyltetrahydrofolate. Increased levels of thyroxine or thyroid powder increase methylenetetrahydrofolate reductase levels above normal which is also consistent with the elevated FlGlu excretion and the reduced histidine oxidation rates (9). The studies reported here deal with effect of hypo and hyperthyroidism on methylenetetrahydrofolate reductase and methionine synthetase levels in kidney and liver.

The earlier studies on the effects of hypothyroidism were made in rats on diets adequate in vitamin B_{12} but low in methionine. Under these conditions hypothyroidism normalized folate metabolism (as measured by histidine oxidation) similar to that produced by feeding methionine. Experiments are reported here on the effect of thiouracil on folate function in rats with severe vitamin B_{12} deficiency produced by feeding pectin. The feeding of 10% of pectin to rats on a low vitamin B_{12} diet has been shown to produce a rapid depletion of vitamin B_{12} as measured by an elevated excretion of methylmalonate (12).

Methods

A low methionine diet (0.3%) was used which contained 20% of isolated soy protein and which was supplemented with adequate amounts of the known vitamin including 5 mg per kg of folic acid (13). Vitamin B_{12} when used, was added at a level of 50 or 100 μg per kg diet as indicated, thiouracil at 1 g per kg, thyroid powder at 3 g per kg and pectin at 100 g per kg. Histidine oxidation was measured by I.P. injection of 0.5 μ Ci [2 - ^{14}C] histidine per animal together with 1.0 μmol of cold histidine per gram of body weight, and placing the animal in a 1.5 L respiration chamber through which air was aspirated at a rate of 300 ml per minute. The expired $^{14}CO_2$ was trapped in 30 ml of a 1:2 mixture of ethonalamine and ethyleneglycol-monomethyl ether for two hours. Radio activity of respired CO_2 was measured using the scintilating fluid of Jeffay and Alvarez (14). Formate oxidation was measured similarly by injecting 0.04 μ Ci ^{14}C formate together with 1 μmol cold formate per g body weight and collecting respired carbon dioxide for 1 hour. Methylmalonate was measured in urine acidified during the collection by the diazotization method of Giorgio and Plaut (15) modified for use with the autoanalyser (16). Hepatic and kidney 5,10-methylenetetrahydrofolate reductase was measured by the method of Kutzback and Stokstad (17) and methionine synthetase by the method of Sauer and Jaenicke (18). Nitrous oxide treatment, where indicated, was produced by placing a rat in a 1.5 liter metabolic chamber and passing 300 ml per minute of a mixture of 50% oxygen and 50% nitrous oxide for

5 hours. This has been shown by Deacon et al (19) to produce maximum inhibition of methionine synthetase.

Results and Discussion

The compartive effects of hyper and hypothyroid states on methylenetetrahydrofolate reductase and methionine synthetase levels in liver and kidney are shown in table I. The effect of thyroidectomy and thiouracil in reducing methylenetetrahydrofolate reductase in liver is similar to that previously reported (9). However the effects in kidney are not parallel

Table I

COMPARISON OF EFFECTS OF THYROIDECTOMY, THIOURACIL AND THYROID POWDER ON HISTIDINE OXIDATION AND ON LEVELS OF METHYLENETRAHYDROFOLATE REDUCTASE AND METHIONINE SYNTHETASE IN LIVER AND KIDNEY IN RATS.

Time on diets: 21-25 days

Grp. #	Treatment No. of animals ( )	Histid. Oxid. % 2h	Relative Enzyme Levels MethyleneTHF Red. Liver	Kidney	Meth. Synthetase Liver	Kidney
1	None (5)	4.5	100 (519)*	100 (617)*	100 (381)*	100 (554)*
2	Thyroidcetomy(6)	29.8	48	46	164	160
3	Thiouracil(5)	29.4	73	38	297	341
4	Thyroid Powder(4)	0.5	189	101	130	127
5	Thyroidectomy +Thyr. Powd.(4)	0.9	214	82	139	141
6	Thiouracil +Thyr. Powd.(3)	1.1	196	40	147	196

* Absolute levels of these enzymes in the control group. (Basal diet containing 50 µg B_{12}/kg and 0.3% methionine in soy protein). Enzyme levels expressed as n mol of product formed per h per g of fresh tissue.

with those in liver. The addition of thyroid powder doubled the levels of the enzyme in the liver but had almost no effect on the kidney levels of this enzyme. Thus the addition of thyroid powder increased methylenetetrahydrofolate reductase levels twofold in the livers of normal (Group 4), thyroidectomized (Group 5), and thiouracil fed animals (Group 6). However, in the kidney, thyroid powder produced no change in the reductase level in the normal animal (Group 4). It approximately normalized the reductase level (82%) in the thyroidectomized animal (Group 5) but had no effect in normalizing the depressed reductase level in the thiouracil fed animal (40%) (Group 6). Thus thiouracil may have some other effect than the simple disturbance of thyroxine production.

Both thyroidectomy and thiouracil increase methionine synthetase in the liver. The increase induced by thiouracil is ca two times that produced by thyroidectomy. This larger increase in methionine synthetase produced by thiouracil, compared to thyroidectomy, has also been observed in previously published studies (9). The effects of the various treatments on methionine synthetase levels in kidney are very similar to those in liver and the effects are essentially parallel. Thus the effect of thyroxine on methionine synthetase levels is the same in liver and in kidney but has different effects on methylenetetrahydrofolate reductase in these two tissues.

The next experiments deal with the effect of thiouracil on folic acid function in severe vitamin B_{12} deficiency. The previous experiments described in table I, were carried out using a diet adequate in vitamin B_{12}(50 μg/kg diet) but very low in methionine (0.3%). Vitamin B_{12} deficiency is produced slowly in the rat by simple dietary exclusion but develops more rapidly when 10% pectin is added to the diet (12). The overall purpose of this study was to determine whether thiouracil, which reduces methylenetetrahydrofolate reductase, can provide adequate folate function as measured by histidine oxidation on a diet which produced severe vitamin B_{12} deficiency.

The next experiment, the results of which are presented in table II, deal with the assessment of vitamin B_{12} deficiency as measured by excretion of methylmalonate and the oxidation of formate to respiratory CO_2. Methylmalonate, which is normally metabolized by the vitamin B_{12} dependent enzyme methylmalonylCoA mutase to succinate is excreted in elevated amounts in vitamin B_{12} deficiency in humans (20) and in rats (21). The oxidation of I.P. injected formate was also studied because the oxidation rate of this to respiratory carbon dioxide is reduced in vitamin B_{12} deficiency but the decrease is not as great as with histidine (22,23). The results of this experiment, presented in table II, show that feeding pectin markedly increased the excretion of methylmalonate. The excretion of methylmalonate on the vitamin B_{12} supplemented diet (Group 1) was 18 µmol/day which increased to 35 µmol on the vitamin B_{12} free diet, and

Table II

COMPARISON OF EFFECTS OF SEVERE VITAMIN B_{12} DEFICIENCY PRODUCED BY PECTIN FEEDING AND NITROUS OXIDE ON METABOLISM OF METHYLMALONATE, FORMATE, AND HISTIDINE.

Means ± S.D.; Values of n for each observation ( )

Diet Supp.	Body Wt.	Methyl-malonate Excr. µmol/day	Formate Oxid. 1 h $- N_2O$	Histidine Oxidation 2 hours $- N_2O$ %	$+N_2O$ %
	49 days[1]	53 days	61 days	52 days	56 days
$+B_{12}$[2] -Pectin	254 ± 24(6)	18 ± 10(6)	38.1 ± 3.9(6)	6.1 ±2.7(6)	-
$-B_{12}$ -Pectin	248 ± 12(6)	35 ± 12(6)	26.4 ± 3.2(6)	1.20 ± .74(6)	0.93 ± .16(4)
$-B_{12}$ +Pectin[3]	142 ± 6(6)	930 ±216(6)	6.2 ± 1.1(6)	0.25 ± .27(6)	0.12 ±0.03(4)

[1] Days on diet when measurements were made

[2] 100 µg vitmain B_{12} per kg diet

[3] 100 g pectin per kg diet

which increased further to 930 µmol when 10% pectin was added. This is similar to the increase observed by Cullen and Oace (12). The histidine oxidation rates were 6.1% with the $+B_{12}$ group, 1.20% with the $-B_{12}$ group and 0.25% with the $-B_{12}$ +Pectin group. These correlate with the methylmalonate excretion levels. It has been previously reported that nitrous oxide, which inactivates methionine synthetase (19), decreases histidine oxidation (24). In previous studies (24,25) nitrous oxide decreased histidine oxidation to _ca_ 0.5% in the presence of vitamin B_{12} in the diet. It will be noted that the oxidation rate on the severely deficient $-B_{12}$ +Pectin group was 0.25% which is lower than that with nitrous oxide. This indicates that the functional folate deficiency, as measured by histidine oxidation, is more severe than that which can be produced by nitrous oxide on diets with or without vitamin B_{12}. Exposure of the $-B_{12}$ +Pectin rats to nitrous oxide further decreased this to 0.12%. Formate oxidation provides another criteria of efficiency or folate function. The oxidation of formate is decreased markedly from 35% on a folate adequate diet to _ca_ 7% on a folate deficient diet (22). However, vitamin B_{12} deficiency, in the presence of dietary folate, decreased formate oxidation to only 12.5% (22). In the experiments described in table II histidine oxidation was decreased from 38% on the $+B_{12}$ diet to 26% on the $-B_{12}$ diet and then further reduced to 6.2% on the $-B_{12}$ +Pectin group. This provides further evidence of the effect of severe vitamin B_{12} deficiency in disturbing folate function.

The methionine synthetase levels for this experiment are given in table III. The levels of this enzyme on the $+B_{12}$ group were 674 n mol/h x g tissue at 82 days. This decreased to 306 n mol on the $-B_{12}$ group and decreased further to 91 n mol in the $-B_{12}$ +Pectin group. This level of methionine synthetase was lower than that reported earlier for rats receiving vitamin B_{12} but treated with nitrous oxide. This provides further evidence that the severity of vitmain B_{12} deficiency produced by feeding pectin is more severe (as measured by methionine synthetase activity) than that produced by nitrous oxide. The next experiment, reported in table IV, describes the results of adding thiouracil to the $-B_{12}$ +Pectin group. The histidine oxidation rates at 42 days show that the oxidation rate for the

Table III

COMPARISON OF EFFECTS OF VITAMIN B_{12} DEFICIENCY AND NITROUS OXIDE TREATMENT ON HEPATIC METHIONINE SYNTHETASE LEVELS AND ON HISTIDINE OXIDATION

Exp. No.	Diet and Supplements	Days on Diet	Histidine Oxid. 2h %	Methionine Synthetase n mol[1]
I	$+B_{12}$[2]	63	6.1 ± 2.7(6)	916 ± 103(4)
		82	-	674 ± 62(4)
	$-B_{12}$	63	1.6 ± 0.7(6)	307 ± 101(4)
		82	-	306 ± 121(4)
	$-B_{12}$	63	0.25 ± 0.27(6)	175 ± 34(4)
	+Pectin[3]	82	-	91 ± 35(4)
II[4]	$+B_{12}$[2] $-N_2O$	20	2.00 ± 0.38(6)	1160 ± 44(5)
	$+B_{12}$ $+N_2O$	20	0.38 ± 0.10(6)	298 ± 22(5)

[1] n mol / h x g liver tissue

[2] 100 μg vitamin B_{12} /kg diet

[3] 100 g pectin/kg diet

[4] Data previously published [Stokstad et al. Seventh Inter. Symposium. Pterdines and Folic Acid Derivatives.. 1982 (25)].

$-B_{12}$ +Pectin group (Group 1) was 2.6% which is very similar to that reported in table II. The rate for $+B_{12}$ +Pectin group was 3.6% while that for the $-B_{12}$ +Pectin +Thiouracil group was 10.4%. This level of histidine oxidation is comparable to that produced by a diet containing vitamin B_{12} and 0.7% added methionine (25). Previous work has shown that hypothyroidism produced either by thyroidectomy or by feeding thyroid inhibitors such as thiouracil increases the proportion of tetrahydrofolate (8) which can function in the degradation of FIGlu and in the oxidation of formate to carbon dioxide. It would thus appear that the reduction of methylenetetrahydrofolate reductase levels by thiouracil is able to decrease the

the formation of methyltetrahydrofolate to the point where a normal proportion of tetrahydrofolate is maintained even when the conversion of methyltetrahydrofolate to tetrahydrofolate by methionine synthetase is reduced to a very low level.

Table IV

EFFECT OF THIOURACIL ON HISTIDINE OXIDATION IN SEVERE VITAMIN B_{12} DEFICIENCY

Diet Containing 10% Pectin

Group	Diet Supplement	Histidine oxidation % in 2 h	
		21 days	42 days
1	$-B_{12}$	0.27 ± 0.10(4)[1]	0.26 ± 0.32(4)
2	$+B_{12}$[2]	1.42 ± 0.40(4)	3.64 ± 1.48(4)
3	$-B_{12}$ +Thiouracil[3]	10.30 ± 5.30(4)	10.40 ± 4.60(4)
4	+B +Thiouracil	32.00 ± 5.00(3)	34.80 ± 9.80(4)

[1] ±Mean ± S.D., number of animals ( ).

[2] 100 µg vitamin B_{12}/kg diet

[3] 1 g thiouracil/kg diet

Acknowledgement

This work was supported in part by U.S. Public Health Service Grant No. AM-08171 from the National Institutes of Health.

References

1. Betheil, J.J., Lardy, H.A.: J. Nutr. 37, 495-509 (1949).
2. Register, U.D., Ruegamer, W.R., Elvehjem, C.A.: J. Biol. Chem. 177, 129-134 (1949).
3. Lewis, U.J., Tappan, D.U., Register, U.D., Elvehjem, C.A.: Proc. Soc. Exp. Biol. Med. 74, 568-571 (1950).

4. Stokstad, E.L.R., Webb, R.S., Shah, E.: Proc. Soc. Exp. Biol. Med. 123, 752-754 (1966).

5. O'Dell, A., Stokstad, E.L.R.: J. Nutr. 92, 127-132 (1967).

6. Batra, K.K., Buehring, K.U., Stokstad, E.L.R.: Soc. Exp. Biol. Med. 72-79 (1974).

7. Brown, D.D., Silva, O.L., Gardiner, R.C., Silverman, M.J.: Biol. Chem. 235, 2058-2062 (1960).

8. Chan, M. M-S, Stokstad, E.L.R.: Biochim. Biophys. Acta. 632, 244-253 (1980).

9. Stokstad, E.L.R., Chan, M. M-S, Watson, J., Brody, T., Jaenicke, L.: Proceedings of the Third European Symposium on Vitamin B-12 and Intrinsic Factor. Zurich, Switzerland, (B. Zagalek, W. Friedrich eds) Walter de Gruyter, Berlin, March 1979, pp. 1139-1144.

10. Kutzbach, C., Stokstad, E.L.R.: Biochim. Biophys. Acta. 139, 217-220 (1967).

11. Kutzbach, C., Stokstad, E.L.R.: Biochim. Biophys. Acta. 250, 459-477. (1971).

12. Cullen, R., Oace, S.M.: J. Nutr. 108, 640-647 (1978).

13. Thenen, S.W., Stokstad, E.L.R.: J. Nutr. 103, 363-370 (1973).

14. Jeffay, H., Alvarez, J.: Anal. Chem. 33, 612-615 (1961).

15. Giorgio, A.J., Plaut, G.W.E.: J. Clin. Lab. Med. 66, 667-676 (1965).

16. Gawthorne, J.M., Watson, J., Stokstad, E.L.R.: Analytical Biochem. 42, 555-559 (1971).

17. Kutzbach, C., Stokstad, E.L.R.: Biochim. Biophys. Acta. 139, 217-220 (1967).

18. Sauer, H.J., Jaenicke, L.: Klin. Wochenschr. 50, 986-990 (1972).

19. Deacon, R., Perry, J., Lumb, M., Chanarin, I., Mint, B., Halsy, M., Nunn, J.: European J. Biochem. 104, 419-422 (1978).

20. Barness, L.A., Young, D.G., Nocho, R.: Science 140, 76-77 (1963).

21. Williams, D.L., Spray, G.H., Newman, G.E., O'Brien, J.R.P.: Br. J. Nutr. 23, 343-352 (1969).

22. Chiao, F., Stokstad, E.L.R.:Proc. Exper. Biol. and Med. 155, 433-437 (1977).

23. Waydhas, C., Sies, H., Stokstad, E.L.R.: FEBS Letters 103, 366-369 (1979).

24. Brody, T., Watson, J., Stokstad, E.L.R.: Biochemistry 21, 276-282 (1982).

25. Stokstad, E.L.R., Chan, M. M-S, Watson, J.E.: Seventh Inter. Sympos. Pterdines and Folic Acid Derivative. 1982.

INTRACELLULAR FOLATE CONCENTRATION AND GROWTH RATE IN FOLATE-DEPLETED K562 HUMAN LEUKEMIA CELLS

David Watkins, Thomas E. Shapiro and Bernard A. Cooper
Department of Physiology, McGill University
Montreal, Canada H3G 1Y6

Folate deficiency in humans results in anemia characterized by specific morphological changes of hemopoietic cells of the bone marrow (1). There is evidence of impaired erythroid maturation, with ineffective erythropoiesis and intramedullary death of cells, and cells with abnormal DNA content (2, 3, 4). It has frequently been assumed, on the basis of apparent decreases in incorporation of deoxyuridine into DNA thymine in folate-deficient bone marrow cells (5) and reports of temporary remission of folate deficiency anemia upon treatment with thymine (6), that deficient folate-dependent synthesis of thymidylate from deoxyuridylate is the cause of the hematologic lesions. However, attempts to demonstrate directly that thymidylate synthesis is deficient in folate-depleted cells have not been successful (7, 8).

We have studied the effects of folate deficiency on cell line K562, a human leukemia cell line derived from cells of a patient with chronic myelogenous leukemia (9). This cell line is believed to represent an early pluripotent cell with capacity to undergo limited differentiation along erythrocytic, granulocytic and monocytic lines (10). This cell represented the best available model for a human erythroid cell in long-term culture. We have investigated the relationship between intracellular folate concentration and growth of cell line K562 when it was cultured in folate-free medium, and which of the several folate-dependent processes involved in cellular metabolism was limiting to growth of folate-depleted K562 cells in culture.

Biochemical and Clinical Aspects of Pteridines, Vol. 2

Materials and Methods

Cell culture. Tissue culture media were purchased from Grand Island Biological Company; fetal calf serum was purchased from Flow Laboratories.

Human leukemia cell line K562 was obtained from Dr. A.K. Sullivan of the Haematology Division, Royal Victoria Hospital, Montreal, and the McGill Cancer Center. Cells were maintained in suspension culture in Eagle's minimum essential medium plus non-essential amino acids supplemented with 10% (v/v) fetal calf serum. This medium contained 2.3 μM Pte Glu.

On day 0 of folate depletion experiments, cells were subcultured into folate-free medium. This medium was identical in formulation with maintenance medium except that it lacked PteGlu and was supplemented with 1 μM cyanocobalamin and 10% (v/v) fetal calf serum which had previously been dialyzed 24 hours at 4° against a ten-fold volume of 0.9% sodium chloride with two changes to remove low molecular weight components, including folates. Cultures were split to relieve cell crowding and medium was replaced at three-day intervals.

To determine the effects of various materials - Pte Glu, DL-5-CHO-H_4PteGlu, DL-5-CH_3-H_4 PteGlu, thymidine, inosine, glycine and serine - on growth of folate-depleted K562 cells, cells which had been grown for 12 days in folate-free medium were subcultured into folate-free medium supplemented with one or more of the above materials. Cell numbers were determined at the time of subculture and three days afterward, and growth rate (average rate of increase of logarithm of cell number, doublings/day) was calculated.

Determination of plating efficiency. (We are grateful to Dr. J.F. Prchal for his assistance in these studies.) Plating efficiency was determined by plating cells at low density in complete medium supplemented with 0.88% methylcellulose to immobilize cells. The number of colonies present 10-14 days after plating was determined, and expressed as a percentage of the number of cells originally plated.

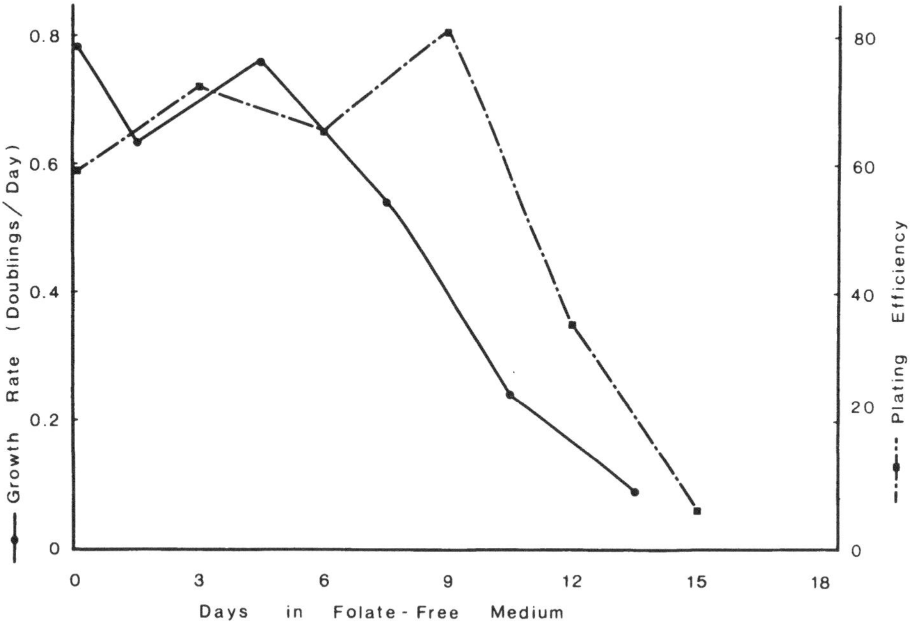

Fig. 1. Effect of folate-depletion on growth and plating efficiency of cell line K562.

Measurement of intracellular folate. Cells were removed from suspension by centrifugation and washed twice with phosphate-buffered saline. The cell pellet was suspended in 0.1 M potassium phosphate, pH 6, containing 1.5 mg/ml ascorbic acid, vortexed, boiled 10 minutes and centrifuged to remove precipitate. The supernatant was incubated for 90 minutes at 37° with 10% (v/v) charcoal-treated human serum to convert folate polyglutamates to monoglutamate forms (11).

Folate in cell extracts was assayed microbiologically with *Lactobacillus casei* ATCC 7469 and *Pediococcus cerevisiae* ATCC 8081 using aseptic addition of samples to sterile medium already innoculated with assay organisms (12, 13).

Intracellular folate content has been expressed in units of µmol of folate/litre of cell water. Mean intracellular water volume was determined by measurement of ^3H_2O space with correction for extracellular space with [^{14}C]-polyethylene glycol (14), and comparison of these data with cell number determined by hemacytometer counting. The mean intracellular water volume determined by this method for K562 cells was 1.65 ± 0.77 pL/cell.

Results

Effect of folate depletion on growth rate and plating efficiency. Cell line K562 maintained in medium containing 2.3 µM PteGlu proliferated logarithmically at a rate of 0.78 ± 0.15 doublings/day. This was equivalent to a population doubling time of approximately 31 hours. When cells were subcultured into folate-free medium, growth rate did not change significantly during the first 6 days (figure 1). After 6 days the growth rate began to decrease, and after 12 days in folate-free medium little or no increase in cell number was observed.

Plating efficiency of cells which had been grown for various periods in folate-free medium was determined. Plating efficiency did not decrease

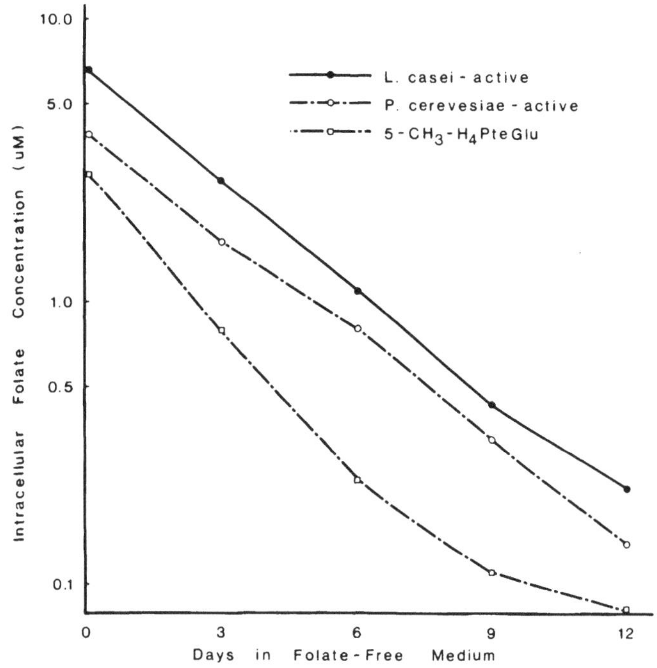

Fig. 2. Effect of folate depletion on intracellular folate concentration of cell line K562. Total intracellular folate was determined by microbiological assay with *L. casei*. Fully-reduced non-methylated intracellular folate was determined by by microbiological assay with *P. cerevisiae*. Intracellular 5-CH_3-H_4 PteGlu was calculated from the difference between *L. casei*-active and *P. cerevisiae*-active folate.

significantly compared with that of cells grown in medium containing 2.3 μM PteGlu during the first 9 days of folate depletion (figure 1). Plating efficiency decreased rapidly after 9 days, and after 15 days in folate-free medium few colonies developed.

Effect of folate depletion on intracellular folate content. The total intracellular folate concentration, determined by assay with *L. casei*, decreased progressively during the folate depletion process (figure 2). Decrease in folate concentration appeared to be exponential, with a calculated $t_{\frac{1}{2}}$ of 58 hours. This decrease occurred during the first 6 days of folate depletion during which no significant decrease in either growth rate or plating efficiency was observed.

The intracellular concentration of fully-reduced non-methylated folate, determined by assay with *P. cerevisiae*, also decreased exponentially in parallel with total intracellular folate concentration, with a $t_{\frac{1}{2}}$ of 61 hours. The intracellular concentration of 5-CH_3-H_4PteGlu, calculated from the difference between *L. casei*-active and *P. cerevisiae*-active folate concentration, decreased exponentially with a $t_{\frac{1}{2}}$ of 46 hours during the first 9 days of folate depletion; after 9 days in folate-free medium the intracellular concentration of 5-CH_3-H_4PteGlu was too low to be measured accurately.

Comparison of $t_{\frac{1}{2}}$ for loss of intracellular folate with the population doubling times of cells in folate-free medium disclosed no evidence for loss of folate from cells during the depletion process. The entire decrease in intracellular folate concentration could be accounted for by dilution of the initial folate content in an increasing volume of cells during proliferation.

Relationship between intracellular folate concentration and growth rate. The observation that growth rate did not decrease significantly during the initial stages of folate depletion although the intracellular folate concentration was decresing suggested that cells grown in medium containing 2.3 μM PteGlu contained intracellular folate in excess of that needed to sustain optimal growth rate. This was confirmed by

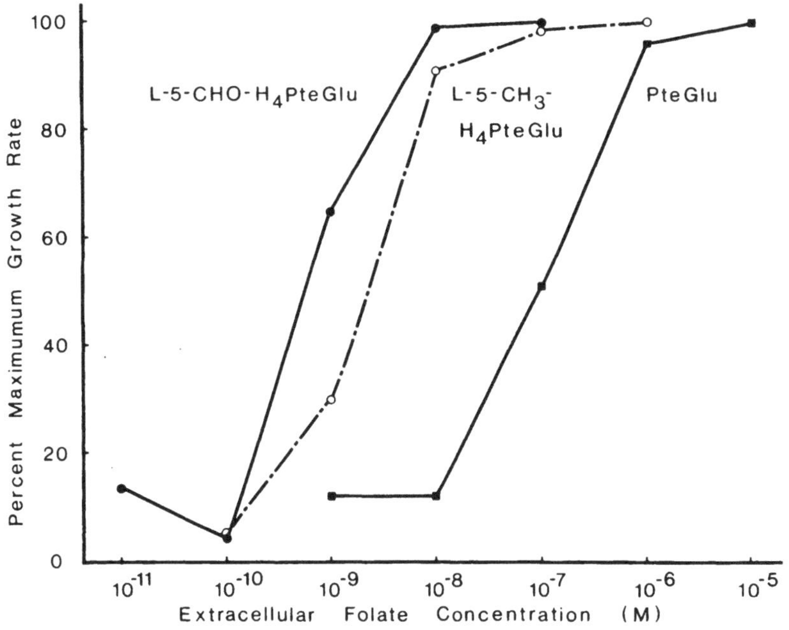

Fig. 3. Effects of PteGlu, 5-CHO-H_4 Pte Glu and 5-CH_3-H_4 PteGlu on growth rate of folate-depleted K562 cells.

comparison of growth rate data with measurements of intracellular folate concentration from the same cells. Cells grown in medium containing 2.3 µM PteGlu contained 6.6 ± 1.5 µM (mean ± SD) L. Casei-active folate, of which 3.9 ± 1.1 µM was also active for P. cerevisiae. Growth rate was independent of intracellular folate concentration when the concentration of L. casei-active folate was greater than 1.5 µM or when the concentration of P. cerevisiae-active folate was greater than 1.0 µM.

The relationship between the logarithm of intracellular folate concentration and growth rate for cells containing less than 1.5 µM L. casei-active folate is shown in figure 3. Growth rate was proportional to the logarithms of intracellular L. casei-active folate ($r=0.807$, $p<0.001$) and intracellular P. cerevisiae-active folate ($r=0.848$, $p<0.001$) concentration. No correlation was observed between growth rate and the logarithm of intracellular concentration of 5-CH_3-H_4 PteGlu ($r=0.141$, $p=0.30$).

Folate requirements of folate-depleted cells. Cells which had been grown in folate-free medium for 12 days were subcultured into folate-free medium supplemented with different concentrations of PteGlu, DL-5-CHO-H_4 PteGlu or DL-5-CH_3-H_4 PteGlu. Media containing DL-5-CH_3-H_4 PteGlu were also supplemented with 50 µg/ml sodium ascorbate. Results of these experiments are shown in figure 4. PteGlu at 100 µM or greater supported optimal growth rate of folate-depleted K562 cells. Optimal growth rate was supported by a 100-fold lower concentration of either DL-5-CHO-H_4 PteGlu or DL-5-CH_3-H_4 PteGlu, assuming that growth was supported only by the biologically-active optical isomer of the reduced folates.

Effects of products of folate metabolism on growth of folate-depleted cells. Cells which had been grown for 12 days in folate-free medium were subcultured into folate-free medium supplemented with products of folate metabolism (Table 1). Thymidine at either 1 µM or 40 µM did not significantly increase growth rate over that of unsupplemented cultures. Thymidine at concentrations greater than 40 µM inhibited growth of cells. Inosine at a concentration of 100 µM increased growth rate to

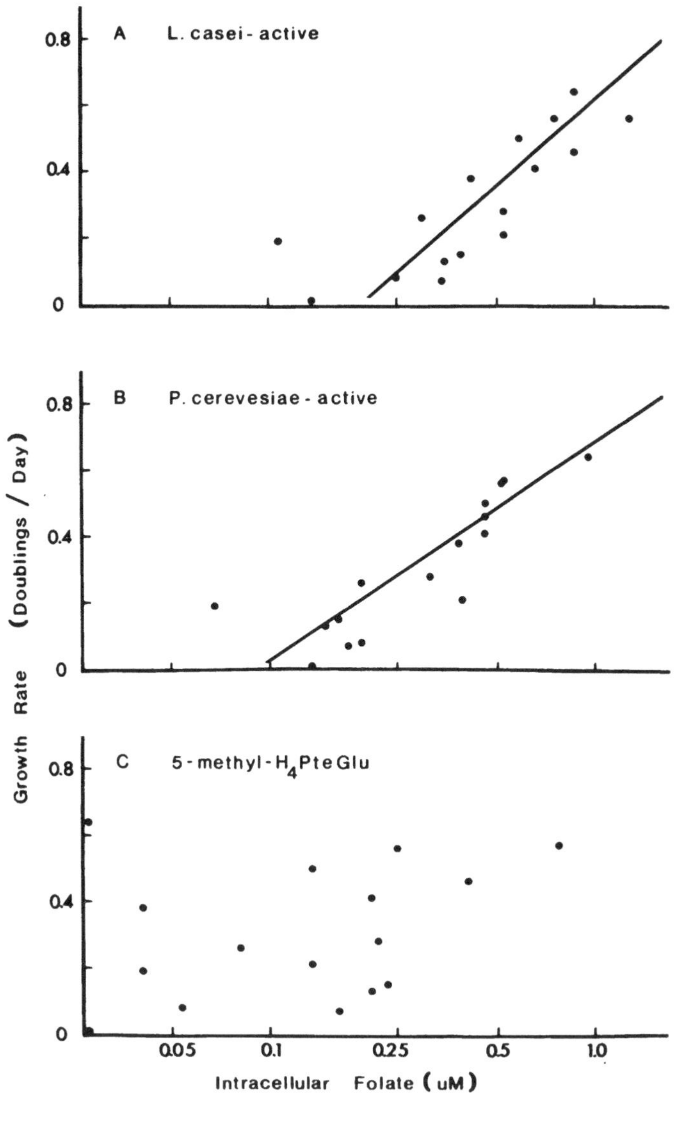

Fig. 4. Growth rate of cell line K562 in folate-free medium as a function of intracellular concentration of: A. L. casei-active folate; B. P. cervisiae-active folate; and C. $5\text{-}CH_3\text{-}H_4$ PteGlu (calculated as the difference between L. casei-active and P. cerevisiae-active folate).

78% of that of cultures supplemented with 0.2 µM DL-5-CHO-H_4 PteGlu (maximum folate-dependent growth rate, figure 4). When both thymidine and inosine were both present, growth rate was further increased to 82% of the maximum folate-dependent growth rate.

Subculture of folate-depleted K562 cells in medium containing 1 mM glycine resulted in a significantly decreased growth rate compared to unsupplemented cultures. Supplementation of cultures with 1 mM serine resulted in an increase in growth rate compared to unsupplemented cultures which was significant at the 0.10 level.

Discussion

Growth rate of human leukemia cell line K562 decreased when the cells were grown for an extended period of time in folate-free medium. That this decreased growth rate was the result of folate-depletion was supported by the increased growth rate which was observed when folate was added back to the cultures (figure 4). The concentration of PteGlu which was required to support optimal growth rate of these cells was 1 µM. This value was similar to the reported folate requirement of several murine leukemia cell lines in culture (15, 16, 17). Both 5-CHO-H_4 PteGlu and 5-CH_3-H_4 PteGlu supported optimal growth rate at 100-fold lower concentration than did PteGlu. This was most likely the result of more effective transport of the reduced folate derivatives into the cell (18).

When folate-replete K562 cells were subcultured into folate-free medium, there was an initial period when neither growth rate nor plating efficiency decreased, although intracellular folate concentration was decreasing. This suggested that cells grown in medium containing 2.3 µM PteGlu contained intracellular folate in excess of that necessary to support optimal growth rate. This was confirmed by comparison of measurements of intracellular folate concentration and growth rate of cells growing iin folate-free medium, which showed that growth rate was

Table 1

EFFECT OF PRODUCTS OF FOLATE METABOLISM ON GROWTH RATE OF FOLATE-DEPLETED K562 CELLS

Supplement	Number of Cultures	Growth Rate (Doublings/Day) Mean ± SD
None	15	0.27 ± 0.08
0.2 μM DL-5-CHO-H_4PteGlu	16	0.90 ± 0.11 *
1 μM Thymidine	6	0.26 ± 0.10
100 μM Inosine	9	0.71 ± 0.08 *
1 μM Thymidine + 100 μM Inosine	10	0.74 ± 0.12 *
1 mM Glycine	6	0.12 ± 0.04 *
1 mM Serine	9	0.33 ± 0.15

* Growth rate significantly different from that of unsupplemented culture (p 0.01) by Student t-test.

independent of intracellular folate concentration when this was greater than 1.5 μM.

Growth rate was proportional to total intracellular folate concentration when this was less than 1.5 μM, and to the intracellular concentration of fully-reduced non-methylated folate when this was less than 1.0 μM (figure 3). No correlation was found between the intracellular concentration of 5-CH_3-H_4 PteGlu and growth rate. Cell line K562, like several other transformed cell lines in culture, has a growth requirement for methionine which cannot be replaced by its precursor homocysteine (18, 19). In these experiments, cells were grown in medium containing excess methionine and it is likely that cellular methionine requirements were met under these conditions from exogenous sources.

Folate coenzymes are required for thymidylate synthesis, de novo purine biosynthesis and interconversion of serine and glycine in addition to methionine synthesis. To determine whether any of these processes is limiting to growth rate of folate-deficient K562 cells, the ability of thymidine, inosine, glycine and serine to replace folate was tested. Thymidine alone had no effect on growth rate. Inosine at 100 μM, on the other hand, increased growth rate to 78% of its maximum level in folate-supplemented cells. This suggested that the primary metabolic lesion in folate-depleted cells was one of de novo purine biosynthesis. When the purine deficiency had been corrected by addition of 100 μM inosine, the addition of thymidine further increased growth rate. Thus thymidine deficiency may represent a secondary cause of growth restriction in folate-deficient K562 cells.

Supplementation of cultures with 1 mM glycine resulted in a decrease in growth rate. This result was in contrast to experiments with Chinese hamster cells which required glycine to grow in folate-free medium (20). It is consistent however with observations that addition of glycine to bone marrow cell cultures aggravated abnormalities of DNA synthesis (21). It is possible that excess glycine prevents the transfer of one-carbon units from serine to H_4PteGlu by serine hydroxymethyltransferase. Addition of serine to cultures of folate-depleted K562 cells resulted in a small increase in growth rate which was significant at the 0.10 level; this result is also consistent with an effect on serine hydroxymethyltransferase.

The observed decrease in intracellular folate concentration in K562 cells grown in folate-free medium could be entirely accounted for by dilution of the folate present at the start of folate depletion in a larger cell volume during proliferation. Thus, there is no evidence for loss of folate from cells such as was observed in confluent cultured human fibroblasts (14). This behavior is similar to that of murine erythroleukemia cells in culture (17).

There was a lag of approximately 3 days between the first significant decrease in growth rate and the decrease in plating efficiency in cells subcultured into folate-free medium. Cells may first become reversibly arrested, retaining the capacity to resume proliferation when conditions are once again favorable. Subsequently, cells become irreversibly arrested and ultimately die. It is not clear to what extent the early decrease in growth rate of folate-depleted cells represents an increase in cell cycle transit time of individual cells or reversible arrest of a proportion of cells. During the later stages of folate depletion it is clear that irreversible arrest of a large proportion of cells occurred.

Summary

Growth rate and plating efficiency of human leukemia cell line K562 decreased when cells were subcultured in folate-free medium. Addition of 1 μM PteGlu or 0.02 μM of either DL-5-CHO-H_4 PteGlu or DL-5-CH_3-H_4 PteGlu to folate-deficient cultures restored optimal growth rate. Growth rate was proportional to total (*L. casei*-active) intracellular folate concentration when this was less than 1.5 μM and to the intracellular concentration of fully-reduced, non-methylated (*P. cerevisiae*- active) folate when this was less than 1.0 μM. No correlation could be detected between intracellular concentration of $5\text{-}CH_3\text{-}H_4$ PteGlu and growth rate. A deficiency of purine biosynthesis appeared to be the primary cause of decreased growth rate in folate-depleted K562 cells. Deficiency of thymidylate synthesis was a secondary cause, but addition of thymidine to folate-deficient K562 cells did not increase growth rate unless purine deficiency was simultaneously corrected. Addition of glycine to cultures of folate-deficient cells inhibited growth, while serine exerted a slight stimulatory effect.

Bibliography

1. Chanarin, I.: The Megaloblastic Anaemias, Second Edition, Blackwell Scientific Publications, London and Edinburgh, 1979.
2. Finch, C.A., Coleman, D.H., Motulsky, A.G., Donohue, D.M., Reiff, R.H.: Blood 11, 807-820 (1956).
3. Yoshida, Y., Todo, A., Shirakawa, S., Wakisaka, G., Uchino, H.: Blood 31, 292-303 (1968).
4. Messner, H., Fliedner, T.M., Cronkite, E.P.: Ser. Haemat. 2, 44-64 (1969).
5. Metz, J., Kelly, A., Swett, V.C., Waxman, S., Herbert, V.: Brit. J. Haematol. 14, 575-592 (1968).
6. Vilter, R.W., Horrigan, D., Mueller, J.F., Jarrold, T., Vilter, C.F., Hawkins, V., Seaman, A.: Blood 5, 695-717 (1950).
7. Ganeshaguru, K., Hoffbrand, A.V.: Brit. J. Haematol.: 40, 29-41 (1978).
8. Iwata, N., Omine, M., Yamauchi, H., Maekawa, T.: Blood 60, 918-923 (1982)
9. Lozzio, C.B., Lozzio, B.B.: Blood 45: 321-334 (1975).
10. Lozzio, B.B., Lozzio, C.B., Bamberger, E.G., Feliu, A.S.: Proc. Soc. Exp. Biol. Med. 166, 546-550 (1981).
11. Rosenblatt, D.S., Cooper, B.A., Lue-Shing, S., Wong, P.W.K., Berlow, S., Narisawa, K., Baumgartner, R.: J. Clin. Invest. 63, 1019-1025 (1979).
12. Baker, H., Frank, O., Hunter, S.H.: Methods Enzymol. 18, 624-629 (1971).
13. Cooper, B.A., Jonas, E.: J. Clin. Pathol. 26, 963-967 (1979).

14. Hilton, J.G., Cooper, B.A., Rosenblatt, D.S.: J. Biol. Chem. 254, 8398-8403 (1979).

15. Chello, P.L., Bertino, J.R.: Cancer Res. 33, 1898-1904 (1973).

16. Moran, R.G., Werkheiser, W.C., Zakrzewski, S.F.: J. Biol. Chem. 251, 3569-3575 (1976).

17. Steinberg, S.E., Fonda, S., Campbell, C.L., Hillman, R.S.: J. Cell. Physiol. 114, 252-256 (1983).

18. Watkins, D., Cooper, B.A.: Biochem. J. (In press).

19. Kano, Y., Sakamoto, S., Kasahara, T., Kusumoto, K., Hida, K., Suda, K., Ozawa, K., Miura, Y., Takaku, F.: Cancer Res. 42, 3090-3092 (1982).

20. McBurney, M.W., Whitmore, G.F.: Cell 2, 173-182 (1974).

21. Waxman, S., Corcino, J., Herbert, V.: Clin. Res. 17, 347 (1969).

QUEUOSINE IN TRANSFER-RNA IN RELATION TO DIFFERENTIATION AND PTERIDINE METABOLISM

Helga Kersten, Elfriede Schachner, Gerlinde Dess
Institut für Physiologische Chemie der Universität Erlangen-Nürnberg
D-8520 Erlangen

Annerose Anders
Genetisches Institut der Universität Giessen
D-6300 Giessen

Susumu Nishimura, Nobuko Shindo-Okada
Biology Division, National Cancer Center Research Institute,
Tsukiji 5-1-1 Chuo-ku, Tokyo 104 Japan

Introduction

Transfer-RNAs play a key role in protein synthesis. The anticodon of a tRNA specifies the amino acid that is attached to the 3′terminal adenosine residue and decodes a corresponding codon in messenger RNA on the ribosome during protein synthesis. Besides their role in protein synthesis, tRNAs also play an important role in the regulation of cell metabolism. These regulatory functions involve alterations of tRNA modifications. The modified nucleosides occur at well defined positions in specific tRNAs (Fig. 1a) (summarizing references in 1,2,3,4). Queuosine (Q) (7-(((4,5-cis-dihydroxy-2-cyclopenten-1-yl)-amino)-methyl)-7-deazaguanosine) is unique in that its purine skeleton is modified to a 7-deazastructure (Fig. 1b). Queuosine occurs at position 34 in the anticodon of $tRNA^{Tyr}$, $tRNA^{His}$, $tRNA^{Asn}$ and $tRNA^{Asp}$ of eubacteria and eukaryotes, except yeast. In eukaryotes D-mannose or D-galactose can be attached by β-glycoside-linkage to the oxygen atom on carbon-4 of the cyclopentenediol (6).

Biochemical and Clinical Aspects of Pteridines, Vol. 2

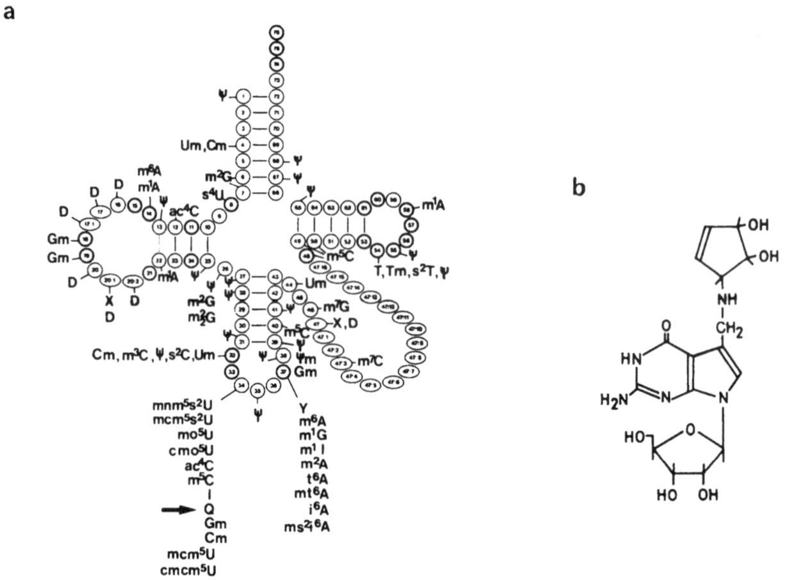

Fig. 1: a) General cloverleaf structure of tRNA and positions of modified nucleosides (positions and abbreviations of modifications in ref. 4).
b) Structure of queuosine.

Eubacteria synthesize the base queuine de novo (for review see ref. 7) whereas mammalian cells are supplied with queuine by nutrition or the intestinal flora (8). The slime mould Dictyostelium discoideum which grows in its natural environment on E. coli, is supplied with the modified base by the bacteria probably after degradation of Q-containing tRNAs. The mechanism of queuine biosynthesis in eubacteria is not fully understood. It is assumed that during the conversion of guanine to queuine the carbon at position 8 is lost together with N-7 of the purine skeleton (7,9). The biosynthesis of queuine is suggested to share common pathways with that of pteridines, folate and riboflavine. In E. coli, a precursor of queuine and in higher organisms queuine is inserted into the polynucleotide chain of tRNA by exchange with guanine at position 34 by tRNA: guanine transglycosylases (E.C.2.4.2). These enzymes have been purified from E. coli (10) and from rabbit reticulocytes or erythrocytes (11). The eukaryotic enzyme was found to be subject to inhibition by pteridines in a competitive manner (12).

$$\text{queuine} + (\text{G34})\text{tRNA} \underset{\text{pteridines}}{\overset{\text{tRNA transglycosylase}}{\rightleftharpoons}} (\text{Q})\text{tRNA} + \text{guanine}$$

tRNAs from exponentially growing E. coli are almost completely modified with respect to Q whereas tRNAs from stationary cells contain small, but measurable amounts of tRNA with G at position 34 in place of Q. Especially tRNAs from eukaryotes exhibit variable degree of Q content. The extent of Q modification in specific tRNAs changes during development of Drosophila (12) and during erythroid differentiation of leukemia cells (13). Furthermore tRNAs with a G in place of Q accumulate in fetal and regenerating rat liver and in a variety of animal tumors (14).

Since the biosynthesis of queuine and the biosynthesis of pteridines in eubacteria share common pathways, and pteridines inhibit the insertion of queuine into tRNA of eukaryotes it is possible that alterations of Q modification in tRNA are caused by changes in the metabolism of pteridines.

In order to elucidate the biological significance of alterations of Q modification in tRNA in relation to pteridine metabolism and differentiation we use E. coli and, as a simple model system of differentiation, the slime mould Dictyostelium discoideum. Furthermore Xiphophorine fishes serve as animal tumor models to study differentiation of their melanophore system in relation to Q modification of tRNA and pteridines.

Results and Discussion

Biopterin and the synthesis of (Q)tRNA in E. coli.

The tRNA:guanine transglycosylase was purified from *E. coli* according to Okada et al. (10). The *E. coli* transglycosylase exchanges in corresponding tRNAs the guanine residue by all precursors of queuine (14). Since guanine itself serves as a precursor in the biosynthesis of queuine a guanine at position 34 of corresponding tRNAs can be exchanged by radioactive labeled guanine, added to the tRNA-transglycosylase reaction assay. The kinetics of the transglycosylase reaction in the absence or presence of added biopterin show that the *E. coli* enzyme is competitively inhibited by biopterin. The calculated K_m of the enzyme preparation,used in these experiments, is 3×10^{-7}M, the K_i value is $2\text{-}5 \times 10^{-4}$M.

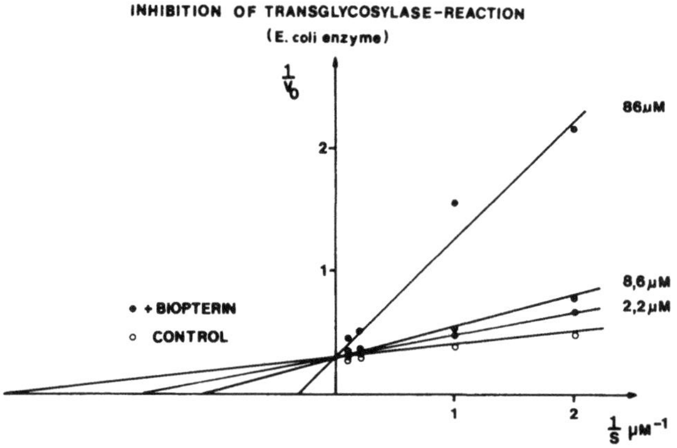

Fig. 2: Inhibition of the *E. coli* tRNA transglycosylase by biopterin. Assay conditions according to ref. (14). Biopterin was obtained from Serva, Heidelberg.

In order to establish as to whether biopterin has any effect on the biosynthesis of queuine or its precursors or the insertion of ´precursor´Q (pre Q) into tRNA, E. coli cells were grown either aerobically or anaerobically in glucose-salt medium in the presence or absence of biopterin (5×10^{-5}M) and kept in the stationary phase for about 6-8 hours. During this time the synthesis of tRNA proceeds and small amounts of tRNAs with G in place of Q, referred to as (Q^{-})tRNA, accumulate. The accumulation of (Q^{-})tRNA during the stationary phase is greatly enhanced in the presence of added biopterin under aerobic conditions, suggesting that biopterin exerts an inhibitory effect on the tRNA transglycosylase only in its oxidized state. Biopterin does not inhibit the insertion of queuine into tRNA during exponential growth. It is possible that this is caused by the fact that,during growth, biopterin is present predominantly in its reduced states as dihydro- or tetrahydrobiopterin.

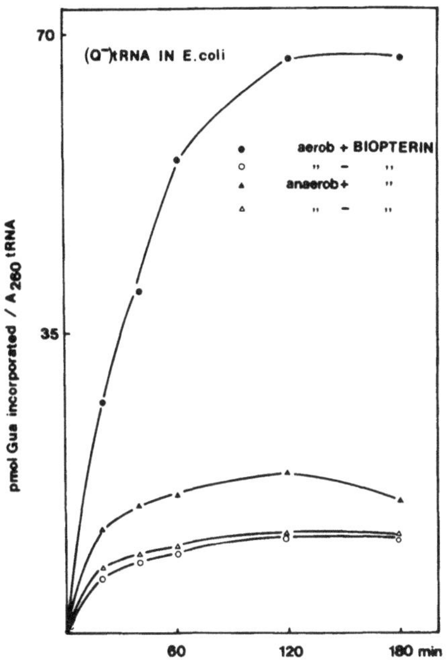

Fig.3: The accumulation of Q-lacking tRNA during stationary phase in E. coli, grown aerobically in the presence of added biopterin.
E. coli cells were grown in glucose-salt medium, with 5×10^{-5} M or without biopterin under aerobic or anaerobic conditions. The cells were harvested during the stationary phase. The tRNA was isolated from 5 g (wet weight) cells by conventional methods and purified by G-150 Sephadex chromatography. The amount of Q-lacking tRNA was determined by exchange of Gua34 with [^{3}H]Gua in the transglycosylase test described in ref.(14).

Alteration of Q Modification of tRNA in Dictyostelium discoideum.

The slime mould Dictyostelium discoideum is a simple multicellular eukaryote. Wild type strains of D. discoideum grow as single amoebae on bacteria as natural food supply, while axenic strains grow vegetatively on a partial or fully defined medium. Developmental transition from unicellular to multicellular stages is induced by starvation and proceeds when a certain amount of cells is placed on agar or filter supports in the presence of phosphate buffer. After an early preaggregation stage, during which the cells become responsive to cAMP,and form cAMP binding sites on the cell surface, the cells aggregate into mounds, the mounds form slugs which subsequently differentiate into mature fruiting bodies comprising spores at the top of a vacuolized stalk (for summarizing references see 15).

In a well defined amino acid medium the cells can be grown in the total absence of queuine. Under these conditions the tRNAs of the Q-family are lacking Q (16). The addition of queuine to the medium at 10^{-7}M restores Q-deficiency in tRNA completely. The presence or absence of queuine has no effect on the growth rate of the amoebae but clearly accelerates the overall developmental cycle. During early development, the preaggregation stage, the level of cAMP to its binding sites increases, in addition characteristic changes in the synthesis of early regulated proteins are observed in response to queuine (17).

Here we show an important survival advantage of cells having queuine or (Q)tRNA. Amoebae that were grown in the absence or presence of queuine and starved during the preaggregation period in suspension culture for several hours differ: Those that comprise queuine or (Q)tRNA respectively, remain aggregation competent and differentiate, but those that are lacking the modified base will not undergo further development (Fig.4 a,b).

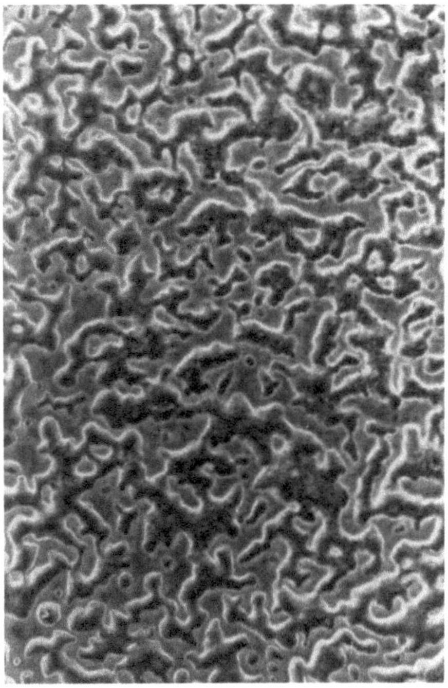

a) preaggregation stage

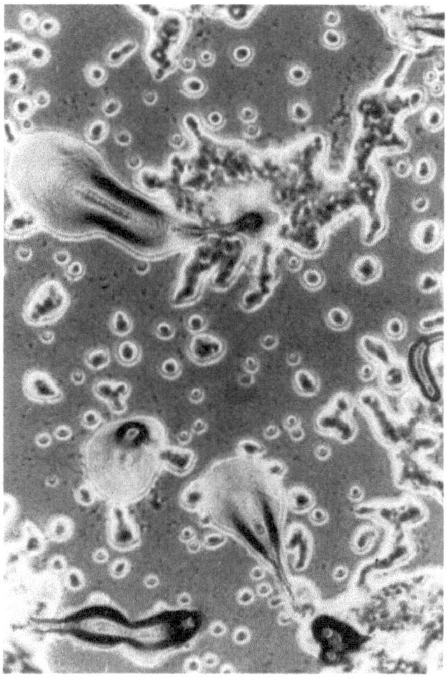

b) culmination stage

Fig. 4: The influence of queuine on development of Dictyostelium discoideum.
The cells were grown in: a) defined queuine free medium, b) defined medium, containing 1×10^{-7} M queuine to a density of 3.5×10^{6}, harvested by centrifugation at 800 x g, washed two times and suspended at a density of 1×10^{7} in 0.017 M phosphate buffer at pH 6.7 a) queuine-free and b) with 1×10^{-7} M queuine and shaken for 24 hours. Aggregation competence was monitored by plating several 50 µl samples onto agar plates. Photographs were taken after further 24 hours of development.

The primary target that is affected by queuine or (Q)tRNA is presently unknown. Further investigations (Schachner, Aschhoff and Kersten unpublished) show that the presence or absence of queuine in Dictyostelium discoideum causes specific changes in the pattern of lactic acid oxidoreductases. These changes might be related to the observed alterations of glucose metabolism in D. discoideum grown either in axenic medium (insufficiently supplied with queuine), or on E. coli (sufficiently supplied with queuine): Axenic strains accumulate large amounts of glycogen (see ref. 15). It was therefore of interest to find out whether the presence or absence of queuine in developing cells causes alterations in biological oxidation reactions. It is known that during development the rate of oxygen consumption decreases (15). Here we show an influence of queuine on endogenous O_2 consumption during preaggregation, as measured with the oxygen electrode. In myxamoebae that were grown in the presence of queuine, the 50% saturation level is reached faster than in cells that were grown in the absence of queuine. This indicates that the decrease in the rate of oxygen consumption is less pronounced during early development in queuine treated cells.

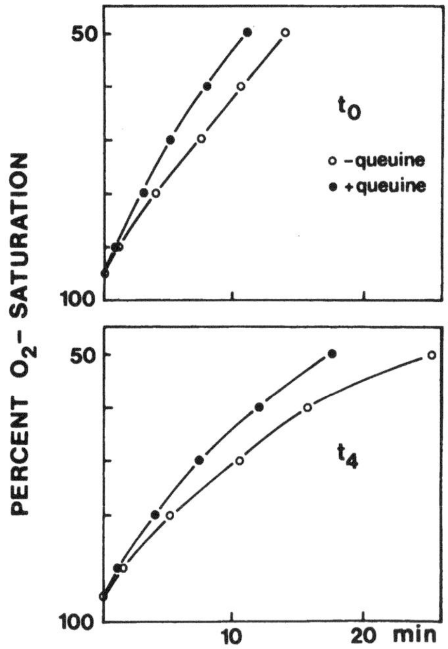

Fig. 5: The influence of queuine on endogenous O consumption immediately and 4 hours after the onset of development in D. discoideum. Conditions of growth and development are described in the legend of Fig 4.

Pteridines and Queuosine-deficiency in tRNA of Skin and Melanoma of Xiphophorine Fishes.

Many alterations of tRNA modification have been observed in tumor tissues and have been suggested as being related to metabolic control mechanisms of neoplastic transformation (18). An incomplete modification of G34 to Q in tRNAs of tumor tissues from various sources has been described (10,19). Also increased levels of pteridines have been found in tumor tissues which might inhibit the insertion of queuine into tRNAs (20).

To investigate the Q-content of tRNA in relation to differentiation, neoplastic transformation and pteridines we use *Xiphophorine* fishes as a model. *Xiphophorus*, a genus of small viviparous freshwater fish from Central America lives in brooks, rivers, ponds and pools. Neoplasia appears to develop only very exceptionally in the wild populations. If, however, interpopulational and interspecific crossings are performed between platys and swordtails the progeny spontaneously develops neoplasms, e.g. melanoma, pterinophoroma, neuroblastoma, thyroid carcinoma, kidney carcinoma, or reticulosarcoma, depending on the genotype (21 and references therein). This crossing conditioned type of neoplasia is known since about 50 years. From formal genetics it has been shown that a specific oncogene, ´Tu`, is responsible for the neoplastic transformation in *Xiphophorus*.

The oncogene ´Tu` is probably identical with the highly conserved c-src gene (21), discovered first in the Rous sarcoma virus. It is now well established that also cells of vertebrates contain the c-src gene. This gene codes for a phosphoprotein (M_f 60.000) ($pp60^{c-src}$) phosphorylating specifically tyrosine residues in particular proteins. The transformation of a normal cell to a tumor cell was suggested to be related to a single base change in this particular gene (22, 23).

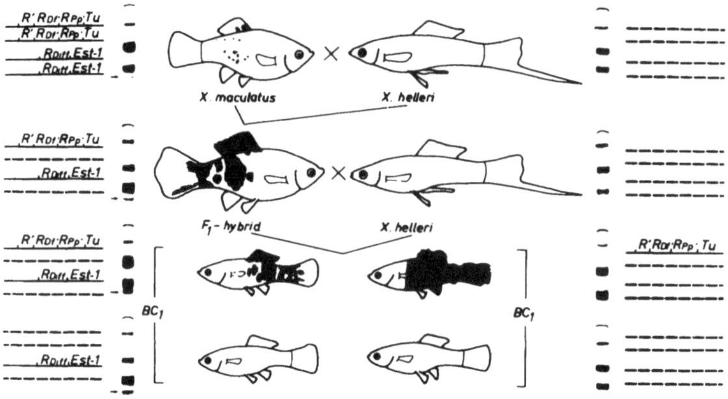

Fig. 6: Crossing scheme which displays the genetic conditions for the ´spontaneous` development of spots, benign melanoma, and malignant melanoma. ——— chromosomes of X. maculatus; ---- chromosomes of X. helleri; Tu = tumor gene; R = regulatory genes, controlling Tu; R_{Diff} = regulatory gene controlling differentiation of normal and neoplastically transformed melanoblasts (for details see ref. 21).

Genetic experiments with Xiphophorine fish showed that ´Tu` is normally repressed by cell type-specific systems of regulating genes ´R`, which suppress the development of the different neoplasms in the fish. In all genotypes, so far studied, most of the R-genes are linked to ´Tu`.

Neoplastic transformation in Xiphophorine fish is well established in the melanophore system: Crosses of spotted X. maculatus (platyfish) with a nonspotted X. helleri (swordtail) result in F_1-hybrids that develop ´benign melanoma` instead of spots. Backcrosses of the $F_1$-hybrids using the swordtail as the recurrent parent result offsprings ($BC_1$), 50% of which exhibit neither spots nor melanoma while 25% develop ´benign melanoma` (like F_1), and 25% develop ´malignant melanoma` (Fig. 6).

The cytological, fine structural and biological data suggest that a specific gene - the regulatory gene ´R_{Diff}`, non-linked to ´Tu` - promotes the final differentiation of normal melanoblasts as well as transformed melanoblasts. Malignant melanoma develop in the ´Tu`-carrying hybrids if ´$R_{Diff}$`-containing chromosomes are substituted by ´R_{Diff}`-lacking ones. If ´$R_{Diff}$` is lacking, the majority of the transformed cells persist in the stage of poorly differentiated, continously dividing TA-melanoblasts and T-melanocytes, and only few cells differentiate to the final stage, the T-melanophores. These are not able to divide any more and are removed by macrophages after they have reached a certain age. Therefore ´R_{Diff}` is suggested to be responsible for the development of benign instead of malignant melanoma (further details see 21 and references therein).

Normal skin of X. maculatus and X. helleri and skin and melanoma of their offsprings have been used to investigate Q modification of tRNA in relation to differentiation of the melanophore system and its neoplastic transformation. Furthermore different genotypes of Xiphophorus, containing variable amounts and pattern of pteridines were chosen to study the relationship between Q modification and pteridines.

In normal skin of Xiphophorus, relatively high amounts of tRNAs with G in place of Q occur, the amount of (Q^-)tRNA in X. helleri (´R_{Diff}-`) is higher than in X. maculatus (´$R_{Diff}$+`) (see Fig. 7). Whether the product of ´R_{Diff}` influences the insertion of Q into a particular tRNA species of the Q-family is presently being investigated. Interestingly we observe that malignant melanoma contain variable, but always higher amounts of Q-lacking tRNA than benign melanoma (Fig. 7b).

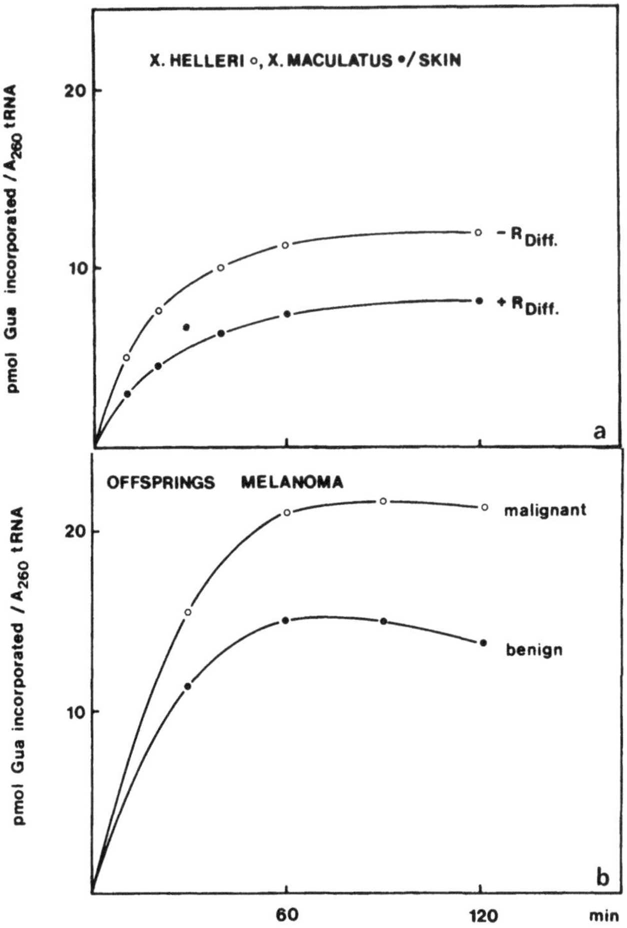

Fig. 7: Guanylation of (Q^-)tRNA from skin (a) and from malignant and benign melanoma (b) of <u>Xiphophorine</u> parents and offsprings. The figures show the kinetics of the exchange of G34 of tRNA by (^3H) Gua, a reaction used to evaluate the amount of (Q^-)tRNA (10).

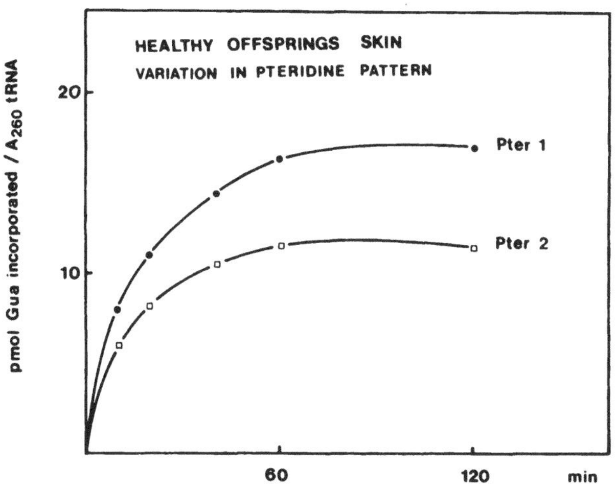

Fig. 8: Guanylation of (Q^-)tRNA from the skin of Xiphophorine offsprings varying in amount and patterns of pteridines. Further details are described in the legend of Fig. 7 and, with respect to pteridines in ref. 24.

Analysis of tRNA isolated from offsprings of different genotypes of X. helleri and X. maculatus showed that the levels of (Q^-)tRNA vary with the pattern of pteridines in their skin. Offsprings, designated in Fig. 8 as Pter 1 have high amounts of drosopterines, whereas those designated as Pter 2 have predominantly sepiapterin, isoxanthopterin and biopterin in their skin (24).

Concluding Remarks

From the results presented here it is evident that queuine and pteridines in eubacteria not only share common biosynthetic pathways but, that at least, biopterin can interfere with the insertion of ´queuine precursors` into E. coli tRNA, however, only under aerobic conditions. The inhibitory effect of pteridines on Q modification in tRNA probably depends on the pteridine level and the redox-state of the cell.

Since alterations in the Q-content of tRNA are characteristic for cells undergoing differentiation or neoplastic transformation, pteridines might be involved in these processes via their inhibitory effect on Q modification in tRNA. Here we show that a lack of (Q)tRNA or of queuine prevents differentiation in Dictyostelium discoideum when the cells are exposed to unfavourable conditions.

From the analysis of Q in tRNAs of the skin and melanoma of different genotypes of Xiphophorine fishes, varying in the types and patterns of pteridines, a relationship between Q modification in tRNA and the metabolism of pteridines emerges which might be important for the differentiation of the melanophore system in Xiphophorus.

The differences in the functional properties of Q-containing and Q-lacking tRNAs are presently being investigated. The (Q)tRNAs are suggested to prefer codons NAU over NAC, whereas the Q-lacking tRNAs read NAC and NAU equally well (7). This can be an important mechanism in the regulation of translation. For eukaryotic $tRNA^{Tyr}$ it has been shown that the Q-lacking species reads a terminator codon, probably UAG (25). Therefore the Q-lacking and Q-containing tRNAs might select mRNAs for translation by a regulatory mechanism, hitherto established for bacteria and designated as termination transcription control (26).

Acknowledgements

This work was supported by the Deutsche Forschungsgemeinschaft, Grant Ke 98/17/6, and SFB 118 and 103.

References

1. Cortese, R., Landsberg, R., von der Haar, R.A., Umbarger, H.E., Ames, B.N: Proc. Natl. Acad. Sci. 71, 1857 - 1861 (1974).

2. Dirheimer, G., Keith, G., Sibler, A.P., Martin, R.P.. In: Transfer RNA, structure, properties and recognition (P.R. Schimmel, D. Söll and J.N. Abelson, eds.) 19 - 41 Cold Spring Harbor Lab. 1979.

3. Kersten, H.. In: Biochemistry of S-adenosylmethionine and related compounds (E.Usdin, R.T. Borchardt and C.R. Creveling, eds.) 357 - 369 Scientific and Medical Division Macmillan Press Ldt. London, 1982.

4. Sprinzl, M., Gauss, D.H.,: Nucl. Acids Res. 10, r1-r55 (1982)

5. Kasai, H., Ohashi, F., Harada, S., Nishimura, S., Oppenheimer, N.J, Crain, P.F., Liev, I.G., van Minden, D.L., McCloskey, J., Biochem. 14, 4198 - 4208 (1975).

6. Kasai, H., Nakanishi, K., Macfarlane, R.D., Torgenson, D.F., Ohashi, Z., McCloskey, J.A., Gross, H.J., Nishimura, S.: J. Am. Chem. Soc. 98, 5044 - 5046 (1976).

7. Nishimura, S.: Progress in Nucleic Acids Res. and Mol. Biol. in press.

8. Reyniers, J.P., Pleasants, J.R., Wostman, B.S., Katze, J.R. Farkas, W.R.: J. Biol. Chem. 256, 11591 (1981).

9. Harada, F., Yamaizumu, K., Nishimura, S.: Biochem. Biophys. Res. Comm. 49, 1605 - 1609 (1972).

10. Okada, N., Nishimura, S.: J. Biol. Chem. 254, 3062 - 3066 (1979).

11. Howes, N.K., Farkas, W.R.: Proc. Natl. Acad. Sci. 76, 3271-3275 (1979).

12. Jacobson, K.B., Farkas, W.R., Katze, J.R.: Nucl. Acids. Res. 9, 2351-2366 (1981).

13. Shindo-Okada, N., Terada, M., Nishimura, S.: Eur. J. Biochem. 115, 423 - 428 (1981).

14. Okada, N., Shindo-Okada, N., Sato, S., Itoh, Y.H., Oda, K.J. Nishimura, S.: Proc. Natl. Acad. Sci. 75, 4247 - 4251 (1978).

15. Loomis, W.F.: In Dictyostelium discoideum, a Developmental System Academic Press, New York, 1975.

16. Ott, G., Kersten, H., Nishimura, S.: FEBS Lett. 146, 311 - 314 (1982).

17. Schachner, E., Kersten, H.: J. Microbiol. 1983 in press.

18. Borek, E.: TIBS 2, 3 - 6 (1977).

19. Kersten, H.: Recent Results in Cancer Res. 84, 255 - 263 (1983).

20. Biochemical and Clinical Aspects of Pteridines. (H. Wachter, H. Ch. Curtius and W. Pfleiderer, eds.) Walter de Gruyter, Berlin-New York, 1982.

21. Anders, F.. In: 33. Colloquium-Mosbach, Biochemistry of Differentiation and Morphogenesis (R. Jaenicke, ed.) 91- 115 Springer-Verlag Berlin-Heidelberg, 1982.

22. Tabia, C.J., Bradley, C.J., Bargmann, R.A., Weinberg, A.G., Papageorge, E.M., Scolnick, Dhar, R., Lowry, D.R., Chang, E.H.: Nature 300, 143-149 (1982).

23. Reddy, E.P., Reynolds, R.K., Santos, E., Barbacia, M.: Nature 300, 149-154 (1982).

24. Henze, M., Rempeters, G., Anders, F.:Comp. Biochem. Physiol. 56B, 35 - 46. (1977).

25. Bienz, M., Kubli, E.: Nature 294, 188 -190 (1981)

26. Yanofsky, C. In: Molecular Mechanisms in the Control of Gene Expression (Nierlich, D. P., Rutter, W. J., Fox, C. F. eds.) 75 - 86 Academic Press New York, 1976

PTERINS AND FOLATES AS EXTRACELLULAR SIGNALS FOR CHEMOTAXIS AND DIFFERENTIATION IN THE CELLULAR SLIME MOLDS

René J.W. De Wit and Theo M. Konijn

Cell Biology and Morphogenesis Unit, Zoological Laboratory
Kaiserstraat 63, 2311 GP Leiden, The Netherlands

1. Developmental signals in the cellular slime molds

The biochemical basis of regulatory and developmental processes receives much attention recently. The cellular slime molds provide a convenient eukaryotic model system for the study of these processes. In contrast to vertebrates the slime molds have a short and very simple life cycle. The cells grow rapidly on simple axenic media (1,2) or together with bacteria on which the amoeboid cells feed by phagocytosis. Development can easily be synchroneously induced by removal of nutrients. During the following interphase there is no obvious morphological sign of the molecular differentiation process. In the well-studied species Dictyostelium discoideum, however, a variety of developmental changes has been reported during this phase, such as synthesis of enzymes for the production and inactivation of communication signals, enhancement of the number of chemoreceptors and of contact sites A mediating cell to cell interactions and oscillatory secretion of signal compounds (e.g. chemoattractants). The secretion of such an attractant or acrasin (3) induces the formation of cell streams and consequently of tight cell aggregation. These are clear morphological markers of the termination of the interphase. After aggregation is completed, the cells form a migrating pseudoplasmodium, in which a sorting out of 2 cell types proceeds. This is reflected in the final stage, i.e. the fruiting body, consisting of dead stalk cells bearing spores, embedded in a

Biochemical and Clinical Aspects of Pteridines, Vol. 2

small slime droplet. Under favorable conditions spores will germinate and yield amoeboid cells, thus completing the life cycle. Other species such as D.lacteum and D.minutum show an even more simple pattern of development, without oscillations, cell streams and migration of the pseudoplasmodium (4-6). The first studied messenger molecule was the acrasin of D.discoideum, identified as cAMP (3). This compound only attracts cells of four related species (7), and mainly after longer starvation periods (8). This increase in responsiveness corresponds to a rise in the number of cAMP binding proteins (chemoreceptors) present on the cell surface (9,10). A higher activity of cAMP inactivating enzymes coincides with it (11). Several forms of this enzyme exist: extracellular phosphodiesterase (e-PDE), membrane bound (m-PDE) and intracellular PDE. The amount of a specific inhibitor of e-PDE decreases when the aggregation competent phase is reached (12). Earlier in development cells start to secrete cAMP in an oscillatory manner (13, 14). Artificial pulses of cAMP, given to cells before the autonomous oscillations, accelerate differentiation as shown by precocious (dis)appearance of developmentally regulated molecules (15-17) and shortening of the interphase (15,18). Thus, cAMP not only mediates chemotaxis and aggregation, but also induces development. In addition, cAMP is an inducer of stalk cell formation, but not of spore formation (19). During aggregation cAMP is not secreted solely by the center. Instead, the cAMP signal is relayed by surrounding cells that have detected cAMP (20).

In contrast to the acrasins which are species specific and highly active in the aggregative phase another chemotactic system was discovered which was mainly active during the feeding or vegetative stage of all species sofar tested (21-23). The compound was secreted by bacteria (19) and identified as folic acid (21). This chemotactic system was viewed as a food seeking mechanism (21,22). Also repellents, which are secreted by vegetative amoebae are thought to be part of this mechanism, by spreading the cells over larger territories (24,25).

Folic acid is an essential component of growth media for <u>D.discoideum</u> (2,26). Uptake in post-vegetative cells could not be demonstrated, although the pterin moiety was taken up after cleavage of the C_9-N_{10} bond (unpublished). Folate is known to be internalized by mammalian cells (27).
The active part of folic acid in chemotaxis seemed to be the pteridine moiety (22), though recent reports suggest that folate and pterin mediate chemotaxis via separate receptors (28). This may suggest distinct functions. Besides being chemoattractants, folic acid and pterin mimic many of the developmental responses to cAMP. Pulses of folic acid and pterin induce secretion of cAMP, oscillations of cAMP (29,30) and precocious appearance of PDE (17,31), cell aggregation (32) and possibly markers of aggregation competence such as cAMP receptors and contact sites A. The amount of PDE inhibitor is decreased by folic acid and pterin as by cAMP (17,31). Folate degrading enzymes were identified as deaminases (33-35). The localization is similar to that of PDE (11,17,34) as is the developmental time course of the activity (17,34). At least 3 forms exist, having different affinity for folic acid, thermostability, and specificity for folic acid and pterin (31,36). Folate and pterin deaminases are also induced by pulses of cAMP, folic acid and pterin (31,36).

2. A pterin derivative as the acrasin of D.lacteum (37)

During cell aggregation of D.lacteum a dialyzable chemoattractant was secreted. This compound was purified by HPLC on 3 systems. The amount of pure acrasin that could be obtained was about 0.1 nmol per isolation. In addition, the compound was labile and degraded into a still chemotactically active product within 1-2 weeks at -20°C in the dark. The degradation product was identified as pterin by 1. UV spectrum at pH 6.0. 2. Chemotactic activity. 3. Comigration with pterin on 5 different HPLC systems. Purification of the acrasin was rather problematic. Therefore attention was focussed on the product of enzymatic degradation of the acrasin by D.lacteum. It is known that attractants undergo rapid inactivation by extracellular media. It was therefore expected that the product of enzymatic degradation would be present in higher amounts in the material that was secreted by the cells. Indeed five-fold higher amounts were present. Furthermore this compound did not show the chemical lability of the acrasin. UV spectra of this compound resembled those of pterins (Fig.1). When incubating pterin derivatives with the extracellular medium of D.lacteum, all were converted to their respective lumazin derivatives. No other reaction occurred. Thus, the acrasin which is thought to be related to pterin (because of the product of chemical degradation and the UV spectrum of its enzymatic degradation product) should also be deaminated by the medium. This was checked by measuring pK_a values of the acrasin and the inactivated product (38). The only difference was the absence of a pK_a value of 1.0 in the inactivated compound, which might reflect absence of the 2-NH_2 group. Based on two biological criteria further evidence for a structural relationship between the acrasin and pterin was obtained. 1. 6-methyl-pterin, which is a good substrate for the deaminase present in the medium of the cells, competed with the acrasin for deamination. 2. 6-amino-pterin, which is an antagonist for chemotaxis towards pterins, also antagonized the response to the acrasin.

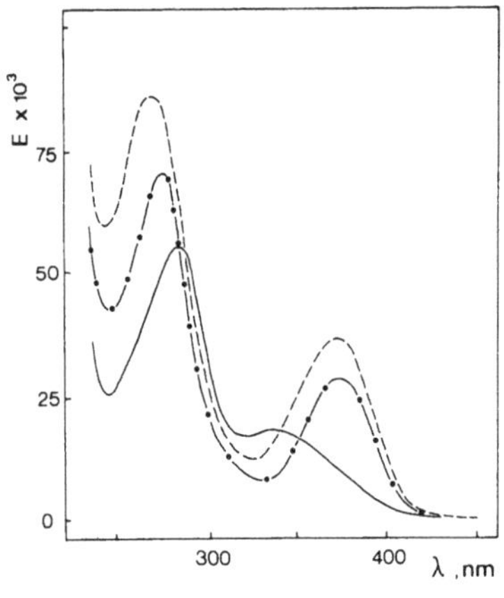

Fig. 1. UV-spectra of the deaminated acrasin of D.lacteum.
—— pH 1.0;
--- pH 5.0;
.-.- pH 13.0
(from ref. 37, with permission).

Testing 30 pterin analogs for their degradation by the deaminase of D.lacteum and their chemotactic activity revealed that modifications in the pyrimidine ring are not tolerated. This was also found for the pyrazine ring with the exception of C_6, in which position modifications are allowed.
Summarizing, we propose that the acrasin, which induces cell aggregation in D.lacteum, is related to pterin, with probably an extra R-group at C_6. The sensitivity of D.lacteum for the acrasin increases 100-1000 fold during development up till 10^{-7}-10^{-8} M and enzymatic inactivation occurs by deamination at C_2.

3. A folic acid derivative as the acrasin of D.minutum (39)

The approach to isolation, purification and characterization of the acrasin of this slime mold species was similar to that of the identification of the acrasin of D.lacteum. After purification on HPLC a preparation was obtained with A_{254} 5 x 10^{-3}, while the concentration of chemoattractant was still 1000-fold the threshold concentration. This suggested the compound either to be very active, or to possess a low UV extinction. The structural relationship with folic acid was suggested by three criteria. 1. Folic acid inhibited degradation of the acrasin by the enzyme(s) present in medium of D.minutum. 2. Methotrexate, an antagonist for chemotaxis towards folates in D.minutum (Fig.2), also antagonized the chemotactic response to the acrasin. 3. High concentrations of attractants are able to delay the optimal chemotactic response, also towards simultaneously tested structurally related attractants, but not to attractants that act via another type of receptor. Thus it was observed that folic acid delayed the reponse to the acrasin. Additional information came from the characterization of the folate degrading enzyme secreted by D.minutum. This enzyme (FAS) cleaved the C_9-N_{10} bond, yielding pterin-6-carboxaldehyde and p-aminobenzoyl-glutamic acid (40,41). Incubation of

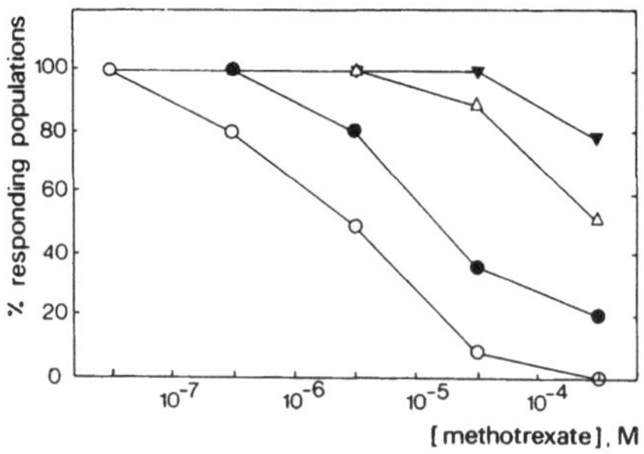

Fig. 2. Competitive antagonism of methotrexate to folic acid induced chemotaxis in D.minutum. Folic acid concentrations: o 5 x 10^{-8}M; ● 5 x 10^{-7}M; Δ 5 x 10^{-6}M; ▼ 5 x 10^{-5}M (from ref. 39, with permission).

the acrasin with FAS rendered the product inactive for D.minutum; in contrast, the chemotactic activity for D.lacteum increased. This is probably due to formation of pterin-6-aldehyde, which is more active than the D.minutum acrasin for D.lacteum. When the D.minutum acrasin was incubated with the deaminase secreted by D.lacteum no change in chemotactic activity for D.minutum was observed, but the chromatographic properties (as judged by reverse phase HPLC) changed. This change was comparable to that when folic acid was deaminated.
The biological activity of eight folate derivatives revealed that N_3 and/or C_4-O, N_5 and/or N_8, N_{10} and the p-aminobenzoyl moiety were necessary for interaction with FAS and the chemoreceptor, whereas C_2-NH_2, N, and the glutamic acid residue were ineffective.
Judged by chromatographic properties of the acrasin and the activity of the analogs the acrasin is proposed to be a folic acid derivative, modified in the glutamic acid moiety, since both charged carboxylic acid functions are absent. The sensitivity for the acrasin increases 10-fold during the interphase up to 10^{-9}-10^{-8}M.

4. The action of folic acid and pteridines during the interphase in *D.discoideum*

Already in 1975 *D.discoideum* was reported to secrete folate and pteridin-like compounds (22). In the *Streptococcus faecalis* assay, vegetative amoebae were found to secrete folic acid in higher amounts than in the aggregative stage. In contrast, as judged by fluorescence, aggregative cells secreted more of folate and pteridine derivatives. More recent investigations confirm the latter observation. Several pterin analogs are found in the medium in increasing amounts as development proceeds (42,43).

Folic acid and pterin degrading enzymes also increase in activity during development (17,31,33,34,49). Both the pattern of secretion of pteridines and of the activity of inactivating enzymes suggest a role for those pteridines during the interphase, i.e. *after* the vegetative stage. The developmental time course of the chemotactic activity of folic acid and pterin is in agreement with this hypothesis. The chemotactic activity of both compounds is constant or increases slightly during the first hours of starvation, followed by a sudden decrease (8). In the aggregation competent phase (after 6-8 h starvation) the activity is at least 10^5 fold less than after 2-3 h of development (unpublished).

In elucidating the molecular basis for the action of folic acid and pterin, attention was focussed on the receptor proteins on the cell surface. Until recently one folate binding protein was detected. This protein bound folic acid and its chemotactically inactive derivative 2-deamino-folic acid with equal affinity (45-47). This might be explained by deamino-folic acid being a chemotactic antagonist of folic acid (45) which was, however, excluded (28). The discrepancy was resolved by the finding of another cell surface binding protein, which was more specific for folic acid (48). When comparing the binding affinity of eleven folate derivates for this binding protein to their chemotactic activity, a high correlation was obtained (unpublished). The previously studied folate binding protein showed an entirely different pattern of binding affinity for the folate analogs. The number of folate chemotaxis receptors per cell during development showed a clear similarity to the chemotactic activity of folic acid (fig.3).

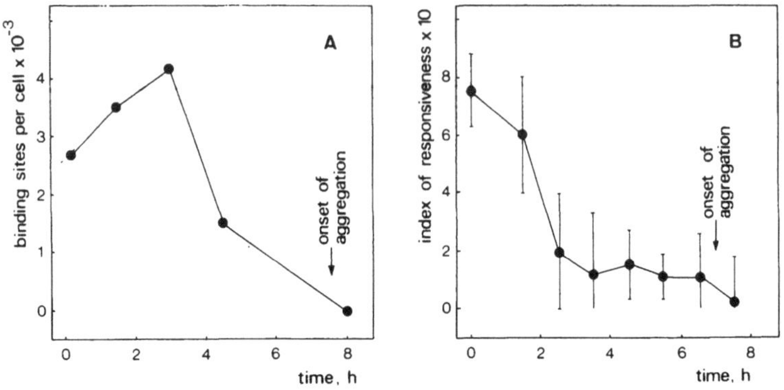

Fig.3. Developmental time course of folic acid receptor levels (A) and relative chemotactic responsiveness to folic acid (B) in D.discoideum. (The data in panel B are from ref. 8, with permission).

A slight increase during the first 3 h was observed, followed by a decline resulting in less than 1% of the original amount after 8 h (onset of aggregation). Also the number of cAMP receptors during development corresponded to the chemotactic activity of cAMP (8,9).

As reported before pterin and folic acid should act via separate chemoreceptors, since a chemotaxis antagonist inhibited the response to pterin but not to folic acid (28). In addition, methotrexate, an antagonist of cAMP secretion triggered by folic acid, did not inhibit pterin induced cAMP secretion (49). The low binding affinity of pterin for the folic acid chemoreceptor (unpublished) confirms the hypothesis of pterin acting via a specific pterin receptor. It will be of interest to examine the binding of pterin to D.discoideum and competition with folic acid for the supposed pterin receptor. Also of importance will be the identification of the natural ligands for folate and pterin receptors and elucidation of their role in development. A separate function for folic acid and pterin is not only suggested by distinct receptors, but also by the presence of an inactivating enzyme(s) specific for folic acid. This folate-specific deaminase is found in the medium and membrane-bound, whereas a pterin-folate aspecific deaminase is present only in the extracellular fraction (31). Extracellular and membrane-bound folate C_9-N_{10} hydrolase activity was also observed in D.discoideum (40, unpublished). An observation which is of interest with respect to the role of folic acid is the fact that folic acid induces a rapid increase in the number of cAMP receptors (50). Thus (a derivative of) folic acid may contribute to the elevation of the sensitivity of the cAMP signalling system and therefore promote differentiation and aggregation.

5. Signal transduction in Dictyostelium and Polysphondylium

Upon administration of a pulse of cAMP or folic acid a number of short-term responses has been observed. The pseudopod formation in the direction of the source of attractant is visible already after about 5 s (16,51). Three biochemical responses are known to proceed on the same time scale as the first visible chemotactic reaction. 1. Protein methylation, which peaks on 3 s for a M_r46,000 protein. Methylation maxima for other proteins were observed after 15, 60 and 150-180 s (52,53). 2. Accumulation of cGMP, maximal 10 s after stimulation (54,55). 3. Equilibrium binding of folic acid and cAMP to their receptors is reached within 10-15 s at 0^oC (unpublished). Evidence has been presented for a messenger function of cGMP in chemotaxis (56) and PDE induction (57). Slower effects are: cAMP accumulation and pulsatile secretion resulting in a relay of the signal (58), Ca^{2+} influx (59), proton release (60), phospholipid methylation (52), lysophosphatidylcholine demethylation (not by folic acid, 61), myosin dephosphorylation (62), vesicle formation (63) and a change in optical density of a cell suspension (light-scattering, 14,64). All these responses peak after 1-3 min. The question arises, which of these changes are involved in chemotaxis, which in the relay secretion of signal molecules and which in changing the pattern of genetic expression, as reflected by the induction or disappearance of various proteins. cAMP, folic acid and pterin seem to behave similarly (17,31) with respect to induction of PDE, folate deaminase and pterin-folate deaminase and the suppression of the PDE inhibitor; also the precocious appearance of autonomous oscillation in cAMP levels can be observed by administration of pulses of both folic acid and cAMP (29). The phase of these oscillations can be shifted in the same manner by both compounds (65). These observations suggest that the intracellular transduction of signals of cAMP, folic acid and pterin proceeds by the same pathway. In contrast, these signals are detected by distinct chemoreceptors, and are degraded by specific (except the pterin-folate deaminase) enzymes (Fig.4).

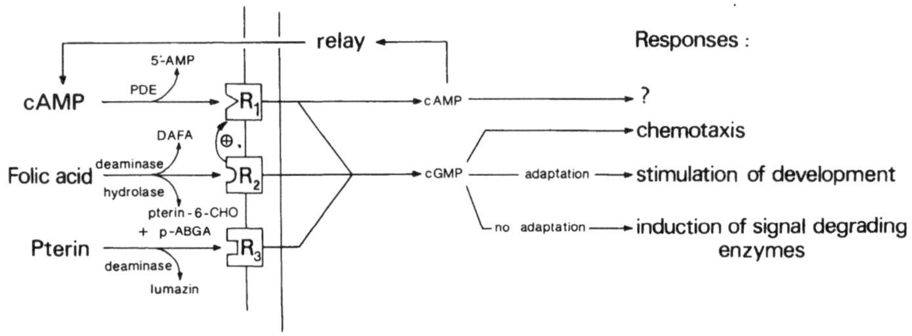

Fig.4. Schematic presentation of the interactions between the components of the signal transduction system. Stimulation of cAMP elicited responses by folic acid is arbitrarily drawn at the level of the receptors for both compounds. It is not clear yet whether pterin exerts a similar effect.

Recently, an asymmetry in this system has been reported, since folic acid, in contrast to cAMP, triggered a very weak relay secretion of cAMP and the signal was not subjected to adaptation (66). Perhaps even this small response was due to folic acid enhanced sensitivity to cAMP. Since D.discoideum cells continuously secrete a small amount of cAMP, which is not detected because of adaptation, a slight elevation in cAMP concentration or the sensitivity to cAMP will induce cAMP relay secretion. By this positive feedback mechanism the signal is then amplified. Thus without addition of exogenous cAMP, the cAMP relay system can be stimulated. This was shown by inhibition of extracellular PDE activity (67) and folic acid enhanced sensitivity for cAMP might have the same effect (66). In this respect the folic acid induced binding of cAMP to its receptor is of much interest (50). It is not clear as yet, how many of the other responses to folic acid are in fact mediated by increased sensitivity of the cAMP detection mechanism. Extracellular cAMP is unlikely to be involved in the action of folic acid in (post-)vegetative cells, since the sensitivity for cAMP is about 100-fold lower than in later stages and the cAMP relay mechanism is not yet functional. Thus chemotaxis and light scattering (the latter of which is thought to be linked to the chemotactic response, 68) are direct responses to folic acid. The responses on a time-scale of hours, such as induction and suppression of proteins and acceleration of development to aggregation competence may be mediated by cAMP relay. The synergistic effect of folic acid on chemotaxis of aggregative amoebae towards low (10^{-9}-10^{-8}M) concentrations of cAMP may also reflect the increased sensitivity to cAMP (19). However, the possibility exists that the chemotactic response to cAMP was potentiated by folic acid stimulated random motility. The maximal motility is reached at about 10^{-7}M cAMP; at lower concentrations of cAMP folic acid might still stimulate the motility. Folic acid increases motility in aggregative cells, but chemotaxis is not observed in that stage (8).

Acceleration of development till aggregation competence is mediated by pulses of folic acid and cAMP. Continuous fluxes of these signals, instead, inhibit development and m-PDE activity (17,69), whereas induction and suppression of the extracellular proteins mentioned above shows a similar pattern, when pulses or fluxes are administered (17,70). This suggests that the two processes are affected by the extracellular signals via at least partially separate routes (in Fig.4 the separation between both pathways is arbitrarily placed after cGMP). Despite other possible explanations, the inhibitory effect of fluxes may originate in adaptation to the constant signal, which interferes with detection and relay of small endogenous pulses, which in turn are necessary to start the signal transduction pathway towards developmental changes.
The detection of signals for induction and suppression of extracellular proteins should not be subjected to adaptation. This detection without adaptation allows cells to respond to high and constant signals by enhancing the activity of inactivating enzymes and thus lowering this background signal. Adaptation would have made this impossible.

6. Summary and conclusions

This paper dealt with the role of endogenous folates and pterins, signal degradation mechanisms, detection of the signal compounds by the cells and the transduction of the signal across the membrane to chemotactic movement and developmental changes.
cAMP mediates cell aggregation in several Dictyostelium species. In addition, physiological pulses of cAMP stimulate development and induce (or suppress) developmentally regulated proteins. In cAMP insensitive species a derivative of pterin (D.lacteum) and a folate derivative (D.minutum) are the chemotactic signals leading to aggregation. Compounds, that are related to pterin and folic acid, are secreted and detected

by specific cell surface receptors during development in D.discoideum. Two types of responses are then observed. 1. Chemotaxis and 2. acceleration of development and changes in the pattern of developmentally regulated proteins. The effect of folic acid on development in D.discoideum may be mediated by the cAMP signalling system.

Also in higher organisms, pteridines are markers for a change in metabolic state (71) and might be signals for changing genetic expression (72). As mentioned before, the cellular slime molds may be viewed as a model system for eukaryote chemotaxis or differentiation. A high correlation exists between the responses of D.discoideum and leukocytes, with respect to morphology (73), intracellular cAMP (58,74,75) and cGMP (54,55,64), calcium influx (59), methylation of phospholipids (52) and several proteins (52,53) and exocytosis of vesicles (63). Both the time scale of these fast responses and the sensitivity for the signalling agents are comparable in D.discoideum and leukocytes, mast cells, platelets and macrophages. N-formylmethionyl peptides, which are the active compounds in leukocytes and macrophages, have a striking similarity to the acrasin of the slime mold Polysphondylium violaceum. This compound also is a N-terminally blocked oligopeptide (76).

Acknowledgements

We wish to thank Peter Van Haastert for the stimulating discussions.

1. Sussman, R.R., Sussman, M.: Biochem. Biophys. Res. Commun. 29, 53-55 (1967).
2. Franke, J., Kessin, R.: Proc. Natl. Acad. Sci. U.S.A. 74, 2157-2161 (1977).
3. Bonner, J.T., Barkley, D.S., Hall, E.M., Konijn, T.M., Mason, J.W., O'Keefe, G., Wolfe, P.B.: Dev. Biol. 20, 72-87 (1969).
4. Raper, K.B.: Mycol. 33, 633-649 (1941).
5. Raman, R.K.: J. Theor. Biol. 64, 43-69 (1977).
6. Gerisch, G.: Roux' Archiv. für Entwicklungs Mechanik 155, 342-357 (1964).
7. Konijn, T.M.: Adv. Cycl. Nucl. Res. 1, 17-33 (1972).
8. Varnum, B., Soll, D.R.: Differentiation 18, 151-160 (1981).
9. Henderson, E.J.: J. Biol. Chem. 250, 4730-4736 (1975).
10. Malchow, D., Gerisch, G.: Proc. Natl. Acad. Sci. U.S.A. 71, 2423-2427 (1974).
11. Pannbacker, R.G., Bravard, L.J.: Sci. 175, 1014-1015 (1972).
12. Riedel, V., Gerisch, G.: Biochem. Biophys. Res. Commun. 42, 119-124 (1971).
13. Gerisch, G., Hess, B.: Proc. Natl. Acad. Sci. U.S.A. 71, 2118-2122 (1974).
14. Gerisch, G., Wick, U.: Biochem. Biophys. Res. Commun. 65, 364-370 (1975).
15. Gerisch, G., Fromm, H., Huesgen, A., Wick, U.: Nature 255, 547-549 (1975).
16. Gerisch, G., Malchow, D., Huesgen, A., Nanjundiah, V., Roos, W., Wick, U., Hülser, D.: Dev. Biol., ICN-UCLA Symposia on Molecular and Cellular Biology, Vol. 2, eds. McMahon, D. and Fox, C.F. W.A. Benjamin Inc., New York. pp 76-88 (1975).
17. Bernstein, R.L., Rossier, C., Van Driel, R., Brunner, M., Gerisch, G.: Cell Differ. 10, 79-86 (1981).
18. Darmon, M., Brachet, P., Pereira Da Silva, L.H.: Proc. Natl. Acad. Sci. U.S.A. 72, 3163-3166 (1975).
19. Bonner, J.T., Hall, E.M., Sachsenmaier, W., Walker, B.K.: J. Bacteriol. 102, 682-687 (1970).

20. Tomchik, K.J., Devreotes, P.N.: Sci. 212, 443-445 (1981).
21. Pan, P., Hall, E.M., Bonner, J.T.: Nature New Biol. 237, 181-182 (1972).
22. Pan, P., Hall, E.M., Bonner, J.T.: J. Bacteriol. 122, 185-191 (1975).
23. Konijn, T.M.: In: Primitive Sensory and Communication systems, ed. Carlile, M.J. Academic Press, New York. pp 102-153 (1975).
24. Kakebeeke, P.I.J., De Wit, R.J.W., Kohtz, S.D., Konijn, T.M.: Exp. Cell Res. 124, 429-433 (1979).
25. Keating, M.T., Bonner, J.T.: J. Bacteriol. 130, 144-147 (1977).
26. Watts, D.J., Guest, J.R.: J. Gen. Microbiol. 86, 333-342 (1975).
27. Huennekens, F.M., Vitols, K.S., Henderson, G.B.: Adv. Enzymol. 47, 313-346 (1978).
28. Van Haastert, P.J.M., De Wit, R.J.W., Konijn, T.M.: Exp. Cell Res. 140, 453-456 (1982).
29. Wurster, B., Schubiger, K., Wick, U., Gerisch, G.: FEBS Lett. 76, 141-144 (1977).
30. Wurster, B., Schubiger, K.: J. Cell Sci. 27, 105-114 (1977).
31. Wurster, B., Bek, F., Butz, U.: J. Bacteriol. 148, 183-192 (1981).
32. Wurster, B., Schubiger, K., Brachet, P.: Cell Differ. 8, 235-242 (1979).
33. Bernstein, R.L., Van Driel, R.: Eur. J. Cell Biol. 22, 234 (1980) abstr.
34. Kakebeeke, P.I.J., De Wit, R.J.W., Konijn, T.M.: J. Bacteriol. 143, 307-312 (1980).
35. Pan, P., Wurster, B.: J. Bacteriol. 136, 955-959 (1978).
36. Bernstein, R.L., Tabler, M., Vestweber, D., Van Driel, R.: Biochim. Biophys. Acta 677, 295-302 (1981).
37. Van Haastert, P.J.M., De Wit, R.J.W., Grijpma, Y., Konijn, T.M.: Proc. Natl. Acad. Sci. U.S.A. 79, 6270-6274 (1982).

38. De Wit, R.J.W.: Anal. Biochem. 123, 285-290 (1982).
39. De Wit, R.J.W., Konijn, T.M.: Cell Differ. 12, 205-210 (1983).
40. De Wit, R.J.W., Van Der Velden, R.J., Konijn, T.M.: J. Bacteriol. 154, in press (1983).
41. Kakebeeke, P.I.J., De Wit, R.J.W., Konijn, T.M.: FEBS Lett. 115, 216-220 (1980).
42. Tatischeff, I., Klein, R.: FEBS Lett. 138, 265-269 (1982).
43. Gerisch, G., Blank, G., Fuchs, D., Hansen, A., Reibnegger, G., Schweiger, M., Wachter, H.: In: Biochemical and Clinical Aspects of Pteridines, eds. Wachter, H., Curtius, Ch. and Pfleiderer, W. W. De Gruyter, Berlin-New York. pp 253-256 (1982).
44. Bernstein, R.L., Van Driel, R.: FEBS Lett. 119, 249-253 (1980).
45. Wurster, B., Butz, U.: Eur. J. Biochem. 109, 613-618 (1980).
46. Van Driel, R.: Eur. J. Biochem. 115, 391-395 (1981).
47. Nandini-Kishore, S.G., Frazier, W.A.: Proc. Natl. Acad. Sci. U.S.A. 78, 7299-7303 (1981).
48. De Wit, R.J.W.: FEBS Lett. 150, 445-448 (1982).
49. Devreotes, P.N.: In: The development of <u>Dictyostelium discoideum</u>, ed. Loomis, W.F., Academic Press, New York. pp 117-168 (1982).
50. Kawai, S.: FEBS Lett. 109, 27-30 (1980).
51. Gerisch, G., Hülser, D., Malchow, D., Wick, U.: Phil. Trans. R. Soc. Lond. 272, 181-192 (1975).
52. Mato, J.M., Marin-Cao, D.: Proc. Natl. Acad. Sci. U.S.A. 76, 6106-6109 (1979).
53. Van Waarde, A.: FEBS Lett. 149, 266-270 (1982).
54. Mato, J.M., Krens, F.A., Van Haastert, P.J.M., Konijn, T.M.: Proc. Natl. Acad. Sci. U.S.A. 74, 2348-2351 (1977).
55. Mato, J.M., Van Haastert, P.J.M., Krens, F.A., Rhijnsburger, E.H., Dobbe, F.C.P.M., Konijn, T.M.: FEBS Lett. 79, 331-336 (1977).
56. Van Haastert, P.J.M.: J. Biol. Chem. in press (1983).

57. Van Haastert, P.J.M., Pasveer, F.J., Van Der Meer, R.C., Van Der Heijden, P.R., Van Walsum, H., Konijn, T.M.: J. Bacteriol. 152, 232-238 (1982).
58. Dinauer, M.C., MacKay, S.A., Devreotes, P.N.: J. Cell Biol. 86, 537-544 (1980).
59. Wick, U., Malchow, D., Gerisch, G.: Cell Biol. Intern. Reports 2, 71-79 (1978).
60. Malchow, D., Nanjundiah, V., Wurster, B., Eckstein, F., Gerisch, G.: Biochim. Biophys. Acta 538, 473-480 (1978).
61. Alemany, S., Garciá Gil, M., Mato, J.M.: Proc. Natl. Acad. Sci. U.S.A. 77, 6996-6999 (1980).
62. Malchow, D., Böhme, R., Rahmsdorf, H.J.: Eur. J. Biochem. 117, 213-218 (1981).
63. Maeda, Y., Gerisch, G.: Exp. Cell Res. 110, 119-126 (1977).
64. Wurster, B., Bozzaro, S., Gerisch, G.: Cell Biol. Intern. Reports 2, 61-69 (1978).
65. Gerisch, G., Maeda, Y., Malchow, D.: In: Development and Differentiation in the Cellular Slime Molds, eds. Cappuccinelli, P. and Ashworth, J.M. Elsevier/North Holland, New York. pp 105-124 (1977).
66. Devreotes, P.N.: Dev. Biol. 95, 154-162 (1983).
67. Van Haastert, P.J.M., Bijleveld, W., Konijn, T.M.: Dev. Biol. 94, 240-245 (1982).
68. Mato, J.M., Van Haastert, P.J.M., Krens, F.A., Konijn,T.M.: Cell Biol. Intern. Reports 2, 163-170 (1978).
69. Washington, A.C., Wynne, V.: Tex. J. Sci. 31, 177-185 (1979).
70. Yeh, R.P., Chan, F.K., Coukell, M.B.: Dev. Biol. 66, 361-374 (1978).
71. Wachter, H., Hansen, A., Reider, E., Schweiger, M.: Naturwissenschaften 67, 610-611 (1980).
72. Wachter, H., Fuchs, D., Hansen, A., Reibnegger, G.: In: Biochemical and Clinical Aspects of Pteridines, eds. Wachter, H., Curtius, Ch. and Pfleiderer, W. W.De Gruyter, Berlin-New York. pp 277-281 (1982).
73. Zigmond, S.H.: J. Cell Biol. 75, 606-616 (1977).

74. Shaffer, B.M.: Nature 255, 549-552 (1975).

75. Roos, W., Nanjundiah, V., Malchow, D., Gerisch, G.: FEBS Lett. 53, 139-142 (1975).

76. Shimomura, O., Suthers, H.L.B., Bonner, J.T.: Proc. Natl. Acad. Sci. U.S.A. 79, 7376-7379 (1982).

Author Index

SUBJECT INDEX

Bei Fragen zur Produktsicherheit wenden Sie sich bitte an:
If you have any questions regarding product safety,
please contact:

Walter de Gruyter GmbH
Genthiner Straße 13
10785 Berlin
productsafety@degruyterbrill.com